Mastering Statistics Using JASP

Second Edition

James MacDougall, Ph.D.

Copyright 2024 by James MacDougall, Ph.D.
All Rights Reserved

No part of this publication may be reproduced, stored in a retrieval system, or transmitted in any form or by any means without the prior written permission of the author.

Almost all data analyses in this text use an open-source program called *JASP*, which is being developed by members of the Department of Psychological Methods of the University of Amsterdam under the direction of Eric-Jan Wagenmakers. It is used with permission of jasp-stats.org. Any errors in the description or use of the program for this text are inadvertent and the responsibility of the author of this text.

Two computational modules involving inference concerning population proportions are available in *Interactive Statistics* by Larry Green of Lake Tahoe Community College and are used with permission of *LibreTexts*. Any errors in the description or use of the program for this text are inadvertent and the responsibility of the author of this text.

The *JASP* logo on the back cover of the book is used with permission of jasp-stats.org.

The cover design and artwork are by Talha Mujaddidi.

Published by Amazon KDP.

ISBN: 979-8-218-39719-7

Contents

Preface to the Second Edition: i

Introduction Mastering Statistics: 1

Chapter 1 Basic Concepts: 3

Chapter 2 Describing Data: 15

Chapter 3 Standard Scores & Normal Distributions: 31

Chapter 4 Getting Started with *JASP*: 43

Chapter 5 Pearson Correlation: 57

Chapter 6 Linear Regression: 73

Chapter 7 Partial Correlation: 89

Chapter 8 Basic Probability: 97

Chapter 9 Sampling Distributions & Statistical Inference: 109

Chapter 10 Single-Sample Inference: 125

Chapter 11 Two-Sample Inference: 135

Chapter 12 One-Way ANOVA: 151

Chapter 13 Two-Way ANOVA: 163

Chapter 14 Factorial ANOVA: 175

Chapter 15 Repeated Measures ANOVA: 185

Chapter 16 Chi Square: 197

Appendix: 209

Preface to the Second Edition

The first edition of *Mastering Statistics Using JASP* was a formal product of an informal textbook that I developed over many years of teaching statistics at a small liberal arts college. The impetus for writing and publishing an actual textbook was the discovery of Jeffreys's Amazing Statistics Program, aka *JASP*. Here was a highly sophisticated statistical analysis program with an easy-to-use graphical interface that was free! Without cost, every student could have a copy on his or her computer. No more expensive short-term licensing fees or trips to the college's computer lab. An inexpensive companion textbook seemed like a natural. Enter Amazon's Kindle Direct Publishing (KDP) and the task of turning a college-printed text into an actual book became remarkably straightforward. So, I did it!

Then I discovered one of the pitfalls in writing a statistics textbook. When you finally see it through to publication, you begin thinking of all of the changes and additions you would like to make! The first edition reflected some of the limitations and biases that I had acquired teaching a 14-week semester course built around the need to teach students the technical details of using the *Statistical Package for the Social Sciences* (*SPSS*) in the college's computer lab. But because *JASP* is so easy to use and always available, it makes it possible to teach a broader range of more sophisticated topics and to give students the opportunity to solve more sophisticated statistical problems.

The second edition includes changes that range from minor improvements in statistical formatting to the inclusion of two entirely new chapters on analysis of variance. The end-of-chapter problems have been expanded for almost all of the chapters. Here is a chapter by chapter list of the more important changes.

Chapter 1, Basic Concepts is largely unchanged except for the inclusion of a more complete discussion of types of variables and the statistics that can be performed on them.

Chapter 2, Describing Data, already long, introduces some new statistics and graphs that are available in *JASP* and explains more completely the algebraic summation conventions used in subsequent chapters.

Chapter 3, Standard Scores & Normal Distributions remains unchanged except for minor formatting improvements and addition of more end-of-chapter normal curve problems.

Chapter 4, Getting Started with *JASP* is completely rewritten and greatly expanded. It now includes a much more thorough discussion of the construction and organization of data files using either a separate spreadsheet program or *JASP*'s revised Data Editor. The description of the statistical analysis interface was expanded and there are now separate sections devoted to explaining (a) how *JASP* output tables and figures can be modified for different purposes, (b) how data selection is accomplished using *JASP*'s point and click interface, (c) how new variables can be created from existing variables, and (d) how one customizes the *JASP* interface and graphical and tabular output. In all, this chapter is a much more complete introduction to *JASP* and should enable most students to teach themselves how to use *JASP* with minimal instructor assistance.

Chapter 5, Pearson Correlation was already a rather thorough introduction to Pearson *r*, but the discussion of using *JASP* to analyze correlational data was expanded a bit and a new section added to introduce other correlational statistics and their use in later chapters as measures of effect size.

Chapter 6, Linear Regression is a fairly lengthy and complete introduction to multivariate regression and remains largely unchanged, save for some minor improvements in figures and tables, and the inclusion of additional regression problems at the end of the chapter.

Chapter 7, Partial Correlation remains unchanged except for some minor formatting improvements.

Chapter 8, Basic Probability was shortened a bit in a couple of places to allow an expanded discussion of the binominal theorem.

Chapter 9, Sampling Distributions & Statistical Inference is a long chapter that covers all of the bases on the logic of hypothesis testing, estimation, inferential error, power, and bias and precision. Students have complained that it was too long and there were too many graphs, so the section on power was stripped of several overly complicated graphs and shortened considerably.

Chapter 10, Single-Sample Inference is an introduction to hypothesis testing and estimation with sample means and proportions. It remains unchanged except for some formatting improvements and the addition of a couple of new end-of-chapter problems.

Chapter 11, Two-Sample Inference extends the logic and procedures developed in Chapter 10 to two-sample hypothesis testing and estimation. The discussion of assumptions has been tightened a bit and there is a new section on Cohen's d and r_{pb}^2 as effect size statistics. JASP's Raincloud plot is introduced as an interesting way to graphically summarize independent t test results. A correlated t test example with clear assumption violations introduces the non-parametric Wilcoxon test for matched pairs.

Chapter 12, One-Way ANOVA is a very complete introduction to the logic of the F ratio statistic and its use in testing hypotheses about multiple sample means. There is a new section on the meaning of η^2 and ω^2 and their use as effect size statistics.

Chapter 13, Two-Way ANOVA introduces the concept of an interaction between independent variables and the use of the F ratio to evaluate main effects and interactions. It is largely unchanged except for some upgraded figures and a more complete discussion of the uses and interpretation of the various effect size statistics available with JASP.

Chapter 14, Factorial ANOVA is an entirely new chapter that gives a more formal presentation of the concept of partitioning variance and introduces more of the terminology commonly used in the analysis of variance. It explains how first- and higher-order interactions are evaluated and how they modify interpretation of main effects. The logical process of carrying out a complete three-way ANOVA with JASP is described, including the interpretation of post-hoc evaluation of simple effects and the construction of interaction plots.

Chapter 15, Repeated Measures ANOVA is another new chapter that introduces the student to an important analysis of variance procedure that is widely used in psychology, biology, and medicine. It explains the general logic of partitioning variance into between-subject and within-subject components and the advantage of removing individual difference variation from the error estimates in the F ratio. The randomized blocks design is explained as an alternative to a pure repeated measures design. There is a focused discussion of the mathematical assumptions underlying repeated measures ANOVA and the consequences of violating the sphericity assumption. A mixed model repeated measures sample experiment is analyzed using JASP, including the use of Mauchly's W in detecting sphericity violations and Greenhouse-Geisser in correcting for them. The issues involved in deciding which effect size statistics to use in repeated measures designs is explored and an alternative approach to dealing with sphericity violations is explained.

Chapter 16, Chi-Square introduces the chi-square statistic and explains how it is used to solve goodness-of-fit and contingency table problems. Except for some minor formatting clean-up, it is unchanged.

Introduction
Mastering Statistics

Mastering Statistics?

Yes, the purpose of this text is to help you master *applied statistics*. You won't become a statistician *per se*, but if you apply yourself, you will develop a clear conceptual understanding of the logic underlying descriptive and inferential statistics. You will also develop the data analysis skills to analyze and interpret the types of data found in modern research. Ideally, at the end of this text, if someone should ask you what multiple linear regression is, you will be able to explain what it is, where it's useful in research, its assumptions and limitations, how it's carried out in *JASP*, and how a final regression model is interpreted.

To achieve these goals, I have stripped away much of the by-hand calculating normally associated with introductory statistics textbooks. This will free you to concentrate on mastering the ideas of statistics. Almost everything of importance in statistics can be learned in terms of pictures and words, and it is entirely possible to develop a clear, intuitive understanding of statistical methods without a background in mathematics. If you come to have this level of conceptual understanding of statistics, you will never really forget it. If you try to learn the material by brute force to get through the course, you will forget most of what you have "learned" in a few months.

OK, that sounds really encouraging, but I must admit that for purposes of completeness, there are mathematical formulas throughout the text that are used to define specific statistics and to illustrate important ideas. You won't need to remember the formulas, but a *little* knowledge of algebra would help in understanding some ideas.

This textbook is unlike other introductory textbooks in another important respect. The three chapters on linear correlation and regression (5–7) are based on a multiple variable approach, rather than the traditional bivariate (two variable) approach seen in most other introductory textbooks. The reason for this is simple. Two-variable research doesn't exist in the real world. All modern research is multivariate (many variables) in nature and students need to have the conceptual background to understand and participate in such research. I believe that the multivariate approach is no more difficult to understand than the traditional approach and I have put my theory into practice in writing those chapters.

If that sounds a bit obscure, don't worry about it. This book is designed to make it un-obscure.

What's In the Book?

Chapters 1–3 teach you the basic vocabulary of statistics and introduce some important concepts and statistics that get more fully developed in later chapters. They're an important foundation for everything that follows. For the most part, the material is straightforward and easy to learn!

Chapter 4 is your introduction to *JASP*, which you will use for all the different types of data analyses that you learn in the course. *JASP*, which stands for *Jeffreys's Amazing Statistical Program*, is a free, open-source and extremely powerful program that you will download to your computer—to keep forever! If you're curious about *JASP*, skip ahead to Chapter 4, download the program, and start playing around with it. It's easy to learn and almost all the figures and tables in the text were generated using *JASP*.

Chapters 5–7 cover Pearson correlation, multiple linear regression, and partial correlation. Most of the ideas in these chapters are relatively intuitive and should present no problems if you stay up with the assignments. These are among the most widely used statistical procedures in the sciences and knowing them gives you some very valuable skills.

Chapters 8 and 9 will teach you the essentials of statistical inference, beginning with a brief introduction to probability, followed by a longer chapter that presents the critical core concepts of statistical inference: sampling distributions, hypothesis testing and estimation, Type I and II Error, power, and estimation bias and efficiency.

This is followed by two chapters (10 and 11) that introduce one and two-sample inference, where sample data are used to test hypotheses about and make estimates of population means and proportions.

Chapters 12–15 give a thorough introduction to a very powerful and widely used statistical procedure known as analysis of variance (ANOVA). If your instructor decides to include all four chapters in the course, you will wind up knowing as much about ANOVA as do many Ph.D. researchers.

The final Chapter 16 introduces a very valuable statistical procedure known as Chi-square, which is used to analyze frequency data of all types.

Why Learn Statistics?

As my undergraduate statistics teacher used to say, "It's good for you," but it turns out that there actually are three really good reasons to master basic statistics.

First, you may be in a program of studies like psychology or biology that requires you to pass a stat course to graduate or take other required courses. If this is your primary motivation, then of course you will want to excel because many professors ascribe to the belief that success in statistics is predictive of success in other courses (as well as life).

Second, much of the modern world runs on statistics and you will greatly enhance your chance of getting many jobs in government or industry if you can confidently say that you are proficient in statistics and data analysis. People with statistical skills typically make good salaries and report low levels of job stress!

Third, in case you hadn't noticed, what passes for news these days is full of misleading and erroneous "information" and a good deal of illogical reasoning. Learning statistics will make you a much more critical thinker—able to detect fallacious reasoning, bogus "facts", and unwarranted conclusions. Basically, knowing statistics makes you smarter!

What This Text Is Not

This is a wonderful text, but it has a point of view. It is frankly traditional in its approach and focuses on what are called "classical" or "empirical" judgments about the probability of events. In this approach, values of random samples are used to make judgments about characteristics of populations, the process driven by mathematical treatment of the sample data itself.

This contrasts with what is called Bayesian statistics –named after Thomas Bayes, an 18th Century philosopher, mathematician, and minister. The Bayesian approach incorporates subjective probabilities, prior beliefs, and computer simulations in shaping how data are used to arrive at conclusions about populations. Happily, knowledge of the classical approach is a good place to start in learning Bayesian statistics.

Final Comments

Each chapter is followed by a set of questions and problems that test your understanding of the material and your ability to apply the statistical procedures. The first set of questions in each chapter require a single word or a very short answer and are intended to give you immediate feedback about whether you've mastered the ideas in the chapter. The remaining questions contain data to be appropriately analyzed and summarized. It's to your advantage to solve these problems, as they reinforce your knowledge and highlight things that you need to study further. Plus, it's fun to "play around" with your new statistical package and learn all of the things that you can do with it!

Chapter 1
Basic Concepts

What's In This Chapter?

When people collect information to describe some situation or to discover the causes of some event, they are doing **research**. Research involves measuring the values of **variables**, which are measurable aspects of the world. When variables are measured, they yield **data**. Variables are classified as either **categorical** or **numerical** in character. Categorical variables are such things as class rank, gender, or job type, in which the values of the variable are different names. Categorical variables can have either **unordered** or **ordered** categories.

Numerical variables vary in amount and have numbers as their values. There are three types or "levels" of numerical variables--**ordinal**, **interval**, and **ratio**. Some variables can be measured in an error-free fashion and yield values that are said to be **exact**. More often, however, measurement introduces error into the process, yielding only **approximate values** for the variables. The accuracy of the approximation is indicated by the **real limits** of the number.

Most research is directed at drawing conclusions about a large group of data called a **population** by studying the data for only a portion of the population called a **sample**. Accurate conclusions about a population are possible only if the sample is **representative** of the population. If a sample is selected in such a way that it cannot be representative of the population, it is called a **biased sample**. A good way to select samples is to use **random sampling**, which is any process that ensures that each bit of data in the population has an equal chance of being included in the sample. Random samples are not biased, but they may not be representative because of **sampling error**. Larger samples are more likely to be representative of their populations than smaller samples.

Statistical analysis of sample data involves two different types of procedures—**descriptive statistics** and **inferential statistics**. Descriptive statistics include summarizing data using **frequency distributions** and **histograms**, and numerical indices called **statistics**. **Statistical inference** consists of **parameter estimation**, in which the value of a sample statistic is used to estimate the value of the population parameter, and **hypothesis testing** in which the value of a sample statistic is used to decide whether a population parameter (or parameters) is (are) likely to be equal to a particular value or range of values.

Introduction

Statistics is a branch of mathematics concerned with describing and drawing conclusions about large amounts of numerical information. Although many aspects of statistics are highly theoretical and abstract, this textbook focuses on applying statistics to solve very practical problems. Understanding the ideas in this textbook does not depend upon a sophisticated knowledge of mathematics. All the important concepts presented in the chapters that follow can be understood in terms of words and pictures. You will need to recall a bit of high school algebra to follow the formulas that define specific statistics, but you will not be expected to algebraically derive new equations or to "prove" statistical theorems.

A key feature of this text is the de-emphasis of manual (i.e., hand calculator) computations of statistics. Manual calculation of most statistics is extremely laborious and leaves little time for the more important tasks of understanding key concepts and learning to correctly apply statistical techniques. Moreover, reliance upon manual calculation makes it impossible to teach several statistical procedures that are widely used in the modern world—techniques with which the beginning student should be familiar. In some chapters I will work through a sample problem "by hand" to show the logic of the procedure and give you a feel for what the statistic actually measures.

Almost all statistical analyses presented in this text have been carried out using a **free** data analysis program called *JASP*, which is short for **Jeffreys's Amazing Statistics Program**. It is named in honor of a famous statistician, Sir Harold Jeffreys. *JASP* statistical analyses appear throughout the text and in Chapter 4 you will learn how to use *JASP* to do all your data analyses.

Clearwater College Study

The Student Legislative Council (SLC) of Clearwater College had been squabbling for several weeks about a proposal to increase student activity fees from $400 per year to $700 per year. Proponents of the increase claimed that the higher fee was needed to support a broader range of extracurricular activities, while opponents argued that fees were already too high. To help resolve the issue, the SLC commissioned a survey of student opinion. The Registrar's office supplied the SLC with an alphabetical listing of all full-time students. Every 20th student on the list (135 total) was contacted by telephone and asked the following three questions:

1. What is your class rank? (Freshman, Sophomore, Junior, or Senior)
2. Are you for or against the proposal to raise the student activity fee from $400 to $700? (For or Against)
3. Last semester, the SLC sponsored 22 activities (lectures, films, concerts, dances). How many of those activities did you attend?

Responses to those questions were obtained for 119 students. Sixteen students who were selected for the survey refused to answer the questions. Below are the overall results for each question:

1. Class Rank

Freshman	38	31.9%
Sophomore	33	27.7%
Junior	25	21.0%
Senior	23	19.3%

2. Attitude Toward Proposal

For	63	52.9%
Against	56	47.1%

3. Events Attended

   Ranged from 0 to 21 Events with an average of 11.5 Events

Responses to Questions 2 and 3 were also tabulated separately for each of the classes:

2. Attitude Toward Proposal

	# For Prop.	% For Prop.
Freshman	26/38	68.4%
Sophomore	18/33	54.5%
Junior	10/25	40.0%
Senior	9/23	39.1%

3. Events Attended

	Average
Freshman	16.8 Events
Sophomore	11.4 Events
Junior	8.2 Events
Senior	6.5 Events

It would be nice to relate that the results of the survey helped to resolve the controversy. Unfortunately, both sides claimed that the results supported their position. The pro-increase faction argued that the survey showed that a "majority of students" favored the increase. The against-faction pointed to the strong opposition among Juniors and Seniors as being decisive. The squabbling continued for the balance of the semester.

This first chapter has two objectives. The first is to introduce some of the key concepts that you will need right away in your study of statistics. The second is to give you an overview of the general processes involved in statistical analysis. We'll accomplish both objectives in the context of a specific example—a hypothetical survey of student attitudes at a small college concerning the desirability of raising the student activity fee. Most surveys are more complicated than our example, but it will serve to introduce many important ideas. **Read the description of the "Clearwater College Study" above carefully before proceeding**.

Basic Concepts

Like all disciplines, the field of statistics has its own specialized vocabulary, which the beginning student needs to master. As in the chapter overview, in the presentation below, important terms are presented in boldface type. The meaning of each term is discussed and usually illustrated with an example. Some terms in statistics have very precise and restricted meanings, and it is a good idea to memorize their definitions. Where the precise definition is important, it is stated in bold-faced type.

Research

When people systematically collect information to achieve some end, we say that they are doing **research**. A specific research project is sometimes referred to as a "study." People do research for two general reasons. The first is to accurately describe some situation. The SLC survey is an example of this type of research. Here, the students doing the survey collected information to learn how students at Clearwater College felt about the issue of raising activity fees.

The other reason for doing research is to learn the **cause** or causes for some event of interest. We noted in the SLC survey that Freshmen were much more favorably inclined toward the proposed fee increase than were Juniors and Seniors. Why? One obvious possibility is that Juniors and Seniors have become too busy to attend many social events and, thus, don't feel inclined to pay for even more unattended events. Or perhaps the onus is on the Freshmen. Perhaps the Freshman class includes many party animals whose propensity for extracurricular activities will cause them to be missing by the time their Junior and Senior years roll around. To decide which of these possibilities (if any) is the real cause of the differences between classes would require more information, i.e., more research.

Cases

In our example, the telephone surveyors secured information from 119 students. Technically, statisticians would refer to each student respondent as a **case**. The word "case" is used to mean **the entity on which information is gathered**. In the example, the cases are students, but if we were collecting information on a group of cities, the cases would be individual cities. The survey results are said to be based on 119 cases.

Variables

The questions asked in the survey define what are called **variables**. A variable is **some measurable aspect of the world that takes on different values or states**. In the SLC study, there were three variables: *Class Rank*, *Attitude Toward Proposal*, and *Events Attended*. The variable *Class Rank*, when measured for each case, could result in the values of "Freshman," "Sophomore," "Junior," or "Senior." The variable *Attitude Toward Proposal* can take on the values "For" and "Against." The variable *Events Attended* can take on the values "0," "1," "2," "3," through "21."

Notice that the variables differ in the type of values that they can assume. The values for the first two variables, *Class Rank* and *Attitude Toward Proposal*, are categories rather than numbers. It makes sense, then, that these types of variables are called **categorical variables**. Another name sometimes given to such variables is **qualitative variables**, because "qualitative" means differences in "kind" rather than "degree." We will use the term "categorical" when referring to variables of this type.

There actually are two types of categorical variables. The first is called **unordered categorical variables**. These are variables such as type of pet, or religion, or job, where the categories can be ordered in any fashion. For example, a list of pets that includes dogs, cats, and birds could just as well be ordered cats, birds, and dogs. The second type is called **ordered categorical variables**, where the categories represent some underlying dimension that goes from low to high. In our example, *Class Rank* and *Attitude Toward Proposal* are actually ordered categorical variables because each is a point on one or more dimensions, e.g., juniors have more courses completed than sophomores, are older, etc. Ordered categorical variables are sometimes assigned numbers and analyzed using the correlational and linear regression statistics explained in Chapters 5 and 6.

The third variable, *Events Attended*, yields numbers when measured. Variables of this type are referred to as **numerical variables**. They are also called **quantitative variables**, because "quantitative" denotes differences in amount of something. We will use the term "numerical variables" in this text.

As you might anticipate, numerical variables also divide into types. The "lowest" type is called **ordinal variables**, where the numbers are basically ranks. If a panel of judges ranks the quality of five Bordeaux wines from best (1st) to worst (5th), each wine's score is an ordinal number ranging from 1 to 5. Ordinal numbers are limited in their usefulness in statistics because the distance between numbers often doesn't reflect the real distances in the thing being measured. In our example, the wines ranked 1 and 2 might be very close to one another in quality, but far superior to wine ranked 3. As a result, ordinal variables are very limited in the statistics that meaningfully can be computed on them. This issue will be discussed further in Chapter 3 in the context of what are called percentile ranks.

Numerical variables in which the numbers accurately represent distances in the thing being measured are called **interval variables**. Virtually all of the statistics in this text except for Chapter 16 assume the variables are interval in character. **When the term numerical variable** is used in this text you can assume that it means **interval variable**.

Just to make this discussion complete, I'll note that there is a third type of numerical variable called **ratio variables**. Variables of this sort not only have equal intervals, but a meaningful zero point where the quantity being measured ceases to exist. With such variables anchored at zero it's possible to talk about ratios in the numbers, e.g., 10 is twice as much as five. Ratio level variables are common in the natural sciences, but surprisingly rare in the social sciences. Fortunately, the statistics in this text assume only interval qualities.

The *Events Attended* variable would be considered ratio in nature. Because the numbers accurately reflect distances in the thing being measured and possess a meaningful zero, it makes sense to find the mean number of events attended for each class and to say that "Suzi attended half as many events as Joe."

Data

People are always talking about **data**, saying things like, "Show me the data" or "Let's look at the data." What are "data"? Technically, the data of a study are **the values of the variable or variables obtained for the cases**. In the SLC survey study described above, there were three variables, so each case (student) contributed three bits of data to the entire data set.

Here's a terminological problem. Technically, the word "data" is the plural of the word "datum", and one should say, "The data are..." rather than, "The data is..." For good or ill, however, modern usage now accepts data as a singular noun. Being old fashioned, I'll continue to use data as a plural noun.

Accuracy of Data

To produce data, we need to measure values of one or more variables on several cases. With some variables we can achieve data that are perfectly accurate, but with other variables our measurement procedures introduce error. Consider first error-free data. Categorical variables can in principle be measured in an error-free fashion. In the SLC survey, it was determined that there were 38 Freshmen included in the study. Assuming that a consistent definition was employed in determining class rank, the number "38" is *exactly* the number of Freshmen in the study.

Some numerical variables can also be measured exactly. The variable *Events Attended* can only assume integer values, i.e., "0", "1", "2", and so on. Variables that only change in fixed increments such as integers are called **discrete numerical variables**. The measurement of such variables involves counting the number of discrete increments present for each case. If Joe College attended three SLC-sponsored events, the value "3" is an exact number.

It may be that you are suspicious of Joe's ability to remember exactly the number of events that he attended, so here is a more persuasive example. Consider the variable *Amount of Money in Male Acquaintances' Wallets*. You ask your friend Sam to let you measure the value of this variable for his wallet. A careful inspection of the wallet yields three one-dollar bills and one five-dollar bill, so the value of the variable for this case is $8.00. Sam doesn't have about $8.00; he has exactly $8.00.

Most numerical variables are not discrete, however. Instead, they vary continuously along some dimension and, for this reason, are referred to as **continuous numerical variables**. Because no measuring instrument or operation is capable of infinitely fine measurement, any attempt to measure a continuous variable inevitably introduces error into the process. Measurement error causes the measured value for the variable to be **approximate** rather than exact.

Here's an example. You have taken your pet Doberman pinscher, Big Bird, to the veterinarian for his annual checkup. During the procedure, the vet tells you that Big Bird weighs 97 lb. You know, of course, that the number "97 lb." is an approximate value, because weight is a continuous variable and vets' scales are not perfectly accurate. Big Bird weighs "somewhere around" 97 lb., but where? Let's assume that the vet's scales are not biased to weigh light or heavy and are accurate to the nearest 1 lb. If this is the case, what are the possible true values for Big Bird's weight that would cause the scale to read 97 lb.? It turns out that if Big Bird weighs any amount between 96.5 lb. and 97.5 lb., the unbiased scale will record a value of 97 lb. The value 96.5 lb. is the *lowest* possible amount that Big Bird could weigh and still have the scale record 97 lb. Any value less than 96.5 lb. is closer to 96 lb. and would cause the unbiased scale to record the value 96 lb. On the heavy side, if his true weight were even a little bit more than 97.5 lb., the scale would record a value of 98 lb. (This discussion assumes that Big Bird's weight cannot be *exactly*, to an infinite number of decimals, either 96.5 or 97.5 lb.) **Figure 1** represents this situation graphically.

The values 96.5 lb. and 97.5 lb. are called the **real limits** of the approximate value 97 lb. The real limits of a number define the **range of possible true values for a number containing measurement error**. The real limits of

Figure 1

Real Limits of Big Bird's Weight Measured to One Pound Accuracy.

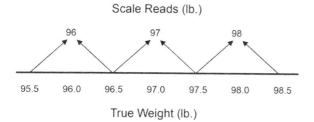

a number are obtained by taking the obtained value of the number plus and minus one-half of the unit of accuracy of measurement (number farthest to the right). Big Bird had been measured to the nearest 1 lb. of accuracy, so the real limits for the obtained value of 97 lb. would be computed as

$$\begin{aligned}\text{Real Limits} &= 97\text{ lb.} \pm .5\text{ lb.}\\ &= 96.5\text{ lb. to } 97.5\text{ lb.}\end{aligned}$$

If the vet's scale had been accurate to the nearest tenth of a pound instead of to a pound, she might have told you that Big Bird weighed 97.2 lb. The ".2" indicates that the reported value is accurate to one decimal place of accuracy. Because measurement is now accurate to one decimal place, the real limits for the approximate value of 97.2 lb. would be the observed value plus and minus one-half the unit of accuracy, i.e.,

$$\begin{aligned}\text{Real Limits} &= 97.2\text{ lb.} \pm .05\text{ lb.}\\ &= 97.15\text{ lb. to } 97.25\text{ lb.}\end{aligned}$$

This abstract discussion of measurement accuracy turns out to have one important implication for statistics students. When you compute things like averages on approximate numbers, it is important to recognize that your computed values are themselves approximate in character. You shouldn't imply greater accuracy in the computed value than was present in the original numbers. For example, suppose that Big Bird has two siblings, Barney and Donatello, who weigh in at the approximate values of 103 and 96 lb., respectively. If you were to use your phone calculator to compute the average weight of your pack of three Dobermans, you would find the result to 98.66666667 lb. Although the computation is numerically correct to seven decimals, this value seems suspiciously precise because the original numbers were only accurate to the nearest 1 lb. For the sake of honesty, you need to "round off" the value of the average to a smaller number of decimals to reflect the accuracy present in the original numbers.

Here is a simple, widely accepted rule for doing so: **You should express computed values to one or at most, two more decimals of accuracy than were present in the original numbers.** Using this rule, you would report the average weight of your Doberman pack to be either 98.7 or 98.67 lb.

What about the situation where you are dealing with exact numbers? Suppose that three of your friends have $3.00, $5.00, and $6.00, respectively, in their wallets? If you used your calculator to compute the average amount of money possessed by your friends, you would arrive at a value of $4.66666667, which is the correct value to seven decimal places of accuracy. Because the three values are exact numbers, you are mathematically justified in expressing the average value to any desired number of decimal places of accuracy. But wouldn't you look silly telling your friends that their average amount of money is $4.66666667? Common sense (and cents!) would suggest that a rounded-off value of $4.67 might be good enough for most purposes. In the social sciences a good rule of thumb might be to express computed values based on exact numbers to two more decimals than were present in the original numbers.

Constants

Some measurable aspects of the world, when measured, yield only one value. These attributes are called **constants** because their values do not change. In the world of the physical sciences and mathematics some attributes that take on only single values are the velocity of light ("c"), the ratio of the circumference of a circle to its diameter ("pi"), and the natural logarithm ("e"). Few (if any) universal constants have been identified in social science research, but often many attributes are constant within a particular research setting. In the case of our survey study of student attitudes, the attribute "college attended" would be the same for all individuals; it would be a constant in this setting with a value of "Clearwater College."

Populations

Almost all research is intended to arrive at conclusions concerning large bodies of data. In their survey study of student attitudes, the SLC wanted to know how Clearwater students "in general" felt about raising the activity fee. The attitudes of all the students at Clearwater College constitute what is technically known as the **population** of interest in the study. We may define a population as **the complete set of data about which we wish to draw conclusions**. Notice that this definition defines populations in terms of variables, not the entities on which the variables are measured. Sometimes, however, we might get a little sloppy and say something like, "The population is all students at Clearwater College." This is all right if it's understood that the population is really the values of some attribute of all of the students that we're talking about.

In our survey study example, the population of student opinions is "real" in the sense that each student could be surveyed, and information collected for the entire population. Some populations, however, are **hypothetical**, either because the population is so large and/or unstable over time that it is impossible to enumerate all the cases comprising the population, or because the population is "created" only by the special conditions of an experiment.

Let's consider the latter situation more fully. Suppose that the SLC at Clearwater College wanted to find a way to change student attitudes about raising activity fees. A psychology major on the Council notes that appeals made by high-status, respected speakers have been found to be effective in changing peoples' attitudes about things. In appreciation for his willingness to speak up, the SLC assigns the psychology major the task of conducting an experiment to determine whether such an appeal would be effective.

Volunteers from introductory psychology courses are recruited for the experiment. One group of 20 Clearwater College students is assembled in a classroom and listens to a speech by Lexi Putellas, the college's very popular soccer player, on the objective need to raise activity fees. At the end of the speech, the students complete the SLC survey. This is technically known as the "experimental condition." Their responses are compared with those of another group of 20 Clearwater students who are assembled in a classroom and merely complete the survey without the benefit of listening to Lexi. This is the "control condition."

There are two groups of students in the experiment, so we need to distinguish between two different hypothetical populations of Clearwater College students. The population associated with the experimental condition might be defined as "the survey responses that would be generated if all Clearwater College students completed the survey after hearing Lexi discuss the need for increased activity fees." The population associated with the control condition of the experiment would then be "the survey responses that would be generated if all Clearwater College students completed the survey without first listening to Lexi discuss the need for increased fees." Clearly, all Clearwater students can't be in both populations—both for practical reasons (time, effort, etc.) and for the conceptual reason that having been exposed to one condition of the experiment would alter how students responded to the other condition. The two populations are, thus, hypothetical.

The existence of hypothetical populations leads to the peculiar situation that one frequently uses sample data to arrive at conclusions about a population that doesn't exist—but could exist. Surprisingly, this mismatch between statistical theory and practical reality doesn't seem to trouble anyone.

Samples

The SLC survey sought to find out what all Clearwater students thought about the activity fee question by querying only a portion of the students concerning their attitudes. Technically, the group of students who completed the survey is called a **sample** of students. A sample is defined as a **subset of a population**, where the term "subset" means a portion of something. The idea is that information gathered from a sample can tell you what the population is like. In the case of the survey study, the proportion of students in the sample who favor raising the activity fee can be used to "guess" what the proportion would be for all students. This process of using sample information to guess about the population is known as **statistical inference** and is described in more detail below.

Not all samples are equally useful for guessing about population characteristics. The "utility" of a sample for this purpose depends on the extent to which the sample is **representative** of the population. A sample is said to be representative **if the values of the variables of interest are distributed in the sample in approximately the same way that they are distributed in the population**. For example, the SLC's sample of 119 students would be considered representative of the population for the variable *Class Rank* if the percentage of cases in each class (Freshman, Sophomore, etc.) was about the same for the sample as for the population.

There are two reasons why a sample might be unrepresentative of the population from which it is drawn. The first has to do with problems in the way that the sample was selected. If the procedures used to select the sample tend to systematically include or exclude cases having particular characteristics and those characteristics have an influence on the variables of interest, then the sample will yield results that are not representative of the population. Such samples are said to be **biased**.

In many survey studies, for example (including the SLC survey), a fairly large proportion of people contacted refuse to answer orally presented questions or to return written questionnaires. People who refuse to participate in such studies often are systematically different in many respects from those who agree to participate. Non-participators may have less free time, be less sociable, or less interested or involved in the topic of the survey. If these characteristics are related to how one responds to the questions contained in the survey, then omission of these cases will make the sample results systematically different from the results that would be obtained for the population.

Obtaining samples from preexisting groups such as business, academic classes, military units, and the like are often referred to as **convenience samples**. Such samples are not inherently biased, but often make it unclear as to which population the results apply.

The second factor that can cause a sample to be unrepresentative is the operation of chance. Even if the selection procedures are unbiased in the sense used above, a small sample "by chance" may over- or under-represent cases having particular values of the variables, and thus be unrepresentative. In the SLC survey study, if only 10 students had been queried concerning their attitudes, it might easily turn out that all or most of the students comprising such a small sample were against the proposal, even if most of the population favored it. The smaller the sample, the more likely it is that the sample will differ from the population, i.e., be unrepresentative.

Random Sampling

How then should a sample be selected to maximize the likelihood that it will be representative of the population? There are a variety of techniques that have been developed to achieve this goal, but the simplest is a procedure called **random sampling**. Samples selected in this manner are called **random samples** and virtually all the procedures discussed in this textbook assume random sampling. A random sample is defined as a sample selected in such a way that **every piece of data in the population has an equal chance of being included in the sample**.

There are a variety of ways to select random samples. The most tried-and-true of these procedures is the "names in a hat" technique. Using this approach, the authors of the SLC study could have written each student's name on a uniform slip of paper, thoroughly mixed the slips of paper up in a large bowl, and without looking at them, drawn out 135 slips. The "names in a hat" technique is convincingly random but is too labor intensive for practical use. A simpler approach would have been to use a computer program that generated random student ID numbers. Students whose ID numbers were generated would be included in the study.

Random sampling has two major advantages over other selection procedures. First, a random sample, by definition, cannot be biased, because the selection procedures are independent of all characteristics of the cases comprising the population. This does not, however, mean that any particular random sample is representative of the population. Recall that chance factors (which operate in random sampling) may lead to un-representativeness, particularly when the sample size is small. But here is where the second advantage of random sampling comes into play. Based on the assumption of random sampling, statisticians have developed procedures for knowing the extent to which random error can influence the value of sample characteristics, i.e., how unrepresentative a sample might be for any particular characteristic.

Where does this discussion of sample representativeness leave us with respect to the SLC survey study? Recall that the SLC selected every 20th student from an alphabetical listing of all students at Clearwater College. Did they achieve a random sample? Technically no, because selection of one member from the list automatically precluded selection of the next 19 members on the list. Is this sample a biased one? Here the answer is less clear. Probably the variables of *Class Rank*, *Attitude Toward Proposal*, and *Events Attended* are not related to the order in which student's names appeared in the list of names. If this is the case, then the initial selection procedure *per se* was not biased

But, recall, that 16 of the students whose names were selected from the list refused to answer the questions. If those 16 students share characteristics in common that might be related to attitudes about raising fees, then the results would be biased. If, for example, most of the 16 refused to answer because they were mistrustful of the SLC's motives in conducting the survey, they are also probably people who would not trust the SLC with more of their money. In that case, most of them probably would have been against the fee increase. Their exclusion from the sample would bias the results in favor of the proposal. Unhappily, we will never know for sure. In the face of such uncertainty, what most statisticians would do is to hope that the 16 failed to respond for reasons unrelated to attitudes concerning the proposal and pretend that the sample is a random sample.

Analyzing Data

There are two stages to the process of statistically analyzing sample data. The first involves summarizing the data in such a way that the important features of the data are clearly shown. Techniques for accomplishing this summary function are collectively referred to as **descriptive statistics**. The term "descriptive statistics" is a little slippery because it has two different but related meanings. One meaning is general and refers to the whole compendium of techniques for describing important features of a set of data. The other meaning is more restricted and refers technically to specific **numerical characteristics of a sample**. For example, the mean of a sample of numbers is technically called a statistic.

The second stage in a statistical analysis is to use the sample data to draw conclusions about the population from which the sample is selected. Procedures for arriving at such conclusions are called **inferential statistics**. There are many different types of inferential statistical procedures and the explanation of the logic and procedures underlying their use will constitute the bulk of this textbook.

Descriptive Statistics

Descriptive statistics are needed to discern general trends or patterns in the data that would not be obvious upon inspection of the "raw data." Imagine, for example, that in the presentation of the Clearwater survey study, we had simply listed the individual responses of the 119 cases. **Table 1** shows what such a tabulation would look like.

Basically, lists of anything are hard to make sense of. To save you this chore, we initially presented the survey results in partially summarized form. We had already begun the process of describing the sample data by applying some basic descriptive statistical techniques. Consider the first two variables, *Class Rank* and *Attitude Toward Proposal*. Each had been presented in the form of a **frequency distribution**, which is a very common descriptive procedure. A frequency distribution is a **table showing how often each value of a variable occurs**. A frequency distribution has at least two columns. The first column always lists the values that the variable can assume, while the second column lists the number of cases having each of the values.

Frequency distributions make it easy to see patterns in the values of the variable. Refer to the description of the SLC survey. Note, for example, that the frequency distribution for the variable *Class Rank* makes it clear that the sample contains almost twice as many Freshmen as Seniors. It would be much tougher to pick that important fact up from a simple listing of the data, e.g., **Table 1**. Because the variable *Class Rank* is categorical in nature,

Table 1

Partial List of the 119 Cases Sampled for Clearwater College Study.

Student Case	Name	Class Rank	Resp.	Events Attend
001	Ackerman, S.	Fresh.	Yes	16
002	Agastino, B	Senior	No	3
003	Ali, J.	Fresh	No	0
004	Armstrong, S.	Soph	Yes	8
005	Baker, J.	Fresh	Yes	11
|	|	|	|	|
|	|	|	|	|
|	|	|	|	|
119	Yali, A.	Junior	No	9

the frequency distribution of different ranks is called a **categorical frequency distribution**.

We could also generate a frequency distribution for the numerical variable, *Events Attended*. Here one would list the values of the variable, usually from low to high, along with the frequencies with which the different values occurred. Such a distribution would be called a **numerical frequency distribution**. One form of this frequency distribution is shown in **Table 2**.

Table 2

Frequency Distribution of the Number of SLC-Sponsored Events Attended by 119 Clearwater College Students.

# of Events	Freq.	# of Events	Freq.
0	3	12	9
1	3	13	3
2	6	14	5
3	3	15	3
4	3	16	4
5	7	17	1
6	13	18	1
7	10	19	2
8	16	20	0
9	4	21	1
10	12	22	0
11	10		

This frequency distribution leaves something to be desired from a technical standpoint, but it serves to show

how a frequency distribution can be constructed with numerical data. Even in its crude state, however, the distribution suggests that most students (74 or 62%) attend between 6 and 12 events. Twenty-five (21%) attended five or fewer events and almost nobody went to 18 or more events (3%). Chapter 2 goes into more detail about constructing numerical frequency distributions of this type.

The next chapter also explains how to construct **graphs** based on frequency distributions. A graph is **a pictorial representation of one or more aspects of the data**. Frequency distribution graphs show visually how often different values of the variable occurred. **Figure 2** is a type of graph called a **histogram** produced by *JASP* for the *Events Attended* data for the SLC study. The graph was generated using the *JASP* histogram chart option under the DESCRIPTIVES - Basic Plots procedure.

Figure 2

JASP Histogram of the Number of SLC Sponsored Events.

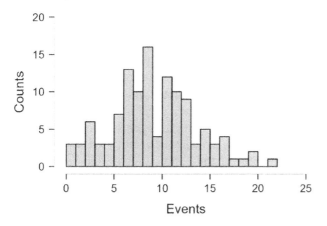

Another way to describe sample data is to compute **statistics** that summarize important features of the data. Again, the term "statistic" used in this sense technically means a numerical characteristic of a sample. In presenting the attendance data, we reported that the "average" number of events attended by the 119 students was 11.5 events. Most of you probably are familiar with this type of average, which is technically called the **arithmetic mean** (the sum of the score values divided by the number of values). The term "arithmetic mean" is used to distinguish it from other types of means, such as the "geometric mean" or "logarithmic mean." Having made this distinction, we'll drop the qualifier. For purposes of this course, the term "mean" will always refer to the arithmetic mean.

The mean is a member of a class of descriptive statistics whose function is to describe the **central tendency** (average or middle) of a set of numerical data. There are other such statistics which you will learn about in the next chapter.

The descriptive procedures discussed thus far—frequency distributions, histograms, and the sample mean—are examples of what are called **univariate descriptive procedures**. The word "univariate" means "one variable," so these are procedures for describing characteristics of variables taken one at a time. A univariate descriptive analysis of the SLC survey data would begin with the first variable, *Class Rank*, describe it as completely as possible, move on to the next variable and describe it, and so on, until all the variables had been described.

Another group of descriptive statistics is concerned with describing the *relationships* between variables. Statistics of this type are called **bivariate** (two-variable) or **multivariate** (multiple variable) **descriptive statistics**. Many of these statistics are used to describe the **correlation** between two or more variables. The term "correlation" means the **tendency for values of one variable to be systematically related to the values of another variable**. In our example we can consider the correlations between *Class Rank* and each of the other two variables. Look back at the table in the description of the Clearwater College Study that shows the number and percentage of "For" responses separately for each class rank. Note that the Freshmen are strongly in favor of the proposal to raise fees, while the Juniors and Seniors are against the proposal. There is, thus, a correlation between the two variables, because values of the variable *Attitude Toward Proposal* are systematically related to values of the variable *Class Rank*. A correlation also exists between *Class Rank* and *Events Attended*, because the mean number of events attended declines systematically as class rank increases. Another term for correlation when categorical variables are involved is **contingent**, as in the number of events attended is contingent on the students class rank.

It is extremely probable that a correlation also exists between *Attitude Toward Proposal* and *Events Attended*, but the necessary frequency distribution was not presented in the discussion of the study.

Inferential Statistics

Most of the time in this course you will be concerned with going beyond simple description of sample data to drawing conclusions from the data. This process of drawing general conclusions from sample data is the business of **inferential statistics**. Said a bit more technically, statistical inference involves **drawing conclusions about characteristics of populations by** studying characteristics of samples drawn from those populations. There are two major divisions within the broad topic of inferential statistics. The first is called **parameter estimation** and the second, **hypothesis testing**. We'll discuss parameter estimation first.

Parameter Estimation

In our example, the SLC wanted to know what proportion of all Clearwater students favored the proposal to raise the activity fee. To find out, they obtained a sample of students and ascertained the percentage of the sample that favored the increase. The sample value of 52.9% provides an estimate of the percentage of the population of all Clearwater students who would favor the increase. The population characteristic being estimated from sample data is called a **parameter** of the population. The term "parameter" is technically defined as a **numerical characteristic of a population**. The phrase, "numerical characteristic" should sound familiar, because it was part of the narrower definition of a "statistic," i.e., a numerical characteristic of a sample. This should lead you to the idea—which is correct—that statistics and parameters must have something in common. Actually, they have a lot in common because they are, in a sense, the same thing. A statistic is to a sample what a parameter is to a population. One can have the mean or percentage for a sample and the mean or percentage for a population. The basic difference is that we know the value of the statistic but must estimate the value for the population. Thus, parameter estimation involves **using the value of the sample statistic to estimate the value of the corresponding population parameter**.

Actually, there is a bit more to parameter estimation than is first implied by this definition. The SLC obtained a sample value of 52.9% of students favoring the fee increase. The authors of the study therefore estimate that 52.9% of all Clearwater students favor the increase. This is known as a **point estimate** of the parameter, because only a single value is specified. The problem with this estimate is that it is almost certainly wrong. Wouldn't you be surprised if the percentage for all students turned out to be exactly 52.9%? What they really are estimating is that the parameter value is "around" 52.9%. But how close is "around 52.9%?" Could the parameter value be as high as 55% or as low as 49%? The answer turns out to be yes, easily. How about 45%? Not so likely. The fact of the matter is, there is always error or inaccuracy in the estimate.

Given this possibility of error, what the SLC needs to do is generate a **range of possible values on either side of the sample value that has a known probability of including the parameter value**. This is known as an **interval estimate**. The logic and procedures underlying interval estimation are relatively complicated and will be developed in later chapters. Right now, just focus on the general concept.

Hypothesis Testing

The other general application of inferential statistics is called hypothesis testing. In the simplest application of hypothesis testing, one uses sample data to decide whether a population parameter is equal to some hypothesized value or range of values. We have an implicit instance of this process in our SLC example. Recall that after looking at the sample percentage of students favoring the fee increase (52.9%), the SLC concluded that a "majority of students" favored the increase.

What they were doing (presumably without realizing it) was testing the hypothesis that 50% or fewer of the population (a minority) favored the proposal. If this sounds a bit strange or backward to you, don't be alarmed—it's how statisticians think! To accept the proposition that the population percentage is greater than 50% (technically called the **Alternative Hypothesis**), they must reject the proposition that the result is a chance outcome when the population value is 50% or less (technically called the **Null Hypothesis).**

The problem with the SLC's conclusion is that there is a high probability of being wrong. Let's see why. Suppose that only 50% of the entire student body favored the proposal. If the population value really were 50%, is it reasonable to believe that a sample of 119 cases could yield a value of 52.9%? The answer turns out to be yes. Such a sample value could readily occur through sampling error ("chance") even though the population percentage was 50%. Thus, the sample value of 52.9% does not provide strong enough evidence to reject the hypothesis that the proposal is favored only by 50% or fewer of the entire student body. This is why pollsters like Roper or Gallop who forecast the percentage of people who believe this or that always indicate that their results are subject to errors of $\pm$ 2% or $\pm$ 3%, sometimes referred to as the "margin of error."

On the other hand, suppose that the sample value had been 68%. Would this value provide strong enough evidence to reject the hypothesis that the population value was 50% or less? Now the answer would be yes, because if the population value really were equal to or less than 50%, it's extremely unlikely that one would ever get a sample value as high as 68% because of sampling error. It makes more sense to believe that the population value is closer to 68%; i.e., that a majority of students favor the proposal.

In more complex applications of hypothesis testing, investigators are concerned with determining **whether two or more populations have different values for some parameter in question.** Recall, for example, the hapless psychology student who wound up conducting the experiment to see whether Lexi's endorsement of raising activity fees would change student attitudes. At the conclusion of the experiment, the student would need to use the appropriate inferential tests to determine whether the proportion of students favoring raising fees in the *hypothetical experimental population* was greater than the proportion in the *hypothetical control population*.

You will learn a wide variety of hypothesis testing procedures in later chapters.

Two Suggestions for You

Many introductory courses in college are of the "survey" type, presenting information successively on a series of relatively independent topics. Introductory psychology courses are good examples of the survey approach. Because the topics are fairly independent, lapses in studying at the beginning of such courses usually don't impair one's ability to learn material presented later in the course. **This is not true with statistics courses!** Statistics is a cumulative discipline in which one's ability to learn material presented later in the course is critically dependent upon learning the material that comes earlier in the course. You need to master the information in one chapter before proceeding to the next chapter. Putting off studying until you have 3-4 chapters to read for the first time is a recipe for disaster.

There is another important difference between a statistics course and many of the other courses that you will take as a freshman or sophomore in college. The subject matter in many elementary courses—particularly in the social sciences and humanities—is relatively non-technical and "common sense" in character. In such courses it is usually possible to "get the idea" or identify "the major points" by reading through the textbook once or twice. **This doesn't work in statistics.** Statistical terminology is precise, and some very important concepts are completely unfamiliar to most students. **You will not succeed by passively reading this textbook; you must plan to *study* it very carefully.** Be prepared to memorize the definitions of key terms, to take notes on your reading, and to test yourself on your ability to explain difficult concepts. Very possibly you might find it useful to form a study group with friends in the class. Often talking with others about statistics helps you to understand it better.

Good Luck!

Whoops! The first blank page of the text. I read somewhere that an author should never allow blank pages. Always put something there, even if it's an inspirational quotation. So here is your first inspirational quotation:

"It had long since come to my attention that people of accomplishment rarely sat back and let things happen to them. They went out and happened to things."

Leonardo da Vinci

Chapter 1, Basic Concepts
Chapter Exercises

> The questions at the end of each chapter are of two types. The first are short answer questions that will give you immediate feedback regarding your mastery of the vocabulary and concepts. Some call for single word answers; others require a short phrase. The second type of questions are designed to challenge your ability to apply what you have learned to answer more open-ended questions or to solve and interpret actual data analysis problems.

Some questions are purely factual and have only one correct answer. Others require greater thought on your part and may have more than one possible answer.

1. Fill in the missing term(s) that makes the statement true.

    a. _____ are values of variables measured on _____.
    b. Computing statistics by hand is a bad idea because _____.
    c. *JASP* stands for _____ and is explained in Chapter _____.
    d. Systematically collecting information is called _____.
    e. Child, teenager, adult are values for a _____ variable.
    f. Your height in inches is an example of a _____ variable of what type? _____.
    g. The entities on which data are collected are called _____.
    h. Which type of numerical variable has the fewest statistical applications? _____
    i. When the term "numerical variable" is used in this text, it means an _____.
    j. Numbers that increase only by integer (whole number) values are called _____.
    k. 17.5 and 18.5 are the _____ of the number _____.
    l. Generally, values of statistics should be expressed to _____ more decimals than were in the original data.
    m. Samples are defined as _____.
    n. _____ are to samples as _____ are to populations.
    o. Selecting volunteers from an introductory psych class for a study is an example of _____.
    p. Calculating the percentage of dogs in a shelter that have ear mites is an example of _____ statistics.
    q. _____ permit causal conclusions about variables.
    r. The "names in a hat" strategy is a way to achieve a _____.
    s. Using sample information to draw conclusions about a population is an example of _____.
    t. Types of cats is an example of a _____ variable.
    u. When a sample is selected in a way that makes it different from the population, it is said to be _____.
    v. A _____ is a table that shows how often values of a variable occur.
    w. A _____ is a pictorial display of a _____.
    x. Another term for correlation when applied to categorical variables is _____.
    y. _____ estimates suck because they are usually _____.
    z. Research shows that males generally rate attractive women as nicer, an example of a _____ study.
    aa. Random factors can cause a sample to be _____.
    bb. _____ is a free data analysis program.
    cc. A statement about a population parameter that is to be tested by sample data is called the _____ Hypothesis.
    dd. It is pitted against the _____ Hypothesis.

2. For each of the following variables, indicate whether the variable is categorical or numerical, and if it is numerical, whether it is ordinal, interval, or ratio. If a variable could be considered as either categorical or numerical, explain why.

    a. Street addresses
    b. Final course grade (e.g., A, B, etc.)
    c. Number correct on a 25-item True-False test
    d. A person's IQ measured by the Wechsler Adult Intelligence Scale
    e. Types of cell phones owned by college students
    f. A telephone number
    g. Ratings of anxiety on an 11-point scale by students prior to taking a major examination in physics

h. Where horses finish in the Kentucky Derby
i. Percentage of games won by the Ohio State football team

3. Give the real limits (if any) for each of the following measurements.

   a. A board measured as 220 cm long
   b. A reaction time of .67 seconds
   c. A score of 19 correct on a 20-item test
   d. A weight of 123.769 kg

4. A college psychology major named Beth hypothesizes that introverted college students are more likely to own pets than extraverted students. She administers a standard Introversion-Extraversion scale to 195 introductory psychology students who volunteer to participate to receive 5 extra credit points on the midterm exam, along with a simple questionnaire concerning pet ownership. She selects the 50 most introverted and 50 most extraverted of the 195 students for comparison. She finds the following: Of the 50 most introverted students, 34 (68%) owned pets within the past three years. Of the 50 most extraverted students, 19 (38%) owned pets during the past three years. Answer the following questions concerning Beth's study. Justify your answers where necessary.

   a. How many cases are involved in Beth's study?
   b. How many variables are involved in Beth's study?
   c. What are the data for Beth's study (be precise)?
   d. Are the data categorical or numerical?
   e. What are the constants for Beth's study?
   f. Define the population(s) to which Beth **ideally** would like to generalize her results.
   g. Define the population(s) that Beth **is actually using** to conduct her research.
   h. What type of **sampling strategy** is Beth using for her study?
   i. What is the **descriptive statistic** that Beth is using?
   j. What is the **inferential question** that Beth must answer (that is, what must Beth decide about the data)?
   k. If you were going to conduct the study, what modifications would you make in Beth's methodology? (Explain the reasons for any).

5. Altruism refers to the willingness to help others without receiving any immediate benefit in return. In a study of altruism in children, 85 children of different ages were given the opportunity to share a cup containing 10 wrapped hard candies with another child from your school "who is in the next room who doesn't have any candy." Here are the data for the study. The dependent variable was the number of pieces of candy that the child placed in an empty cup to be given to the other child.

	Number of Children Sharing Various Amounts of Candy					
	0 Pieces	1 Pieces	2 Pieces	3 Pieces	4 Pieces	5 Pieces
3 – 4 year old Boys ($n = 12$)	5	5	2	0	0	0
3 – 4 year old Girls ($n = 15$)	4	7	2	1	1	0
5 – 6 year old Boys ($n = 13$)	2	5	5	1	0	0
5 – 6 year old Girls ($n = 15$)	1	4	4	3	2	1
7 – 8 year old Boys ($n = 14$)	2	4	6	1	1	0
7 – 8 Year old Girls ($n = 16$)	0	3	5	5	2	1

Summarize these data more completely and describe what you would conclude from the study.

6. Here's a hypothetical study in evolutionary psychology that reflects what a couple of published studies have actually found. Young males ranging in age from 13 to 26 years were asked to answer the following question: "If you could choose, what would be the best age for a girl to be your girlfriend." Participants answered in "Years Old." Here are the data:

Participant's Age	Participant's Answers
13 – 14 Years ($n = 9$)	13 16 14 16 15 15 14 16 13
15 – 16 Years ($n = 12$)	16 16 17 17 18 16 17 15 16 17 18 17
17 – 18 Years ($n = 11$)	17 18 16 17 16 17 18 17 16 18 17
19 – 20 Years ($n = 12$)	18 17 16 17 18 17 18 17 19 17 18 17
21 – 22 Years ($n = 12$)	18 19 18 18 19 18 18 17 20 19 19 18
23 – 24 Years ($n = 9$)	20 19 18 21 20 18 19 21 19
25 – 26 Years ($n = 9$)	21 21 19 21 22 18 20 18 19

You need to break out your phone calculator for this problem and perhaps peek ahead to Chapter 2 to completely analyze the data. If you were an evolutionary psychologist, how would you interpret the results of your analysis?

7. Recall that the concept of correlation meant that the values of one variable are related to the values of another variable. Here's three problems. You need to decide the <u>extent</u> to which the variables are correlated (if at all).

   a. *Consumer Reports* 2024 quality scores of six brands of electric vehicles (EV's) and their cost:

Brand	Rated Quality	Base Cost
A	86	$60 K
B	93	$80 K
C	77	$70 K
D	83	$120 K
E	65	$40 K
F	88	$60 K

   b. Percentage of students of different college levels who overindulge alcohol at Clearwater College.

Class Rank	%-age Drinking Too Much
Fresh.	26%
Soph.	17%
Jr.	13%
Senior	24%

   c. Preference of 15-year old girls ($n = 70$) and boys ($n = 35$) for riding horses or motorcycles.

	Horses	Motorcycles
Girls	55	15
Boys	10	25

8. Recall the hapless psychology major who got to conduct the experiment to determine whether hearing Lexi Putellas speak on the importance of increasing activity fees would change student attitudes about the proposal. Here are the data he collected.

	Experimental Condition ($n = 20$)	Control Condition ($n = 20$)
Voted "Yes"	14	9
Voted "No"	6	11

a. What would be the Null and Alternative Hypotheses that our student researcher is testing?
b. The appropriate inferential test to analyze these data is discussed in Chapter 16. Baring looking up the test and running the analysis, what does your intuition tell you about the likely result?

Chapter 2
Describing Data

What's In This Chapter?

An **experiment** is a research strategy that permits investigators to determine whether changes in an **independent variable** cause changes in a **dependent variable**. Causal conclusions are possible because all other variables are **held constant** or equalized across the values of the independent variable. In contrast, **correlational studies** simply examine existing relationships among variables as they are encountered in the everyday world. Because the operation of other variables is not controlled in correlational studies, definitive conclusions about causality are not possible.

Frequency distributions are tabular summaries of the data for either categorical or numerical variables. Frequency distributions for numerical data can be either ungrouped or grouped. An **ungrouped** frequency distribution lists all the individual values that the variable can assume and the number of times that each value occurred. A **grouped** frequency distribution collapses the values of the variable into ranges or **class intervals** and shows the number of cases falling into each interval. It is generally better to express frequencies in **relative** rather than **absolute** terms. **Comparative** frequency distributions simultaneously display frequencies of values of the variable for two or more samples. A variant of a frequency distribution is a graphical table called a **Stem and Leaf Table**.

Graphs are pictorial displays that summarize important features of data. **Frequency graphs** are pictorial representations of frequency distributions. Frequency graphs based on categorical frequency distributions are called **bar graphs** and show the frequencies of values with the heights of discrete bars. Frequency graphs based on numerical frequency distributions show frequencies either with bars that touch (**histogram**) or with points that are connected by lines (**frequency polygon**). Comparative frequency polygons show the distributions of values of a variable simultaneously for two or more samples on the same axes. **Function graphs** show the relationship between two variables, usually how a dependent variable changes in response to changes in the value of an independent variable.

Statistics are numerical characteristics of samples that describe the **central tendency, variability**, and **distribution symmetry** of the data comprising a sample. Central tendency refers to the middle or midpoint of an ordered numerical distribution of scores, and is quantified using the **mean, median**, and **mode**. The mean should always be given for sample data. In highly **skewed distributions**, the median is a better indicator of the bulk of the score values and should be given to supplement the mean. A graph that displays information about the median is called a **Box Plot**.

Variability refers to the extent to which the scores differ in value from some measure of central tendency. The primary measure of variability associated with the mean is the **standard deviation.** The measure of variability associated with the median is the **interquartile deviation**. The **range** and **lowest** and **highest scores** are sometimes used to indicate variability.

Distribution symmetry refers to the extent to which scores are symmetrically distributed around measures of central tendency. The **coefficient of skew** is a measure of distribution asymmetry.

Introduction

As noted in Chapter 1, the term "descriptive statistics" refers to a set of graphical and numerical procedures used to describe the important features of a set of data. The procedures described in this chapter are all univariate descriptive statistics. Bivariate description is covered later in Chapter 5. Our discussion of univariate descriptive statistics will proceed in the context of a hypothetical experiment conducted to determine how the size of a group influences the effectiveness with which the group can solve a problem. **Before proceeding, read the description of this experiment on the next page very carefully because it will be referred to throughout this chapter.**

The group problem solving study is an example of what scientists call an **experiment**. In an experiment, investigators systematically change the value of one variable and see how the change affects the value of another variable or variables. The variable whose values are being manipulated or changed is called the

Group Problem Solving Experiment

Naval researchers were interested in how group size influences the quality of problem-solving decision making. Each of 560 enlisted "volunteers" was randomly assigned to one of three group size conditions of an experiment: (a) 20 4-person groups, (b) 20 8-person groups or (c) 20 16-person groups. Each of the 60 groups was presented with a test problem called "Lifeboat" and given 40 minutes to try to reach a solution. All groups met in identical conference rooms and received standard instructions. The group discussions were videotaped for subsequent scoring and analysis. Two variables were measured for each group: (a) the total number of ideas generated that were relevant to the problem (*# of Ideas*) and (b) whether or not the group reached the expert-based "correct" solution to the problem (*Solution*: Coded as 0 = "No," 1 = "Partial," 2 = "Yes" for purposes of data entry into *JASP*). Below are the raw data for the three conditions of the study.

	4-Person Groups			8-Person Groups			16-Person Groups	
Group	# of Ideas	Solution	Group	# of Ideas	Solution	Group	# of Ideas	Solution
1	38	1	1	54	2	1	55	2
2	61	0	2	39	2	2	58	1
3	29	0	3	46	2	3	56	1
4	56	2	4	63	2	4	39	1
5	35	1	5	74	2	5	58	2
6	63	2	6	39	0	6	62	2
7	41	2	7	47	1	7	56	2
8	25	0	8	58	2	8	61	1
9	31	0	9	62	2	9	64	2
10	66	1	10	59	2	10	66	2
11	39	2	11	52	2	11	64	1
12	30	0	12	47	1	12	71	2
13	26	0	13	49	2	13	39	1
14	34	1	14	51	2	14	52	1
15	32	0	15	69	2	15	51	2
16	49	2	16	44	1	16	49	1
17	33	1	17	75	2	17	55	2
18	43	2	18	44	0	18	44	1
19	32	0	19	53	2	19	68	2
20	31	2	20	58	2	20	32	1

independent variable. The variable that potentially is influenced by the independent variable is called the **dependent variable**. The independent variable in this study was *Group Size* and the investigators conducted the experiment using three values for the independent variable— 4, 8, and 16 person groups. The three group sizes selected by the investigators define the three **conditions** of the experiment. Two dependent variables were measured: *# of Ideas*, and *Solution*. The investigators chose to measure two dependent variables because they believed that problem solving is a complex process, and no single measure would reflect the process adequately. **Notice that the cases for this experiment were not individual people, but groups of people**.

In addition to active manipulation of the independent variable, an experiment also requires that all other variables that might influence the dependent variable be **held constant**, i.e., **be held the same for each condition of the experiment**. If this condition of constancy of conditions is met, then any changes in the dependent variable that accompany changes in the independent variable must be due exclusively to the effects of the independent variable. In the problem solving example, the investigators held constant instructions to the groups, the testing environment, the time given to work on the problem, and so on. It is the requirement of constancy of other variables that allows researchers to conclude that the independent variable is the true cause of changes in the dependent variable.

This ability to isolate the causal effects of single variables is what distinguishes an experiment from other types of studies. This difference can be made clearer by referring back to the Clearwater College survey study described in Chapter 1. Recall that the SLC survey results revealed that freshmen were strongly in favor of raising the activity fee, while seniors were opposed to the proposal. Can we therefore conclude that increased education "caused" changes in attitudes about the proposal to raise fees? Unfortunately, not, because many variables change concomitantly with changes in class rank. Freshmen differ from Seniors in terms of age, social maturity, work demands, vocational attitudes, relationship involvement, and so on. Moreover, the composition of the two classes is different. The Freshman class contains many students who will never make it to the Senior year because of deficiencies in ability, motivation, personal organization, and the like. Any or all these variables could be causing the difference in attitudes between the two classes. Studies of this type that simply measure the relationship between variables without meeting the requirement of holding other variables constant are called **correlational studies**. Procedures for analyzing data from correlational studies are covered in Chapters 5-7.

If you are on your toes, you may have noticed that when we discussed the fact that the Naval investigators held all other variables the same for each condition of the experiment, we failed to mention one thing that could not be held constant. The volunteers who made up the groups were different from group to group and condition to condition. The failure to have the same people in all conditions of the experiment violates the requirement of constancy of conditions. It might be, for example, that several of the 8-person groups contained unusually enthusiastic individuals, while the 16-person groups were burdened with a disproportionate share of reluctant "volunteers." This is a real possibility and is an unavoidable consequence of doing research with heterogeneous creatures like human beings.

The researchers attempted to handle this potential problem by randomly assigning volunteers to condition. **Random assignment** to condition means that each volunteer had **an equal chance of being in any of the conditions of the experiment and to any group within the condition**. The hope is that random processes will tend to "equalize" the groups with respect to the average level of all person-associated variables that could influence the dependent variable. If the average level of enthusiasm of the participants is the same for each of the three conditions, then any consistent differences in the dependent variables between conditions could not be due to differences in enthusiasm.

Realize that random assignment to condition does not *guarantee* equivalence of the groups. Random processes *can* produce groups that are markedly different with respect to some variable that can influence the dependent variable(s). Indeed, as we will see, one of the most fundamental questions in inferential statistics is whether the difference among experimental groups is sufficiently large that it is unlikely to be due to chance differences in the composition of the samples.

One last point. It is important to understand the difference between random sampling, which was discussed in the last chapter, and random assignment to condition. The former is a technique for maximizing the possibility that a sample will be representative of the population from which it is drawn. The latter is a technique for maximizing the possibility that experimental conditions will be equivalent prior to the introduction of the independent variable. If the original pool of cases selected for the experiment is representative of some particular population, then the randomly assigned samples representing the conditions of the experiment also may be considered representative of that population. In our example of the influence of group size on problem solving efficiency, we don't know whether the naval "volunteers" selected for the experiment were representative of naval enlisted personnel in general, so we are uncertain to whom the results of the experiment might apply.

The rest of this chapter will be concerned with descriptively analyzing the data generated by the group problem solving experiment. We will begin with a discussion of a basic data-summary technique known as a frequency distribution which was described in Chapter 1. We will then discuss different types of graphs commonly used in descriptive statistics and then go on to consider a variety of statistics that are used to summarize different aspects of the data.

Frequency Distributions

Recall that a frequency distribution is a list of values of a variable along with the number of times that each value occurred. We have already encountered frequency distributions in the presentation of the SLC survey study data. As we saw there, frequency distributions may be constructed for both categorical and numerical variables.

Table 1 presents a frequency distribution for the categorical *Solution* variable from the present group decision-making experiment. This frequency distribution is somewhat more complicated than those given in the previous chapter because it displays the data for all three conditions of the experiment in a single table. This format facilitates comparison between the three conditions and is called a **comparative frequency distribution**.

This frequency distribution shows the number of cases in each sample that received a "No," a "Partial," and a "Yes"

Table 1		
Comparative Distribution of Values for the Solution Variable for All Conditions of the Experiment.			
Value	4-Person Groups	8-Person Groups	16-Person Groups
No	8	2	0
Partial	5	3	10
Yes	7	15	10

for the *Solution* variable. Distributions that report the actual number of cases having each value of the variable are sometimes called **absolute frequency distributions**.

A different, and better, way to construct a frequency distribution is to list the *proportion of cases* in each sample having each of the values of the variable. Such a distribution is called a **relative frequency distribution**. To translate absolute frequencies into proportions, you divide the frequency for each value by the total number of cases in the sample. In our comparative frequency distribution, there are three samples representing the three conditions of the experiment, and we would do the conversion separately for each sample. **Table 2** shows the comparative frequency distribution for the *Solution* variable with the data expressed as relative frequencies.

Table 2		
Proportion of Sample Having Different Values of the Solution Variable for All Conditions of the Experiment.			
Value	4-Person Groups	8-Person Groups	16-Person Groups
No	.40	.10	.00
Partial	.25	.15	.50
Yes	.35	.75	.50

There are two very good reasons for expressing frequencies in relative rather than absolute terms. First, if the sample sizes for the different conditions were not equal, the absolute frequencies would be difficult to compare. For example, imagine the 4-person condition had 20 groups and the 8-person condition 30 groups. If both conditions had 10 groups achieve solution, the absolute number 10 would mean something different for each condition. In the 4-person condition, the 10 would mean that one half of the sample reached solution, while in the 8-person condition, the 10 would mean that only one-third of the group achieved solution. By expressing the data directly as proportions, this mental translation is avoided.

The second and even more important reason for expressing the frequency data as a proportion of the sample size is that a proportion can be applied directly to the population. The sample value of .75 correct solutions for the 8-person groups may be taken directly as an estimate that 75% of all possible 8-person groups would reach solution.

Generally, simply plan on always using relative rather than absolute frequency when you construct either frequency distributions on the graphs based on such distributions (see below).

Before going on, look carefully at the proportions in **Table 2**. What conclusions would you come to regarding the effect of group size on the quality of problem solving?

We can also construct frequency distributions of numerical data. These are a bit trickier. Consider the *# of Ideas* data for the 4-person condition. The values of this variable range from a low of 25 ideas to a high of 66 ideas. We might construct our frequency distribution by listing all possible integer values from 25 up to 66 in one column and the absolute or relative frequency of each value in a second column. This is called an **ungrouped frequency distribution**, because all possible values of the variable are shown, rather than grouping the variable into ranges of values, as will be done below. The ungrouped frequency distribution of the *# of Ideas* variable is shown in **Table 3**.

Table 3					
Number of Cases Having Different Values of the #Ideas Variable for the 4-Personl Condition of the Experiment.					
Value	Freq.	Value	Freq.	Value	Freq.
25	1	39	1	53	0
26	1	40	0	54	0
27	0	41	1	55	0
28	0	42	0	56	1
29	1	43	1	57	0
30	1	44	0	58	0
31	2	45	0	59	0
32	2	46	0	60	0
33	1	47	0	61	1
34	1	48	0	62	0
35	1	49	1	63	1
36	0	50	0	64	0
37	0	51	0	65	0
38	1	52	0	66	1

The major problem with this ungrouped frequency distribution is that it is not particularly useful. There are 42 possible values between 25 and 66. Most had a 0 frequency; 16 values occurred once, and two values occurred twice. The purpose of constructing a frequency distribution is to see general trends in the data, i.e., which values occur most often, how "spread out" the values are, and so forth. General trends often are difficult to discern amid the detail of too many 0's and 1's.

To overcome this problem of too much detail, numerical frequency distributions often express the values of the variables in terms of *ranges of values* rather than individual values. These ranges of values are called **class intervals**.

Look at how the same *# of Ideas* data have been displayed in both absolute and relative frequency form in a grouped frequency distribution in **Table 4**. Inspection of this frequency distribution quickly reveals that most values fall between 30 and 39 ideas, with a gradual "trickling off" of values above 39. This type of frequency distribution is called a **grouped frequency distribution** because the values of the variable are grouped into class intervals.

Table 4

Proportion of Sample Having Different Values of the #Ideas Variable for the 4-Person Condition of the Experiment.

Class Interval	Absolute Frequency	Relative Frequency
20 – 24	0	.00
25 – 29	3	.15
30 – 34	7	.35
35 – 39	3	.15
40 – 44	2	.10
45 – 49	1	.05
50 – 54	0	.00
55 – 59	1	.05
60 – 64	2	.10
65 – 69	1	.05

Notice that the distribution in **Table 4** begins with a value of 20 ideas and increases by increments of 5 ideas. The starting point for a grouped frequency distribution is called the **lowest interval** and the size of the class interval is called the **interval width**. Also notice that the class intervals do not overlap, so that a case can appear in only one interval. The value chosen for the starting point is fairly arbitrary. If the data are in the form of integers (whole numbers), most people will select some multiple of 5 or 10 that is immediately below the lowest value in the data set. Even when the data assume decimal values, most people try to begin the distribution with an integer.

The choice of the class interval width is more complex. The purpose of grouping the data is to "smooth out" raggedness arising from random processes in order to see the general pattern of the distribution of scores, i.e., to estimate what the distribution might look like in the population. If you choose too narrow a class interval width, your distribution will not be "smoothed" enough to reveal the important features of the distribution. If the class interval width is made too broad, the resulting distribution is too "crude" to reveal subtle but potentially important features in the distribution.

One other factor enters into the choice of class interval width—sample size. The data for the *# of Ideas* variable includes 20 values. This is a relatively small sample, particularly if you are trying to estimate what the distribution would look like in the population. With small samples and variables that can take on a wide range of values, it is usually necessary to make the class intervals fairly wide to achieve an adequate degree of smoothing. As the sample gets larger, however, random processes start "canceling" each other out and general trends can be seen with narrower class intervals.

JASP produces a very clever variant of a frequency distribution table that also shows visually the distribution of scores in a graph-like fashion. It is called a **stem and leaf table** and is available under the Tables Option for the DESCRIPTIVES procedure (see Chapter 4). **Figure 1** is the stem and leaf table for the 8-Person condition of the experiment.

Figure 1

JASP Stem and Leaf Table of the Number of Ideas Data for the 8-Person Condition

8-Person

Stem	Leaf
3	99
4	446779
5	1234889
6	239
7	45

Note. The decimal point is 1 digit to the right of the |

The stem is the tens portion of the data and the leaf is a list of the units portion of each data value. The first value in the data is 39, the second is also 39, and the third is 44, and so on. From the table, it is obvious that most scores are in the 40's and 50's.

Graphing Data

Recall that a graph is a pictorial way of describing important features of a set of data. In this chapter we will distinguish two general types of graphs—frequency graphs and function graphs. In Chapter 5 we will introduce a type of graph called a scatterplot, which is used to show the correlational relationship between two variables.

Frequency Graphs

A **frequency graph is a pictorial display showing how often different values of a variable occurred**. Frequency graphs are basically pictures of frequency distributions. The first type of frequency graph that we will construct is called a **bar graph** or **bar diagram**. A bar graph is used when the variable being displayed is categorical in nature. In a bar graph the categories of the variable are listed on the X axis of the graph, while the Y axis is scaled for frequency, either absolute or relative. A bar is drawn vertically above each category, with the height of the bar indicating the frequency with which the category occurred in the sample. In a bar graph, the bars are separated from one another by a space. This is done to emphasize the categorical or discontinuous nature of the variable.

As you would expect, a relative frequency bar graph is preferred over an absolute frequency bar graph for exactly the same reasons that relative frequency distributions are preferred. **Figure 2** shows a bar graph generated by *JASP* for the *Solution* data for the 4-Person condition of the experiment. Note that *JASP* uses absolute rather than relative frequency.

Figure 2

JASP Bar Graph of the Solution Data for the 4-Person Condition.

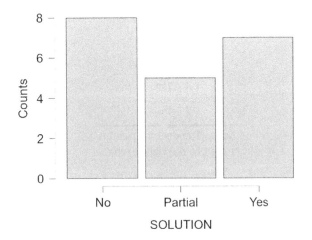

Just as one may construct a comparative frequency distribution, one may also construct a **comparative bar graph**. **Figure 3** shows one way that the *Solution* data for all three conditions could be displayed on the same set of axes. The advantage in doing so, again, is to facilitate comparison of the data for the three conditions of the experiment.

Figure 3

Comparative Bar Graph of Solution Data for All Conditions of the Experiment.

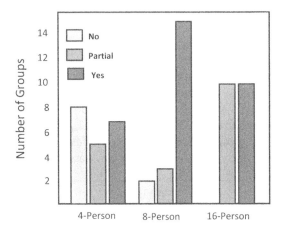

We have seen that bar graphs are used to visualize the distribution of categorical variables. Let's see now how you graphically display the frequency distribution of the values of a numerical variable, using the *# of Ideas* data for our example. There are two types of frequency graphs that can be constructed from a numerical frequency distribution. The first is called a **histogram**, the second a **frequency polygon**. Both graphs show the same information and, in many circumstances, may be used interchangeably. **Figure 4** shows the *JASP* generated histogram for the *# of Ideas* frequency distribution given in **Table 4**. The X axis displays the class intervals.

Figure 4

JASP Histogram of Solution Data for the 4-Person Condition of the Experiment.

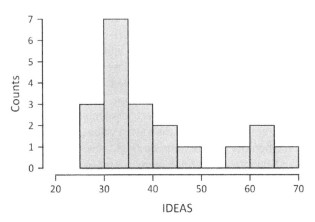

In a histogram, the number or percentage of cases associated with each class interval is also indicated by the height of a bar. But, unlike a bar graph, the **bars touch each other to indicate that the variable is numerical rather than categorical**.

When numerical data are displayed in a histogram (or frequency polygon), it is possible to talk about the "shape" or contour of the distribution. The histogram in **Figure 4** rises to its highest point in the interval of 30-34 and then gradually "trails off" at higher values. The distribution is **asymmetrical**, which means that the values are not distributed in the same pattern on both sides of the distribution's highest point. Technically, distributions that are asymmetrical are described as being **skewed**. The nature of the skewing is defined by the direction in which the frequencies taper off most gradually.

For the data in **Figure 4**, we say that distribution is **positively skewed**, because the extreme, infrequent scores are in the direction of positive (larger) values. Had the situation been reversed, with the bulk of the scores falling at higher values and a few scores trailing off in the lower end of the range of possible values, we would say that the distribution was **negatively skewed**. The presence of a cluster or cases in the 55-70 score range suggests a second "peak" in the distribution. Distributions that have two peaks are said to be **bimodal**.

The other type of frequency graph for numerical data—the frequency polygon—displays the relative frequency of each class interval by the height of a point rather than the height of a bar. The points are positioned at the middle of the class intervals and are connected by straight lines. The "contour" of the lines connecting the points represents the shape of the distribution. **Figure 5** shows a frequency polygon constructed using the same class intervals as **Figure 4**.

As can be seen from comparison of **Figures 4** and **5**, histograms and frequency polygons provide the same information concerning the shape of the distribution. The one advantage that the frequency polygon has over the histogram is that by using points and lines instead of bars to display frequencies, it is possible to display more than one distribution on the axes. This facilitates comparison of different distributions.

Function Graphs

You now know that frequency graphs are used to show how often different values of a variable occurred, i.e., to show the distribution of scores. A **function graph**, by contrast, shows **relationships between variables—how one variable changes in value as a function of changes in value of another variable**. In the context of experimental data, we want to show how a dependent variable changes as a function of an independent variable.

Consider the *# of Ideas* variable for the group problem-solving experiment. An obvious question is whether the number of ideas generated by a group is influenced by the size of the group. One way to graphically display this relationship is to plot the average value of the dependent variable for each of the conditions of the experiment. **Figure 6** shows how such a graph is constructed. The *X* axis is scaled for values of the independent variable and the *Y* axis for values of the dependent variable. The three points represent the mean values of the dependent variable for the conditions of the experiment. The points are connected with straight lines. The vertical brackets are the 90% confidence intervals for the sample means (See Chapter 10!).

Figure 6

JASP Function Graph Showing the Mean Number of Ideas as a Function of Group Size.

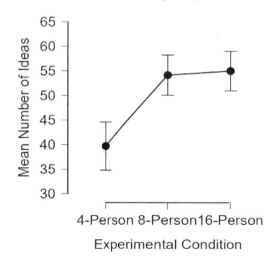

Figure 5

Frequency Polygon of the Number of Ideas Generated for the 4-Persons Condition of the Experiment.

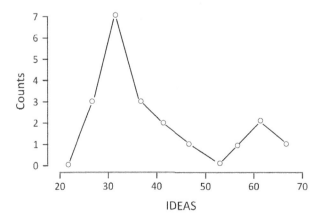

Inspection of this graph reveals that as group size increases from four to eight persons, there is a substantial increase in the number of ideas generated by the group. An additional increase in group size seems to have little additional effect on the number of ideas generated. The use of points connected by lines is a convention that emphasizes that the independent variable displayed on the X axis is numeric in character and is ordered from low to high. In the example, the experimenters chose to use group sizes of 4, 8, and 16 persons—three points along a numeric dimension. They might also have included a condition involving 6-person groups. The line connecting the mean for the 4-person condition and the 8-person condition can be used to estimate what the mean would be for that condition.

Descriptive Statistics – Measures of Central Tendency

Figure 7 shows a table of descriptive statistics produced by the *JASP* DESCRIPTIVES procedure for the *# of Ideas* variable for each of the three conditions of the experiment. In this section we are going to learn about these and a few other statistics.

Figure 7

JASP Descriptive Statistics Table for the Ideas Variable for the Three Experimental Conditions

Descriptive Statistics

	IDEAS		
	4-Person	8-Person	16-Person
Valid	20	20	20
Median	34.500	52.500	56.000
Mean	39.700	54.150	55.000
Std. Deviation	12.645	10.559	10.352
Coefficient of variation	0.319	0.195	0.188
IQR	13.500	13.000	12.000
Skewness	1.011	0.550	-0.648
Minimum	25.000	39.000	32.000
Maximum	66.000	75.000	71.000

The Mean

The term **central tendency** is used to refer to **the middle of a distribution of scores**. There are several statistics that are used to indicate the middle of a distribution; *JASP* routinely computes the most widely used of these statistics. You are already familiar with one of these statistics called the mean. In this section we will discuss the mean in greater detail and introduce two other measures of central tendency—the mode and the median.

The **mean** for a sample is formally defined by an algebraic equation. This is a common practice in statistics because equations are compact and unambiguous. The formula for the mean is:

$$\bar{X} = \frac{\sum X}{n}$$

where: $\sum X$ = sum of the *X* scores and
n = number of scores.

This equation illustrates five important conventions in statistics:

1. Variables are symbolized or represented by capital letters, usually chosen from the end of the alphabet, e.g., *X*, *Y*, and *Z*. When you see the symbol for a variable in an equation, it stands for all the values of that variable in a particular set of data. The particular values of a variable can be indicated by subscripts, e.g., X_1, X_2, and X_3 are the values of the *X* variable for the first three cases of the data set.

2. The Greek capital letter sigma (Σ) is read "the sum of." It means to add up all of the values that follow it. The expression $\sum X$ means to add up all of the values that the variable *X* assumes, e.g., $X_1 + X_2 + X_3$, and so on. Sometimes it is necessary to use parentheses to indicate the order in which the mathematical operations should be carried out. The expressions $\sum(X - 5)$ means subtract 5 from each of the values of the *X* variable and then sum up the remainders. $\sum(X - 5)^2$ means subtract 5 from each of the values of the *X* variable, square the remainder, and then sum up the resulting squares.

3. The symbol "*n*" always stands for the size of the sample. If there are multiple samples, the size of each is indicated by subscripting the *n*, e.g., n_1 and n_2.

4. The symbol $\bar{X}$ stands for the mean of the variable *X*. The mean of any variable is indicated by placing a bar over the letter standing for the variable, e.g., the mean of the *Y* variable would be indicated $\bar{Y}$.

5. Symbols for variables, constants, and statistics are always italicized in print.

The equation is brief and to the point. If you wanted to say the same thing in words, you would need to write, "The sample mean is equal to the sum of the scores divided by the number of scores."

The sample means for the *# of Ideas* variable for the three conditions of the experiment are given in **Figure 7** in the third row of the *JASP* Descriptive Statistics table. The values of the sample means (and many other statistics) are displayed in *JASP* to three decimals. As you will recall from Chapter 1, we might want to round these values off to just one more decimal of accuracy than was present in the original data. In this case, that would be one decimal point of accuracy. Thus, the mean for the 8-Person condition is given by *JASP* as 54.150 ideas, so if you were writing a summary of the results, you could report the value as 54.2 ideas. The mean of the 16-Person condition is 55.000 ideas, so it would be rounded to 55.0 ideas.

The mean has many characteristics to recommend it as a measure of central tendency. First, the mean is the algebraic "balance point" for a distribution of scores. For any set of scores, if you subtract the mean from every score in the set, the sum of all the positive discrepancies will equal the sum of all the negative discrepancies. Said another way, the algebraic sum of the discrepancies will equal zero. Said the easiest way,

$$\sum(X - \bar{X}) = 0$$

To satisfy yourself that this assertion is true, consider the following set of numbers: 5, 10, 4, 9, 6, 11. The value of the mean for this set of numbers is 7.50. The discrepancies are (-2.5, 2.5, -3.5, 1.5, -1.5, 3.5), which, when added up, equal 0. In this sense, the mean really is in the "middle" of a distribution.

Second, because the mean "takes into account" all the scores in a distribution, it is the most **stable** measure of central tendency. For the moment we will define the **sampling stability** of a statistic to be the **tendency for the statistic to be similar in value in successive random samples**. Another name for sampling stability is **precision**. The idea of sampling stability of a statistic is fairly complicated and has many important ramifications in inferential statistics. It will be discussed more fully in Chapter 9.

For our immediate purposes, we can get a feel for what is meant by sampling stability by doing kind of a "mental" sampling experiment. Imagine a jar containing 1000 1" plastic disks. Each disk has an integer number printed on it. The numbers range from 1 to 11, with relatively few 1's and 11's and many more 5's, 6's, and 7's. The mean of this population of 1000 numbers is exactly 6. Suppose that you drew 10 random samples of 15 disks each and computed $\bar{X}$ for each sample, replacing discs as you drew them. Obviously, you would expect that the value of $\bar{X}$ would vary from sample to sample, because two sample means could be the same only if the sum of the numbers was exactly the same for the two samples. This sounds (and is) fairly unlikely for any two random samples. The question is, how much will the sample means vary in value? The answer, surprisingly, is not very much at all. One value might be 6.47, the next 5.42, then 6.20, and so on. It is very unlikely that any sample mean would be as low as 4.00 or as large as 8.00. This reflects the fact that mean is quite stable in repeated random sampling.

Our little thought experiment also points up the advantage of working with stable statistics. Suppose that you didn't know what the mean of the population was and wanted to estimate it from the value of a *single* sample mean. Because the sample mean is stable, it is unlikely that any single sample value will depart terribly far from the population mean and, thus, any particular sample is likely to yield a fairly accurate estimate of the population value.

A third advantage of the sample mean is that it is related to an enormous body of other descriptive and inferential statistics. In substantial part, this is because the mean is an algebraically defined statistic, and algebraic equations are easy to manipulate and combine. The mean, for example, shows up in the definition of a wide variety of descriptive statistics that measure variability, distribution asymmetry, and correlation. It is also integral to most of the major inferential procedures.

The bottom line to this litany of praise for the mean is simple. When describing a set of **numerical data**, always use the mean as a measure of central tendency. There is one circumstance described below when the mean should be supplemented with another measure of central tendency, but you should never abandon it completely in favor of its less meritorious competitors.

Just as we can define a mean for a sample, we can also define the mean for a population. The mean for a population is symbolized μ (lower case Greek "**mu**"), and is defined in a parallel equation:

$$\mu = \frac{\sum X}{N}$$

The Greek symbol μ indicates that you are dealing with a population parameter and the capital *N* is the size of the population. The use of a Greek alphabet symbol for a population parameter is a widely used convention in statistics, although the convention has some exceptions.

A Discursion Into Summation Rules

Summation notation is one of those awkward topics in statistics texts because it's usually introduced before it's really needed. We've already used summation notation beginning with the definition of the sample mean, but that's the simplest application. In later chapters we'll use summation notation to define new statistics and explain important statistical concepts.

Here's a more complex example of a statistical summation for computing what is called the **sums of squares (SS)** introduced in Chapter 12:

$$SS_A = n_a \sum_{a=1}^{A} (\bar{X}_a - \bar{\bar{X}})^2$$

where: $\bar{X}_a$ = the mean of each group,
$\bar{\bar{X}}$ = the mean of all groups,
A = the number of groups, and
n_a = the number of observations in each group.

The" bounded" summation sign means to sum from $a = 1$ to A for the group means, i.e., groups $\bar{X}_1, \bar{X}_2, \bar{X}_3$ and so on up to A groups.

The expression says to subtract the grand mean from each of the group means, squaring the difference, summing up the squared differences for all A groups, and then multiplying the total by the number of observations in each group.

Sometimes double summations are needed as in:

$$SS_{WG} = n_G \sum_{g=1}^{G} \sum_{1=1}^{n} (X_{g,i} - \bar{X}_g)^2$$

Double summations are read from right to left. This expression says to subtract the group mean from every score in the group, squaring the difference, and then sum the squared differences for the group. Do this for each group, sum the groups up, and then multiply the total by the number of scores in each group. You'll get tired of seeing double summation signs.

The Median

A second measure of central tendency is the **median (Md)**. The median is a statistic that is not defined in algebraic terms and, as such, it is not possible to write a compact equation for its definition. A verbal definition is required. The median is **the point (value) in a numerical distribution selected in such a way that half of the scores are numerically larger than that point, and half are numerically smaller**. The median is thus a "balance point" as well. But it is a balance point in terms of the *number* of scores rather than the *values* of the scores.

If it seems to you that the verbal definition of the median offered above is a little "soft," you are correct. Where exactly is that "point" that splits the distribution into half? It turns out that there is some variation in how the median actually can be computed. The simplest way uses the following algorithm. The scores are first sorted numerically from low to high. If there is an odd number of scores, the median is the value of the middle score, i.e., if there were 19 scores, the median would be the value of the 10th score. If there is an even number of scores, the median is taken as the mean of the two middle scores, i.e., if there were 18 scores, the median would be the mean of the values of the 9th and 10th scores. **Table 5** illustrates the computation of the median when there are an even and an odd number of scores in a sample.

Table 5
Location of the Median in an Ordered Distribution of Scores When There is an Even and an Odd Number of Scores.

Even Number of Scores	Odd Number of Scores
32	32
32	32
35	35
37	37
45	45 ---- Md = 45
----- Md = 50.5	56
56	59
59	69
69	87
87	
92	

In our little family of measures of central tendency, the median is a poor relation of the mean. It is somewhat less stable than the mean in random sampling—particularly for small samples—and, because it is not defined in algebraic terms, has many fewer related descriptive and inferential applications. It does, however, have one advantage over the mean. Because it is not influenced by the exact numerical value of every score in a distribution, the median is not influenced by the presence of a small number of scores having extreme values.

This insensitivity to extreme scores can make the median a more informative measure of central tendency than the mean in distributions that are highly skewed. We can get a feel for this by considering the *# of Ideas* data for the 4-person condition. As we saw in **Figure 4**, the distribution of the scores was positively skewed. The mean for this condition is 39.7 ideas, while the *Md* is 34.5 ideas

(**Figure 7**). The appreciably larger value of the sample mean reflects the fact that this statistic is being "pulled up" in value by the relatively small number of more extreme positive values. The sample median falls closer to the interval (30-34) that contains the largest number of cases. In this sense, the median seems more representative of the bulk of the scores.

The median is sufficiently important that it has its own type of graph called a **box plot**. **Figure 8** shows a box plot for all three conditions of the experiment produced by the Customizable Plots option under the *JASP* DESCRIPTIVES procedure.

Figure 8

JASP Box Plot of the Number of Ideas Variable for the Three Conditions of the Experiment.

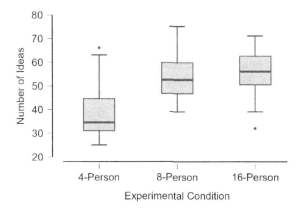

Box plots convey a great deal of information about the data. Relative to the *Y* Axis, the box height indicates the Interquartile Range (*IQR*, see below) for each condition, the solid line inside the box is the value of the median, the brackets are a kind of minimum and maximum, where minimum = $(Q_1 - 1.5 * IQR)$ and maximum = $(Q_3 + 1.5 * IQR)$. The dots are "outliers." The box for the 4-Person condition clearly shows the positive skewing of the data. All of this gets further explained below!!

We said above that the mean should not be replaced by the median; instead, the value of both statistics should be given. Each tells you something slightly different about the data, and comparison of the values of the mean and median is also informative. The fact that the mean is appreciably larger than the median is a tipoff that you are dealing with a positively skewed distribution.

The medians for the *# of Ideas* variable for the three conditions of the experiment are given in the second row of the *JASP* output table. Rounded to one decimal of accuracy, the values are 34.5, 52.5, and 56.0 ideas. For the 8- and 16-Person conditions, the median and the mean are relatively close in value, indicating that both distributions are more symmetrical in shape. *JASP* uses a somewhat more complex procedure for computing the Median, but the idea is the same.

The Mode

A third measure of central tendency is the **mode (*Mo*)**. Like the median, the mode is not an algebraically defined statistic, so a verbal definition must be given. Actually, two definitions are needed, because the term mode can be applied either to data cast in a grouped frequency distribution or to "raw" numerical data (scores not collected into a frequency distribution). When the data are summarized in the form of a grouped frequency distribution, we can define the mode to be **the class interval for numerical data or the category for categorical data having the largest frequency**. We sometimes say the "modal interval" or "modal category." When dealing with raw numerical data, the mode is **that value that occurs most often**.

The problem with both definitions is that when the number of scores is small or the range of possible values is large, the mode often is not a unique value. Several intervals, categories, or scores may occur with equal frequencies greater than one, or in the case of raw data, every score value may occur only once. In either case, the mode is not very informative as a measure of central tendency. For the *# of Ideas* data, the modes were 31 and 32 ideas for the 4-Person condition. There were four modes for the 8-Person condition, and five modes for the 16-Person condition. With respect to the histogram display of the *# of Ideas* data in **Figure 3**, we would say that the modal interval is 30-34.

The mode is a very unstable statistic in random sampling and is associated with no other descriptive or inferential procedures. As such, it should never be used alone as a measure of central tendency with numerical data. At best, if the mode can be meaningfully computed from the data, it could be used to supplement the mean and median. **For categorical data, however, the mode is often used to designate the most frequently occurring category.**

Writing Narrative Summaries

Before going on to introduce other statistics, it's useful to have a brief lesson about how to talk about statistical information. Most students have no difficulty in understanding the definition of statistics like the mean, median, and mode, nor do they have problems in knowing when to apply them. Where they have problems is in knowing how to report the results of statistical analyses. Consider the following "summary" of the information in the histogram in **Figure 4** and the computed values of the mean and median from the *JASP* statistics table:

> "The data for the *# of Ideas* variable for the 4-person condition is displayed in Figure 4. The mean was 39.7 ideas and the median 34.5 ideas."

This type of summary is typical of a lot of first efforts by students. Its major problem is that it just lists information in sentence format; it does not help the reader make any sense of the information. (Naturally, when you read the sentence, you immediately picked up on the fact that the author misused the word "data" by pairing it with the singular verb "is"!) Contrast that effort with this attempt at summarizing the information:

> "The data for the *# of Ideas* variable for the 4-person condition are displayed in histogram form in Figure 4. The distribution is positively skewed, with the modal interval at 30 - 34 ideas, and half of the groups producing fewer than 35 ideas. The skewing of the data is reflected in the fact that the mean was 39.7 ideas, and the median was 34.5 ideas."

Notice that this version explicitly draws the reader's attention to important features of the data. Your professor will be much more impressed with you if you write summaries of this type. Moreover, by forcing yourself to make interpretive statements about the statistics and graphs, you will come to understand the procedures much better.

Descriptive Statistics- Measures of Variability

Some important information is missing from the analysis summary above. Nowhere is there mention of how much **variability** there is in the scores. When we talk about the variability of scores, we are referring to **the extent to which the values of a variable are similar in magnitude**. You can see the need for statistics to quantify variability by comparing Sample A with Sample B in **Table 6**.

Both samples have the same mean (44.0) and approximately the same median (44.0 and 41.0), but they are quite different in the variability of the scores. The nine scores in Sample A are all similar in value (from a low of 40 to a high of 47), so we would say there is little variability. By contrast, the nine scores in Sample B vary markedly in value, from a low of 26 to a high of 71. There is much greater variability in the values of these scores.

Table 6

Two Samples that Have the Same Mean, but Very Different Amounts of Variability in the Scores.

Sample A	Sample B
40 42 45	32 41 42
45 44 43	71 38 29
44 46 47	54 63 26
$\overline{X}$ = 44.0	$\overline{X}$ = 44.0
Md = 44.0	Md = 41.0

The extent of variability in a set of scores must be considered in interpreting measures of central tendency. When variability is low, a measure of central tendency is likely to be "representative of" (close in value) to most of the scores in the sample. If variability is high, however, your measure of central tendency may be quite unrepresentative of most of the scores. This is illustrated in the Sample B data, where the value of the mean (44.0) is similar to only two of the scores (41, 42), and several of the scores (26, 29, 71) are highly discrepant in value.

JASP computes by default only two measures of variability—the **standard deviation** and the **minimum and maximum**. In the sections below, we will consider these, plus three others—the **range**, the **interquartile range** and the **coefficient of variation.**

The Range and Minimum and Maximum

The simplest of all measures of variability is called the **range (R)**. It is simply **the value of the highest score minus the value of the lowest score**. The range is easy to compute, but it provides only a very crude measure of variability—so crude, in fact, as to be almost worthless. Although *JASP* will compute *R* as an option, I recommend that instead of using the range for descriptive purposes, it's better to simply specify the lowest and highest values present in the data set. This communicates more information than *R*. In *JASP* the lowest and highest values are called the minimum and maximum.

For the *# of Ideas* variable, the minimum and maximum are given in the last two rows of the output table shown in **Figure 7**. For the 4-Person condition, Minimum = 25 ideas, and Maximum = 66 ideas, for the 8-Person condition, the values are 39 and 75 ideas, respectively, and for the 16-Person condition, 32 and 71 ideas.

The Standard Deviation

The **standard deviation (S)** is the most commonly used measure of variability. Unfortunately, there is a persistent disagreement between the social and natural sciences in how the standard deviation is defined. We'll compare these two slightly different definitions.

In the social sciences, the standard deviation is usually defined as follows:

$$S = \sqrt{\frac{\sum(X - \bar{X})^2}{n}}$$

where: X = values of the variable,
$\bar{X}$ = mean of the scores, and
n = number of scores.

As you can tell, the standard deviation quantifies variability of scores by finding the average of the squared deviations (differences) of the scores about the sample mean and then taking the square root of this average. The summation portion of the formula is referred to for convenience as the **sums of squares**. The deviations must be squared because (you will recall) the sum of the simple deviations about the mean always equal zero. Unfortunately, squaring the deviations results in squared units of measurement, e.g., # of Ideas². For descriptive purposes, you want to maintain consistency in the units of measurement between the mean and the standard deviation. That is why the square root is taken.

While most students find it easy to interpret the mean, the meaning of the standard deviation is less intuitively obvious. **It's probably best to think of it as the average or mean distance that the scores lie from the mean.** The value of S for Sample A in **Table 6** is 2.0 and for Sample B, 14.7. In words, we might say that in Sample A, the scores are on average about 2 units away from the mean. In Sample B, however, the average distance of the scores from the mean is almost 15 units.

Just as measures of central tendency have their advantages and disadvantages, so also do their associated measures of variability. The standard deviation, like the mean, is a stable statistic across repeated random sampling and sees many applications in other descriptive and inferential statistics. **For numerical data, you should always use the mean and standard deviation together to describe your data.**

The standard deviation shares one disadvantage with the mean. Recall that in highly skewed distributions, the value of the mean may not be representative of the bulk of the scores. A similar situation exists with the standard deviation. **In skewed distributions, the standard deviation will overstate the amount of variability present in *most* of the scores.**

The standard deviation for a population is symbolized σ (lower case Greek "sigma"), and is defined as follows:

$$\sigma = \sqrt{\frac{\sum(X - \mu)^2}{N}}$$

where: X = values of the variable in population,
μ = mean of the population, and
N = number of scores in population.

The Estimated Population Standard Deviation

In most disciplines in the natural sciences and in *JASP* a closely related statistic called the **estimated population standard deviation** ($\hat{s}$) is used for descriptive purposes. This statistic is defined as:

$$\hat{s} = \sqrt{\frac{\sum(X - \bar{X})^2}{n - 1}}$$

where: X = values of the variable,
$\bar{X}$ = mean of the scores, and
n = number of scores.

The $\hat{s}$ is very similar to S, except that the denominator contains $n - 1$ rather than n. It is technically known as the "relatively" unbiased estimate of the population standard deviation and is very important in inferential statistics. As noted above, many social scientists maintained the distinction between S as the actual standard deviation of the sample and $\hat{s}$ as an estimate of the population standard deviation. Those days are mostly over. **Given the fact that most data analysis packages routinely use $\hat{s}$ for descriptive purposes, we'll do so as well.**

You may wonder why a different formula is needed to estimate the population standard deviation. This has to do with the concept of **bias** in estimating population parameters, which will be discussed in Chapter 9. For the moment, simply focus on the descriptive function of either form of the standard deviation. The Descriptive Statistics table produced by *JASP* displays the value of $\hat{s}$ in the fourth row, labeled "Std. Deviation." For the # of Ideas variable for the three conditions it is 12.6, 10.6, and 10.4 ideas.

The Mean Absolute Deviation

For completeness I'm going to include a variant of the standard deviation called the **mean absolute deviation (MAD)**.

MAD is defined as:

$$MAD = \frac{\sum |X - \bar{X}|}{n}$$

where: $|X - \bar{X}|$ = the absolute (unsigned) deviation of each score from the mean.

By taking the absolute deviations of the scores about the mean, all deviations are positive and don't sum to zero. In words, MAD is literally the average deviation of the scores about the mean. Why, this sounds perfect! It's intuitively obvious what it means, so why isn't it used in place of the standard deviation? Actually, some people do recommend that it replace the standard deviation for descriptive purposes. Sadly, however, it's a dead-end statistic. By losing the algebraic quality of the data, MAD can't be related to many other important inferential statistics. JASP computes MAD as an optional statistic, so feel free to use it for descriptive purposes, by don't use it to replace the standard deviation. Generally, MAD will be smaller than S or $\hat{s}$ because squaring large deviations inflates the size of the latter.

The Interquartile Range

There are two other measures of variability yet to be considered. The interquartile range (**IQR**) is appropriate where the median is being given as a measure of central tendency. The definition of *IQR* is based upon the concept of **percentiles**, so before we can offer such a definition, it's necessary to explain the idea of percentiles. The term "percentile" derives from the word "percent," which means "in a 100." Percentiles are **points that divide an ordered distribution of scores into 100 parts**. Recall that the median is a point in a distribution of scores ordered from low to high that is selected so that one-half of the scores lie above the point and one-half lie below. Thus, another name for the median is the **50th percentile**, because **50% of the scores fall at or below the median**. The idea of percentiles is quite general, so one can talk about any percentile. For example, the 25th percentile is that point or score at or below which lie 25% of the scores in an ordered distribution. The 75th percentile is that point or score that separates an ordered distribution into 75% lower values and 25% higher values.

Here's a terminological distinction that confuses everyone, but it is important to make it. A "percentile" refers to the *value of a score* at or below which lies some specified percentage of the sample or population. When you say, "What's the 50th percentile?" you are asking for a score value. There is a related concept called **percentile rank**, which is **the percentage of cases at or below some specified score**. You would say, "My score of 68 has a percentile rank of 72."

With the idea of percentiles in mind, we can then offer a fairly compact definition of *IQR*:

$$IQR = C_{75} - C_{25}$$

where: C_{75} = 75th percentile and
C_{25} = 25th percentile.

The *IQR* statistic quantifies how much variability is present in a distribution of scores by specifying the range of values needed to encompass the middle 50% of the scores. If scores vary a great deal in value, then C_{75} and C_{25} will be far apart, and their difference ($C_{75} - C_{25}$) will be a large number. If the scores are all close together in value, then the difference between C_{75} and C_{25} will be small in value.

JASP computes *IQR* as an option. For the three group size conditions, the values of *IQR* are shown in the sixth row of the Descriptive Statistics table (**Figure 7**) as 13.0, 13.5, and 12.0 ideas. Because *IQR* encompasses the middle 50% of the scores, rather than the average distance of the scores from the middle of the distribution, *IQR* will always be larger than *S* and $\hat{s}$.

With this information in mind, go back and review the information contained in the box plot in **Figure 8**.

The Coefficient of Variation

JASP also computes another measure of variability called the coefficient of variation (**COV**) which is defined as follows:

$$COV = \frac{\hat{s}_X}{\bar{X}}$$

The coefficient of variation is simply the sample standard deviation divided by the sample mean. In words, it is the amount of variation relative to the size of the scores. It is given in line 5 of the Descriptive Statistics table. For the 4-Person Groups, *COV* for the ideas variable is:

$$COV = \frac{12.645 \; Ideas}{39.700 \; Ideas} = .319$$

Because the units of the variables cancel out in the division, *COV* is a "unitless" statistic. It is often expressed as a percentage as in, "the variation was 31.9% of the mean." *COV* is most useful when the means differ greatly in size in such a way as to inflate the size of the standard deviations.

The *COV* is not given as a default statistic in the DESCRIPTIVES procedure but does appear by default in several statistical procedures developed in later chapters.

Descriptive Statistics- A Measure of Skew

Recall that skew refers to the asymmetry of a distribution. The extent of the skewing of the distribution is generally apparent from inspection of a histogram display of the data, but it is useful to have a statistic that precisely quantifies the direction and magnitude of skew.

There are two ways to compute a **coefficient of skew** (*Sk*). *JASP* uses a more complex formula that we don't need to consider. A much simpler formula that is more intuitively obvious is

$$Sk = \frac{3(\bar{X} - Md)}{S}$$

Inspection of this formula makes it clear that in a perfectly symmetrical distribution *Sk* will equal 0. This is because the mean and median will be equal, thus making the numerator equal to 0. When the distribution is positively skewed (scores tail off to the right), the mean will be numerically larger than the median, and *Sk* will have some positive value. In a negatively skewed distribution (scores tail off to the left), the mean will be smaller than the median, and *Sk* will be a negative number. The magnitude of the skew is expressed relative to the size of the standard deviation. Values greater than ± 1.00 indicate a very asymmetrical distribution.

Using the above formula, the value of *Sk* for the *# of Ideas* variable for each of the experimental conditions is as follows: For the 4-Person condition, *Sk* = +1.23, indicating a very substantial degree of positive skewing in the data. This is consistent with the impression gained from the study of the histogram display of these data in **Figure 4**. For the 8-Person condition, *Sk* = +.47, indicating a modest positive skew, and for the 16-Person condition, *Sk* = -.28, indicating only a slight negative skewing of the distribution of scores. The corresponding values computed by *JASP* using a more complex formula are shown in line 7 of the Descriptive Statistics table. They are in general agreement with those computed using the simpler formula.

Having completed our discussion of measures of variability and skew, it might be useful to see how such information would be incorporated into a narrative summary of a complete descriptive analysis. By way of example, let's summarize the data for the *# of Ideas* variable for the 16-Person condition of the experiment.

The distribution of values for the *# of Ideas* variable for the 16-Person groups is shown in a frequency polygon in Figure X (not actually shown). The distribution is sharply peaked at 57.5 ideas and shows a slight negative skewing of values (*Sk* = - .28). The groups' scores ranged from a low of 32 ideas to a high of 71 ideas. The majority (15/20 or 75%) of the 16-Person groups generated 50 or more ideas, and only a minority of groups (3/20 or 15%) produced 40 or fewer ideas. The mean number of ideas for the group was 55.0 with the standard deviation of 10.4 ideas.

"Just as food eaten without appetite is a tedious nourishment, so does study without zeal damage the memory by not assimilating what it absorbs."

Leonardo da Vinci

Chapter 2, Describing Data
Chapter Exercises

Some questions are purely factual and have only one correct answer. Others require greater thought on your part and may have more than one possible answer.

1. Fill in the missing term(s) that makes the statement true.

   a. In an experiment, the _____ is manipulated to see its effect on the _____.
   b. In the "Group Problem Solving Experiment," the cases are _____.
   c. In experiments involving people, _____ is/are "held constant" by _____.
   d. Study of the "Group Problem Solving Experiment" suggests that _____ is the optimal group size.
   e. When constructing a frequency distribution, _____ is better than _____.
   f. The purpose of grouping in a frequency distribution is to _____.
   g. A bar graph shows _____ for _____ variables.
   h. Three types of frequency graphs for numerical data are _____, _____, and _____.
   i. Asymmetry in a frequency distribution is technically called _____.
   j. A distribution that has two "humps" is technically called _____.
   k. _____ plot the effects of an independent variable on the _____.
   l. The best measure of central tendency is the _____.
   m. This is because _____.
   n. Lower case _____ is the symbol for the sample size.
   o. $\bar{Y}$ is the symbol for _____.
   p. The Greek capital letter _____ means "sum."
   q. $\sum_{i=1}^{n}(X_i - \bar{X})^2$ says to _____.
   r. The sample mean can be deceptive in _____ distributions because _____.
   s. The sum of the score discrepancies about the sample mean equals _____.
   t. When a statistic yields similar values across successive random samples, it shows _____, AKA _____.
   u. The median is technically the _____ of a distribution, also called the _____.
   v. For the scores 5,7,8,11,15,16,19, 21 the median is _____.
   w. A _____ is a graph specifically used with the median.
   x. Dispersion of scores about some measure of _____ is referred to as _____.
   y. The most useless measure of variability is the _____.
   z. $\hat{S}$ will always be _____ than S because _____.
   aa. MAD stands for _____ and is computed as _____.
   bb. MAD seems like a perfect measure of variability except that _____.
   cc. IQR stands for _____ and is the difference between _____ and _____.
   dd. If $\bar{X}$ and Md are about equal, the distribution is _____.
   ee. COV is defined as _____ divided by _____.
   ff. For a distribution of scores, Sk = -3.22. We know that the distribution is _____.

2. Forty male and 60 female college students are queried concerning the amount of vigorous physical exercise they get each week. The students responded using the following three categories: 1 = "Little or None", 2 = "A fair amount", and 3 = "A great deal." Here are the data for the 100 students:

   Males: 1 3 3 2 3 2 3 1 3 2 2 3 2 1 2 3 2 2 1 3 3 3
            2 3 3 2 1 1 3 3 3 3 2 3 1 3 2 1 3 2

   Females: 1 2 3 2 1 2 1 3 2 3 1 1 1 2 2 2 3 3 2 3 1
            2 1 3 2 1 3 1 1 1 2 3 1 2 1 2 2 3 1 3 1 1 3
            3 1 3 2 3 1 2 1 1 1 2 3 1 2 2 1 2

   a. Construct a frequency distribution that best summarizes these data. Include an informative title for your distribution. (At this point the highly motivated student will skip ahead to Chapter 4, download JASP and use it to solve problems like this.)

1

b. Now, **neatly** draw the appropriate graph for summarizing the data in this problem. Include an appropriate title for your graph.

3. Participants in a memory experiment read a 2-page story describing Ben Franklin's use of a kite to study the electrical properties of lightning. Then, participants were randomly assigned to one of four retention intervals: 1, 3, 6, or 12 days ($n = 20$ for each interval). At the appropriate interval, participants were brought back into the laboratory where they read 10 statements about the story. Five of the statements were actually true of the story and five were plausibly true, but not actually contained in the story. The dependent variable was the mean number of plausible statements accepted as actually being in the story after each of the retention intervals. Here are the data for each condition:

Descriptive Statistics				
	Plausible Accepted			
	1 Day	3 Days	6 Days	12 Days
Valid	20	20	20	20
Mean	1.112	1.452	2.601	3.846
Std. Deviation	.212	.676	.455	1.011
Minimum	0	1	1	1
Maximum	2	4	5	5

a. Neatly draw the graph that best shows these data. Include an appropriate title for your graph.
b. What inferential question would you pose for this experiment?

4. The Dean of a large Junior College is reviewing the average scores for students from three sections of Introductory Statistics, each taught by a different instructor. All three sections were scheduled at the same time and 35 students were randomly assigned to each section at the start of the semester. Here are the descriptive statistics produced by *JASP* for the three sections. Speculate a bit concerning possible interpretations of the findings.

Descriptive Statistics			
	Prof A	Prof B	Prof C
Valid	25	30	33
Median	72.000	77.500	88.000
Mean	73.840	77.800	86.091
Std. Deviation	15.212	7.757	9.458
IQR	28.000	11.500	10.000
Minimum	49.000	60.000	62.000
Maximum	100.000	92.000	100.000

5. Here are two histograms produced by *JASP*. Describe each distribution as precisely as possible.

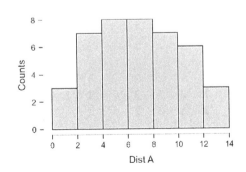
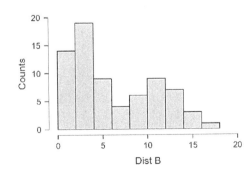

6. Here's another graph produced by *JASP* for 78 cases. Explain what type of graph it is. Then improve the graph by providing an appropriate (made-up) caption and doing whatever else seems appropriate.

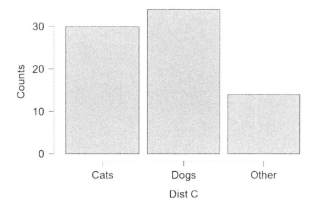

7. Below are hypothetical data that need to be summarized in a Stem and Leaf Table. Complete the table and then comment on what the table shows about the distribution of the scores.

Hypothetical Data Set

114	144	86	161	72
83	135	162	104	131
98	123	175	140	71
118	136	154	102	85
121	96	145	139	140
94	122	138	124	136
150	142	138	88	

Hypothetical Data

Stem	Leaf

Note. The decimal point is 1 digit to the right of the |

8. This Box Plot was produced for another hypothetical data set. Indicate the numeric values of the variable for each of the components of the plot, i.e., show that you know what all of the components mean. (You might need a straightedge for accuracy.)

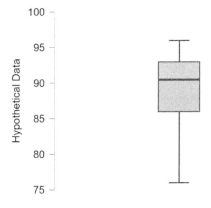

Chapter 3
Standard Scores & Normal Distributions

What's In This Chapter?

A **standard score** is a transformation of a raw score into a form that shows the relative position of the score in a distribution. The use of standard scores gives meaning to raw scores and permits comparison of scores obtained under different conditions.

A **percentile rank** is one type of standard score that shows the proportion or percentage of cases in a distribution that fall at or below a particular raw score. Unfortunately, percentile ranks are **nonlinear transformations** of scores that lose information about the relative distances between scores. As such, percentile ranks cannot meaningfully be subjected to the mathematical operations required for most statistical procedures.

Another type of standard score is the **z score**, which expresses the number of standard deviations that a score lies above or below the mean of a distribution. Because the z score is a linear transformation, z scores may be analyzed statistically. A z score may be transformed into a variety of other standard scores, such a **T scores** and **IQ scores**.

Normal distribution curves are a family of symmetrical, bell-shaped curves whose exact shape is a function of the values of μ and σ. Although no variable in the world is exactly normally distributed, some variables are sufficiently well approximated by a normal curve that it is possible to use proportionate areas under the curve to determine the frequency with which different values of the variable will occur. Rather than having separate normal distributions for all possible values of μ and σ, such approximations are all made using the **unit normal distribution curve**, in which the variable of interest is expressed in terms of z scores.

Some **sampling distributions** also are either exactly or approximately normally distributed, and statisticians use the unit normal curve to determine the likelihood of obtaining different sample outcomes.

Introduction

Two related topics are covered in this chapter. The first deals with a technique for expressing scores in a way that shows their relative standing in a distribution. Such standard scores often are much more meaningful than are the "raw scores" themselves. We can illustrate this with reference to one type of standard score with which most students are familiar. The Scholastic Assessment Test (SAT) is taken by many students as part of the process of applying to college. Examinees receive a score for the verbal test (SAT-V) and the quantitative test (SAT-Q). People with high scores are eagerly recruited by colleges and universities.

Suppose that you score a "720" on the SAT-V. You are very smart and should be pleased with yourself. But what precisely does the 720 mean? It is not the case that you answered 720 questions correctly, because the test does not have that many items. Instead, the 720 is a type of standard score developed by the Educational Testing Service. The entire distribution of raw scores is transformed (mathematically altered) so that the mean score for the distribution is made equal to 500 and the standard deviation of scores equal to 100. Your standardized score of 720 is 220 points or 2.2 standard deviations above the mean. As you will see below, very few people achieve a score as high as yours. The discussion below focuses primarily on a particular standard score called a z score, which is extremely important in statistics and from which most other standard scores can be derived.

The second topic covered in this chapter is a consideration of a class of mathematical functions collectively referred to as **normal distributions**. These mathematical entities are important in statistics for two reasons. First, the frequency distributions of some variables in the real world are rather well approximated by normal distributions (for example, IQ scores). When that is the case, it is relatively simple to estimate how often different values of a variable will occur. Second, many instances of a class of distributions known as **sampling distributions** are either exactly or approximately normally

Table 1		
Scores of 15 Students on the First Two Tests in a Statistics Course.		
Student	Test 1	Test 2
J.M.	42	63
E.S.	25	69
B.S.	49	54
A.C.	41	60
M.D.	44	52
J.H.	38	54
H.C.	42	53
B.B.	46	69
T.B.	44	48
S.C.	40	58
R.H.	50	65
S.S.	47	70
B.A.	39	60
P.C.	48	64
A.S.	42	57
$\bar{X}$ =	42.5	59.7
S_X =	5.9	6.6

distributed. Knowledge of the mathematical properties of such distributions is essential for all of inferential statistics and will be covered in Chapters 9 - 16.

Standard Scores

The data in **Table 1** are the scores attained by 15 students on the first two tests in an introductory statistics course. The first test had 50 points possible, while the second test had 75 total points. Expressed in this raw score manner, the scores have no clear-cut meaning. J.M.'s score of 42 on the first test is just 42 correct answers out of 50 possible. Professors frequently transform such scores into percentages of the total possible points, but still, the percentage itself doesn't really mean much. J.M.'s percentage score of 84% (42/50) could represent sloppy performance on a weak exam or diligent performance on a difficult exam. The traditional guidelines of 90-100 = A, 80-89 = B, etc., are themselves entirely arbitrary.

So, doesn't anything mean anything? Well, theoretically, it is possible to construct an exam so carefully that a percentage score can be interpreted directly as percentage mastery of the material. Unfortunately, this almost never happens in teacher-constructed classroom exams. In fact, the only way that any meaning can be attached to these scores is by expressing each score's standing in the distribution, i.e., by expressing how well each person did compared to all the others who took the test. **This is what standard scores do—express relative position in a distribution**.

Percentile Ranks

You are already familiar with one type of standard score called percentile ranks. Recall from Chapter 2 that the percentile rank of a score is the proportion or percentage of scores in the distribution that are equal to or lower in value than the score. The percentile rank of J.M.'s score is 53 because 8 out of the 15 students (including J.M.) had scores equal to or less than his score of 42 (8/15 = .53 = 53). J.M. should be fairly displeased with his performance because it means that 47% of the students did better than him on the test.

There is a problem with percentile ranks, however. Converting scores to percentile ranks is a **nonlinear transformation**. A nonlinear transformation is **any mathematical transformation that changes the relative distances between scores**. We can see the nature of this problem more clearly by considering the test scores and their percentile ranks of the three students who fared most poorly on the first test. These are shown in **Table 2**.

Table 2		
Test 1 Scores and Percentile Ranks of Scores for Three Students with the Lowest Scores on the Test.		
Student	Test Score	Percentile Rank
B.A.	39 Correct	3/15 = 20.0
J.H.	38 Correct	2/15 = 13.3
E.S.	25 Correct	1/15 = 6.7

Notice that the distance between the three students' test scores is not the same. B.A. got one more answer correct than J.H., but J.H. got 13 more correct than the unfortunate E.S. But when the scores are transformed into percentile ranks, this distance information is lost. Because of the nature of the percentile rank transformation, the distance between the three percentile ranks is necessarily equal, and therefore uninformative about the real differences in magnitude between the students' scores. Because of this loss of information, it makes no sense to perform most arithmetic operations on percentile ranks. If you were to average the percentile ranks, for example, the mean rank would be 13.3, which would suggest that J.H.'s score lay exactly at the mean of the little distribution of three scores. Clearly this is not the case. The mean of the three test scores is 34.0 correct, and J.H.'s score of 38 correct puts him/her appreciably above the mean of this group.

z Scores

Because of this limitation with percentile ranks, another type of standard score is required—one that shows relative position *and* maintains the relative distance relationships among the scores themselves. This need is met with a type of standard score called a **z score**. The z score transformation of a score from a sample of scores is defined as

$$z = \frac{X - \bar{X}}{S}$$

where: X = value of X of interest,
$\bar{X}$ = mean of the distribution, and
S = standard deviation of distribution.

In words, a z score is **the number of standard deviations that a score lies above or below the mean of a distribution**.

You can also compute z scores on population data. Here, the definition for the z score would be:

$$z = \frac{X - \mu}{\sigma}$$

where: X = value of X of interest,
μ = mean of the population, and
σ = standard deviation of population.

A z score is a **linear transformation** because the mathematical operations of addition, subtraction, multiplication, and division maintain the relative distances between scores. What is the effect of this transformation? The numerator of the formula computes the "distance" that a score lies away from the mean of the distribution. This distance is expressed in the measurement units of the variable. This distance is then divided by the standard deviation of the distribution, which is also expressed in the same measurement units. The units in the numerator are "canceled out" by the units in the denominator, leaving a "unitless" number. By way of example, let's look at the z score value of J.M.'s score of 42 on the first test:

$$z = \frac{42 - \bar{X}}{S} = \frac{42 - 42.5}{5.9} = \frac{-.5}{5.9} = -.08$$

How shall J.M.'s score of -.08 be interpreted? First, note that the score is negative. This indicates that J.M.'s score is *below* the group mean. How far below the mean? The value of -.08 indicates that J.M.'s score lies a little less than a tenth of a standard deviation below the mean of the group.

Table 3

Test Scores and z Score Transformations for 15 Students on the First Test in a Statistics Course.

Student	Test 1	Test 1 z Score
J.M.	42	- 0.080
E.S.	25	- 2.985
B.S.	49	+ 1.116
A.C.	41	- 0.251
M.D.	44	+ 0.262
J.H.	38	- 0.763
H.C.	42	- 0.080
B.B.	46	+ 0.604
T.B.	44	+ 0.262
S.C.	40	- 0.422
R.H.	50	+ 1.287
S.S.	47	+ 0.775
B.A.	39	- 0.592
P.C.	48	+ 0.945
A.S.	42	- 0.080
$\bar{X}$ =	42.5	0.00
S_X =	5.9	1.00

A major advantage of z scores is the possibility of comparing scores from different distributions. Back to J.M. On the first test he received a 42, while he got a 63 on the second test. On which test did he do better? The fact that he received a higher raw score on the second test is not very informative, because the second test had many more possible points than the first (75 vs. 50). How about the fact that he got 84% of the questions correct on both tests? Does this mean the performance was the same? Perhaps not, if the two tests were not equally difficult. A possible solution to this problem of comparing scores on two dissimilar tests is to compute the z score for each test and compare the z scores. J.M.'s z score on the second test is computed as

$$z = \frac{63 - \bar{X}}{S} = \frac{63 - 59.7}{6.6} = \frac{+3.3}{6.6} = +.50$$

Good news! J.M. appears to be at least headed in the right direction. His first test score was a fraction of a standard deviation below the group mean, but on the second test his score is one-half of a standard deviation *above* the group mean for the that test. J.M. improved his relative standing on the second test.

Thus, z scores are useful for "standardizing" performance so that comparisons can be made between dissimilar tests. What else do you need to know about z scores? Actually, there are two other important properties we need to mention. First, when all the scores in a distribution are converted to z scores and the mean and standard deviation of the z scores is computed, an interesting thing happens. **The mean of the distribution of z scores will always be 0 and the standard deviation always 1.00**. There is an algebraic proof of this assertion, but we can see it just as clearly by example. **Table 3** contains both the original test scores and their z score transformations for the first test given in **Table 1**.

A second property of z scores concerns the *shape* of the distribution of transformed scores. This is important, so remember it. **Although the *z* score conversion changes the mean and standard deviation of the distribution, it has no effect on the shape of the distribution**. If the distribution of raw scores is skewed, the distribution of z scores will be identically skewed. This is because the z score transformation is a linear transformation. In effect, all you are doing is changing the units on the X axis—the relative position of the scores in the distribution remain unchanged.

Other Standard Scores

One thing about z scores that troubles people is that z scores can be negative. Many people do not like or understand negative numbers. Consequently, z scores themselves are often transformed to create other types of standard scores. Recall the SAT example with which we began this chapter. There we said that this standard score had a mean of 500 and a standard deviation of 100. It turns out that SAT standard scores are derived in principle from z scores. Here is how these two types of standard scores are related:

SAT = (z score) (100) + 500

Notice that if a z score is equal to 0 (indicating that the score is exactly at the mean of the distribution), the equation makes the SAT type score equal to 500. When the z score equals -1.00 (one standard deviation below the mean), the SAT score equals 400. Thus, SAT scores have a mean of 500 and a standard deviation of 100.

It turns out that you can have any type of standard score that you desire. The general relationship to z scores takes the following form:

Score = (z score) (Desired σ) + Desired μ

We can see how this formula works by making up a new standard score for use in J.M.'s statistics class. We'll call it the "*S*" score. We decree that *S* scores shall have a mean of 1000 and a standard deviation of 200. From knowledge of the above general relationship between z scores and other types of standard scores, we can write a specific equation for converting z scores to *S* scores:

S Score = (z score) (200) + 1000

Substituting in J.M.'s z score for the first test yields

S Score = (z score) (200) + 1000

        = (-.08) (200) +1000

        = 984

We haven't changed J.M.'s relatively poor standing among his peers in statistics, but perhaps receiving a large number like 984 will make him feel better about his performance. Certainly, his parents would be impressed to learn that he scored 984 on his first test in statistics!

Some other popular standard scores that relate to z scores are ***T* scores** ($\mu = 50$, $\sigma = 10$), which are used to express the results of many different types of educational and psychological tests and **IQ scores** ($\mu = 100$, $\sigma = 15$ or 16), which are scores used to compare the general intelligence of people.

Normal Distributions

The term "normal distribution" refers to a family of smooth, bell-shaped curves defined by the following general equation:

$$Y = \frac{1}{\sigma\sqrt{2\pi}} e^{-(X-\mu)^2/2\sigma^2}$$

where: Y = value of the function,
            X = any point on the X axis,
            μ = mean of the curve,
            σ = standard deviation of the curve,
            π = "pi," approximately 3.1416, and
            e = base *e* log, approximately 2.7183.

The term "family" is used because the equation defines an infinitely large number of possible normal distributions that all share certain similarities. Let's see why this is so. Notice that the equation contains two components, π and e, that are constants for all curves. If we replace the symbols for these constants with their approximate values and simplify the equation a bit, we can rewrite it as:

$$Y = \frac{1}{2.507\sigma} 2.718^{-(X-\mu)^2/2\sigma^2}$$

Notice that Y is now a function of just three things—the value of X, μ, and σ. The μ and σ are called the **parameters of the equation**. They can be of any value, which is why there are an infinite number of possible normal distribution curves. If we hold the scaling of the X axis constant, changing the value of μ "slides" the curve either up or down the X axis. Changing the value of σ makes the curve either "skinnier" or "fatter." Depending on how the X axis is scaled relative to the value of μ and σ, a normal distribution curve can appear quite "bell shaped." Indeed, most normal distribution curves that you will see in this and other statistics texts are presented in this manner. But normal distributions can also be tall and thin or squat and broad. **Figure 1** shows three perfectly legal normal distribution curves.

Figure 1

Three Normal Distribution Curves Having Different Means and Standard Deviations

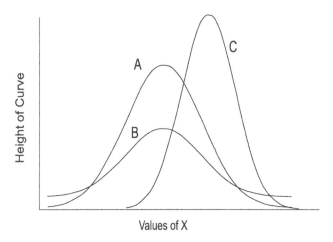

Curve A is the "classic" bell shaped normal distribution curve, having μ = 100 and σ = 15. Curve B is an "Uncle Bud" type of normal distribution, short and squat (μ = 100, σ = 20). Curve C is an "Aunt Gertrude" sort of a curve, tall and skinny, with μ = 110 and σ = 10. All normal distribution curves are at their highest point when X = μ, and fall off symmetrically about μ as X takes on values other than μ. As X gets further and further away from μ in either direction, the curve gets closer and closer to the X axis, but never quite touches it. Technically, we say that the curve is **asymptotic** to the X axis.

Normal Distributions and Real-World Variables

Regardless of their size and shape, it is important to understand that normal distribution curves are mathematical entities that only *approximate* certain real-world distributions. This is, in part, because no variable is measured in a truly continuous fashion and no variable has values that extend to infinity on either side of the mean. But just because the fit isn't perfect doesn't prevent us from making use of mathematical entities. For example, there are no *exactly* spherical objects in the universe, but that doesn't stop us from using the equation for the volume of a sphere to find the volume of such imperfect spheres as ball bearings and planets. The fit is close enough for our purposes. Similarly, statisticians can use knowledge of the mathematical properties of normal distributions to answer questions about the frequency or likelihood of real-world events. Let's see how.

One variable of great interest in the real world is "intelligence," mostly because measures of intelligence are predictive of a large number of important aspects of human behavior, such as how well you are likely to do in school, your level of job success, the effectiveness with which you raise your children, and so on. There are theoretical reasons for believing that intelligence in large, heterogeneous populations should be a normally distributed variable, and virtually all tests of intelligence (IQ tests) have been constructed to yield approximately normally distributed scores.

Figure 2 shows a distribution of IQ scores for a large sample of U.S. adults. The distribution is expressed in the form of a relative frequency histogram having class intervals of two IQ points. The mean of the distribution is approximately 100 and the standard deviation approximately 15. Fitted to the histogram is a normal distribution curve in which μ = 100 and σ = 15. Notice that the smooth normal distribution is a pretty good approximation to the empirically obtained histogram.

Suppose that you were asked to estimate the percentage of adults in the population who have IQ's above

Figure 2

Histogram of Sample IQ Scores Fitted with a Normal Distribution Curve Having μ = 100 and σ = 15.

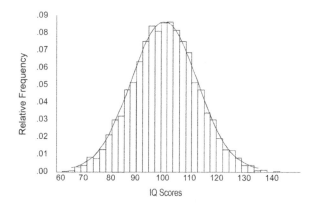

115 **using just the histogram display of the data**. Because the Y axis is already scaled for relative frequency, the obvious way to do this would be to "add up" the relative frequencies for those histogram bars for intervals above 115—in a sense, stack up the bars and measure how high the stack is. The sum of the relative frequencies would then be your estimate of the proportion of the population that have IQ's above 115. For the histogram of IQ scores, this turns out to be .148 or 14.8%.

That approach would work fine, but there's another, easier way to do it. Look at **Figure 3**, where the histogram has been redrawn so that the bars for intervals above 115 are shaded.

Instead of paying attention to the heights of the bars, refocus your attention on the *area* within the bars. There is an important relationship between the area represented by the shaded bars and the total area of the entire histogram. Because the bars are equally wide, the proportion of the total area of the histogram taken up by the shaded bars is exactly the same as the sum of the relative frequencies represented by the heights of the shaded bars. In other words, **one can use the proportion of the area within the histogram to get at relative frequency**.

Let's carry the logic one step further. Notice that because the imposed normal distribution curve is such a good fit to the histogram, the same proportionality holds between the area of the normal curve above 115 and the total area under the normal curve. About 15.9% of the total area of this imposed normal curve lies above a score of 115, which corresponds pretty closely to the value of 14.8% obtained from the histogram itself.

Here is the bottom line. If a variable is approximately normally distributed and you know the μ and σ of the

Figure 3

Histogram of Sample IQ Scores with Bars above 115 Shaded to Indicate Relative Area.

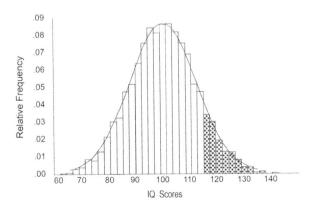

normal distribution curve that approximates the distribution, it is possible to use proportionate area under the normal curve to estimate the frequency with which different values of the variable will occur. You will learn how to solve normal curve area problems like this in the next section.

The Unit Normal Curve

The problem with this approach—at least historically— is that it's not an easy matter mathematically to find areas under a normal distribution curve. It's only been fairly recently that one could punch in the values for μ and σ for a normal curve on a statistics calculator and ask for the proportionate area under the curve beyond some value or between two values.

The historical solution to the problem was to work out the area relationships for one *particular* normal distribution and put these relationships in a table. Thereafter, all normal curve problems could be solved by first translating the scores of the particular problem into the units of the standardized normal distribution and then consulting the table to find proportional areas. The normal distribution which was selected as the standard is the one in which $\mu = 0$ and $\sigma = 1.00$. This is the normal distribution for z scores and is referred to as the **unit normal distribution curve**. **Table A** in the Appendix at the end of the book gives a listing of areas for the unit normal curve for different values of z. **A portion of that table is reproduced in Table 4 on the next page**.

Note that the table is organized into sets of three columns. The first column contains z scores, beginning with 0 and increasing in value by increments of .01. For each value of z, the second column contains the proportion of the area under the unit normal curve between μ and that value of z. The third column contains the proportion of area beyond the z score. Look at the first line in the Table where z = 0. Because a z score of 0 means that you are at the mean of the normal distribution, there is no area of the curve between the μ and the z score, and the Table gives .0000. Conversely, beyond the z score lies one half of the total area under the distribution and the tabled value is .5000. Look down in the right-hand set of columns to where z = .25. Here, the second column indicates that .0987 (9.87%) of the area lies between the μ and z, and the third column indicates that .4013 (40.13%) lies beyond that z score.

The normal distribution curve is symmetrical, therefore areas below μ, indicated by negative z scores, are the same as those given for positive values of z. Thus, if z = -.15, you enter **Table A** for z = .15 and find that .0596 of the area of the curve lies between μ and -.15 and .4404 lies beyond a z of -.15.

Table 4

*Portion of **Table A** in Appendix Showing Areas Under Unit Normal Distribution Curve for Values of z Between 0.00 and 0.29.*

z score	Area µ to z	Area > z	z score	Area µ to z	Area > z
0.00	.0000	.5000	0.15	.0596	.4404
0.01	.0040	.4960	0.16	.0636	.4364
0.02	.0080	.4920	0.17	.0675	.4325
0.03	.0120	.4880	0.18	.0714	.4286
0.04	.0160	.4840	0.19	.0753	.4247
0.05	.0199	.4801	0.20	.0793	.4207
0.06	.0239	.4761	0.21	.0832	.4168
0.07	.0279	.4721	0.22	.0871	.4129
0.08	.0319	.4681	0.23	.0910	.4090
0.09	.0359	.4641	0.24	.0948	.4052
0.10	.0398	.4602	0.25	.0987	.4013
0.11	.0438	.4562	0.26	.1026	.3974
0.12	.0478	.4522	0.27	.1064	.3936
0.13	.0517	.4483	0.28	.1103	.3897
0.14	.0557	.4443	0.29	.1141	.3859

Let's now use **Table A** to solve the problem that was posed initially, i.e., what percentage of the population have IQ's above 115? We will assume that IQ's are approximately normally distributed, with $\mu = 100$ and $\sigma = 15$. To use the unit normal curve table, we need to express the score of interest (115) as a z score relative to its own distribution μ and σ.

The population z score formula yields the following:

$$z = \frac{115 - \mu}{\sigma} = \frac{115 - 100}{15} = \frac{+15}{15} = +1.00$$

With respect to the unit normal curve, a score of 115 corresponds to a z score of +1.00. Entering +1.00 into **Table A** in the Appendix, we find in the third column that .1587 of the area of the unit normal curve lies beyond this value of z. We can translate this proportion into a percentage by multiplying it by 100, yielding 15.9% as the answer to the question. About 16% of the adult population have IQ's above 115.

Table A can be used to answer other types of questions as well. What percentage of the population have IQ's *below* 110? The z score equivalent of 110 is given as

$$z = \frac{110 - \mu}{\sigma} = \frac{110 - 100}{15} = \frac{+10}{15} = +.67$$

Here you need to think a little bit. The question asks for the percentage of the population below the score in question. The area of concern then is *everything* below the z score of +.67. This includes the entire half of the area that lies below μ (.5000), plus the area between μ and z = +.67. The latter, according to **Table A**, is .2486. Adding the .500 to the .2486 yields .7486 as the proportion of the area under the unit normal curve lying below a z score of +.67. Translating the proportion into a percentage gives us our final answer of 74.9% as the percentage of adults with IQ's below 110.

If you are on your toes, you will realize that the value of 74.9 is the *percentile rank* of an IQ score of 110. When a distribution of scores is approximately normally distributed, the unit normal curve may be used to estimate the percentile ranks of scores. **This only works, however, for distributions that are adequately approximated by a normal distribution.** Remember your SAT score of 720? It just so happens that SAT scores are approximately normally distributed, so we can use this approach to learn just how good your verbal SAT score of 720 really is. Your score of 720 translates to a z score of +2.20, i.e., (720 - 500)/100 = +220/100 = +2.20. According to **Table A**, .4861 of the area under the unit normal curve lies between μ and a z score of +2.20. When this is added to the .5000 area below μ, the total proportion of the area below the z score is .9861 or 98.6%. Your percentile rank is about 99. As we said, you did very well indeed on the test.

Not only can you find the percentile rank of scores using Table A, but also you can turn the problem around and find percentiles as well. For example, what is the 90th percentile among IQ scores? Here, you are looking for the IQ score at or below which lies 90% of the IQ scores. **Figure 4** illustrates this situation. The lower 90% of the area is indicated, but the value of the z score that defines it is unknown.

Figure 4

Normal Curve Approximation of Distribution of IQ Scores with Lower 90% of Area Under the Curve Shaded.

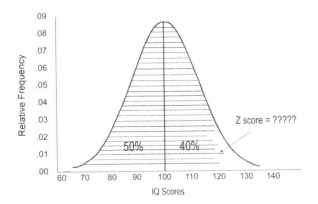

Turning the problem around this way forces you to use **Table A** in a reverse direction. What we are starting with is a percentage value and we want to get an IQ score that corresponds to that percentage. We are going to have to go in the direction: percentage → proportion → z score → IQ score, instead of the usual order: IQ score → z score → proportion → percentage. The 90% translates to a proportion of area of .90. That means we are going to find the z score below which lies .90 of the area under the curve and then translate that z score back into an IQ score. Because 50% of the area under the curve lies below μ, we need to find the z score corresponding to an additional 40% of the area above μ. To do this, run your finger down the *second* column of the table until you find the value closest to .400, then look at the z score that corresponds to that value. The closest value is .3997, which corresponds to a z score of +1.28. Now we need to translate a z score of +1.28 back into an IQ score. There is a formula for doing so, but it is better to think it through in the following way. A z score of +1.28 means that the IQ score is 1.28 standard deviations above the mean of the distribution. The σ for the distribution is 15 IQ points. If the score is 1.28 standard deviations above the mean, then it is 1.28 x 15 IQ points = 19 IQ points above the mean. The mean is equal to 100, so 19 IQ points above the mean equals 119. The 90th percentile is 119 IQ points.

Here is one last type of problem. What percentage of the population have IQs between 80 and 90? Statistics instructors like problems of this type because they discriminate between students who "sort of" understand the use of the normal curve from those who really do. Problems like this are best solved with the aid of a picture, as in **Figure 5**.

Note that what is called for here is the proportionate area that is shaded under the normal curve, i.e., the area between 80 and 90. This is what you do to find the area. First, find the area between an IQ score of 80 and the μ of 100 (indicated by the arrow labeled "A"). This area is obtained for z score equal to (80 - 100)/15 = -20/15 = -1.33, which Table A reports to be .4082. Then find the area between an IQ score of 90 and the μ of 100 (indicated by the arrow labeled "B"). This area is obtained for z score equal to (90 - 100)/15 = -10/15 = -.67, which **Table A** indicates to be .2486. Finally, subtract the **area** for the (90 - μ) range from the area for the (80 - μ) range. What's left is the area between 80 and 90, i.e., .4082 - .2486 = .1596 or about 16% .of the population.

There is one limitation associated with using the normal distribution curve to estimate the frequency of events. **Because the curve is perfectly continuous, the likelihood of any *specific* value of X occurring is infinitely small.** This is an odd situation, but it arises because we are using a continuous mathematical function to approximate real-world variables that are discrete in value. Fortunately, there is a solution to this problem. Suppose, for example, that you wanted to estimate the proportion of adults having IQ's of exactly 100. One way to think of this score is that the value of 100 represents a range from 99.5 to 100.5, i.e., the real limits of the number 100 (Chapter 1). Once we have a range of values (even a very skinny range), we can

Figure 5

Normal Curve Approximation of Distribution of IQ Scores with Area Between 80 and 90 Shaded.

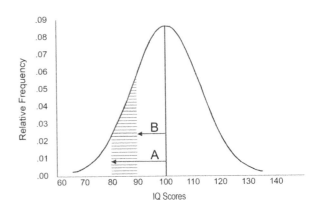

Figure 6

Normal Curve Approximation of Distribution of IQ Scores with Area Corresponding to an IQ Score of 100 Shaded.

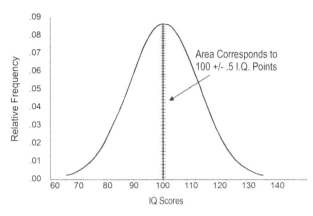

use the normal curve procedure to compute relative frequency. This situation is pictured in **Figure 6**, where the slender shaded area represents a range between 99.5 and 100.5.

With this insight in mind, the solution is straightforward. You first find the z score equivalent for a value of 99.5. This is computed as (99.5 - 100)/15 = -.5/15 = -.03. Consulting **Table A** for a z score of -.03 yields an area of .0120 between the z score and μ. Because the curve is symmetrical, the area for the distance between μ and the upper real limit of 100.5 will be the same, i.e., .0120. The sum of these two areas is .0240, suggesting that about 2.4% of the population will have IQ's of exactly 100.

Normal Distributions as Sampling Distributions

All of the discussion thus far concerning the normal distribution curve has focused on using this mathematical function to approximate real-world distributions. There is a second general application of the normal curve that we will touch upon briefly in this chapter and consider much more fully in Chapter 9. This second application of the normal distribution curve lies in the realm of inferential statistics. We can get a feel for this application by considering a simple problem in inferential decision making, which is called "The (Perhaps) Crooked Coin." **Read the description of this problem at the bottom of the page before proceeding.**

The results of the test described are a bit ambiguous. The 12 out of 16 heads seems to support the idea that the coin is biased toward heads, but you reason that maybe the preponderance of heads just occurred by chance. Faced with the evidence, we have a classical problem in statistical decision making. You have to choose between two alternatives, namely (a) the coin is really "fair" (unbiased) and the preponderance of "heads" was just the result of chance, or (b) the coin is biased to come up "heads." How do you make your decision?

Here is the logic that statisticians use. Let's begin with the conservative assumption that the shady-looking guy is trying to hustle you and that the coin really is fair. This is our "working hypothesis." What we then do is to inquire how likely it is that one would get 12 heads out of 16 flips if the coin were fair. If "chance" would produce 12 heads in 16 flips fairly often in a fair coin, then the results of our little test are not very compelling evidence for the idea that the coin is biased towards heads. If, on the other hand, 12 heads out of 16 flips is very unlikely to occur with a fair coin, then we could accept the second alternative that the coin is biased to come up heads.

The key to this decision-making process is knowing what chance (random factors) would produce if the coin really were fair. One way to find this out would be to take a fair quarter and repeat the 16-flip test a very large number of times and keep track of how often different numbers of heads appeared. One thousand such tests would probably suffice to get an accurate idea of how likely 12 heads is out of 16 flips. This sounds like a lot of work, and it is. Assuming it would take 60 seconds to complete and record the results of each 16-flip test, you would spend 16.7 very busy hours to complete the series of 1,000 tests.

Statisticians have better things to do with their time. Instead, they use mathematics to figure out what would happen across an infinitely large series of such tests. In this case, a mathematical proposition called the "**binomial theorem**" supplies the answer. **Table 5** gives the

---

The (Perhaps) Crooked Coin

One day while lounging in a beach-front tavern, you are approached by a shady looking guy who offers to sell you a "trick quarter" for only $10. The coin's trick, it seems, is to come up "heads" most of the time when flipped. One could imagine a variety of uses for such a coin, so you decide to test it out. If the coin really is biased to come up heads, you will buy it. Your test consists of flipping the quarter 16 times and counting the number of times that it comes up heads. At the end of 16 flips, "heads" has come up 12 times. Should you buy the coin or not?

proportion of times that 0, 1, 2, 3, etc., heads would appear in 16 flips across an infinite number of 16-flip trials. The distribution in the table is known as a **sampling distribution**, because it shows the frequency with which different sample outcomes (values of some statistic) occur. The complete technical definition of a sampling distribution is pretty complex, so we'll leave it for Chapter 9. For the moment, we will define a sampling distribution to be **a relative frequency distribution of the sample outcomes that would occur across an infinite number of random samples**. Note that across an infinitely large number of 16-flip tests of a fair coin, 12 heads would occur .0278 or 2.78% of the time.

This information can also be displayed graphically in the form of a relative frequency histogram, which is shown in **Figure 7**. Did you notice anything familiar about the general shape of the histogram? If you had exclaimed, "Why, it looks like a normal distribution!" you would be correct. The sampling distribution displayed in **Figure 7** can be well approximated by a normal distribution having μ = 8.0 and σ = 2.0. As was the case for our example with IQ scores, statisticians can also apply normal curve approximations to theoretical sampling distributions to answer questions concerning the likelihood of obtaining different sample results. Indeed, as we'll see in later chapters, some sampling distributions are *exactly* normally distributed—no approximation needed.

But back to the (perhaps) crooked coin. Should you conclude that the coin is fair or is it biased towards heads? We can solve the problem in two ways. The first involves summing relative frequencies in **Table 5**. For reasons to be considered later, we need to consider not only the likelihood of 12 heads, but also the likelihood of even more improbable events, i.e., 13, 14, 15, and 16 heads. To find out the likelihood of getting 12 or more heads, we can sum the relative frequencies for 12 through 16 heads, i.e., .0278

Table 5

Proportionate Occurrence of Different Numbers of Heads Across an Infinitely Large Series of 16 Flips of a Fair Coin.

# Heads	Proportion	# Heads	Proportion
0	.00002	9	.1746
1	.0002	10	.1222
2	.0018	11	.0667
3	.0085	12	.0278
4	.0278	13	.0085
5	.0667	14	.0018
6	.1222	15	.0002
7	.1746	16	.00002
8	.1964		

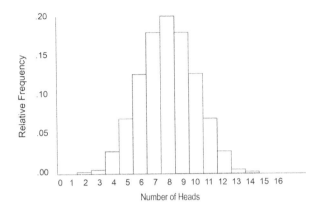

Figure 7

Relative Frequency Histogram of Sampling Distribution Shown in Table 5.

+ .0085 + .0018 + .0002 + .00002. This turns out to be approximately .038 or 3.8%. Your sample result of 12 heads is among the least likely 3.8% of the things that would happen for 16 flips of a fair coin.*

The second approach uses the normal curve approximation to the sampling distribution to compute the same thing. What we want is the relative area under the normal curve approximation *beyond* a value of 11 heads. To get a good approximation, we have to pretend that the exact value of 12 has a real lower limit of 11.5. This is shown in **Figure 8**.

We find the z score equivalent of 11.5 heads to be z = (11.5 - 8)/2 = +3.5/2 = +1.75. According to **Table A**, the area beyond a z score of 1.75 is .0401 or 4%. Notice that this compares favorably to the exact value of 3.8% obtained from the sampling distribution.

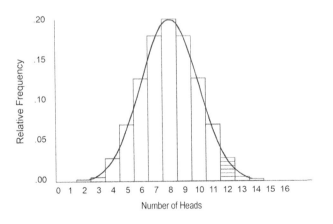

Figure 8

Histogram of Sampling Distribution Shown in Table 5 with Bars for 12 or More Heads Shaded and a Normal Distribution Curve Superimposed.

What both approaches tell us then, is that your sample of 16 flips yielded one of the results that would occur about 4% of the time with a fair coin. So what should you conclude? The coin *could* be fair and this just happened to be one of those rare events. Or, the coin could be biased toward heads and, thus, 12 or more heads out of 16 flips would be an expected outcome. Your choice! In deciding, you may want to reflect on the cost of making a mistake. If the coin is fair and this is just one of those rare outcomes, and you decide to buy it, you are out your $10 investment and maybe more if you use the coin for betting. If, on the other hand, the coin is really a trick coin and you pass up the chance to acquire it, what "opportunities" for using it will you miss? Sadly, in the business of statistical decision making, there are usually no sure bets. We will have more to say about rules that one might use to make such decisions in Chapter 9.

---

* This is not an entirely accurate statement. The 3.8% represents the most unlikely outcomes in the direction of an unusually large number of heads. If one is speaking of unlikely outcomes in general for a fair coin, you have to take into account the fact that an equal discrepancy from the mean in the other direction (4 or fewer heads) will also occur about 3.8% of the time—meaning that a discrepancy of this magnitude is among the least likely 7.6% of things that would happen by chance. But, for purposes of deciding whether to purchase the coin or not, we would use the 3.8% figure.

"Those who educate children well are more to be honored than they who produce them; for these only gave them life, those the art of living well."

Aristotle

Chapter 3, Standard Scores &
   Normal Distributions
Chapter Exercises

Some questions are purely factual and have only one correct answer. Others require greater thought on your part and may have more than one possible answer.

1. Fill in the missing term(s) that makes the statement true.

   a. Standard scores are designed to show _____ in a distribution.
   b. "Percentiles" means _____.
   c. Percentile ranks are poor standard scores because they are _____ and obscure _____.
   d. The mean of a set of percentile ranks is _____ for statistical purposes.
   e. The "unit" of z scores is _____.
   f. If your z score on some test is -2.85 you did _____.
   g. If a set of scores is converted to z scores, the new mean is _____ and the standard deviation is _____.
   h. T Scores have a mean of _____ and a standard deviation of _____.
   i. There are _____ of normal distributions, the exact shape of which depend on _____ and _____.
   j. Table A in the Appendix contains values for the _____.
   k. _____ under the normal curve is used to compute percentage or proportions of variables.
   l. If a fair coin is flipped 16 times, it will come up 8 heads about _____ of the time.
   m. Your answer to l. depends upon a distribution generated by the _____.
   n. Such distributions are called _____.

2. You want to convert z scores into a new type of score called "V Score" having a mean of 100 and a standard deviation of 25. The conversion formula would be _____.

3. Here are scores on the first exam for 11 students in an advanced statistical methods course. Complete the table.

Student	%-age Correct On Test	Percentile Rank *	z score	T-Score
Joe	86			
Sissy	74			
Katy	93			
Carol	86			
Mark	66			
Isra	78			
Noel	98			
Sam	92			
Frank	78			
Ira	86			
Jessie	84			

$\bar{X} = 83.727$
$S = 8.781$

* If scores are tied, assign them the average rank of the scores to compute percentile rank.

4. Normal curve problems. The Wechsler Adult Intelligence Scale (WAIS-IV) has µ = 100 and σ = 15 and in the U.S., the scores for many populations are close to normally distributed. Use **Table A** in the Appendix to answer the following questions.

a. What percentage of people will have scores below 125?
b. What percentage of people have scores above 90?
c. What is the 60th percentile for IQ scores?
d. What percentage of people have scores between 70 and 90?

Use the curve below as a reference figure.

Figure 2

Histogram of Sample IQ Scores Fitted with a Normal Distribution Curve Having $\mu = 100$ and $\sigma = 15$.

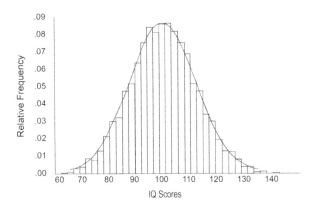

Chapter 4
Getting Started with *JASP*

What's In This Chapter?

In this chapter you will learn how to (a) format data sets in a spreadsheet, (b) do simple analyses using *JASP*, (c) incorporate your analyses in a printable word processing format, and (d) select and modify data in *JASP*.

What Is *JASP*?

The data analysis program used in this course is called *JASP*, which stands for "Jeffreys's Amazing Statistics Program." It is named in honor of the Bayesian statistics pioneer Sir Harold Jeffreys. *JASP* was developed by members of the Department of Psychological Methods of the University of Amsterdam under the direction of Eric-Jan Wagenmakers. It is an easy to use, feature-rich data analysis program that runs on Windows and Macintosh and is available for free by downloading it from the *JASP* web site:

https://jasp-stats.org/download/

At the time this chapter was written, the current version of *JASP* is v 0.18.3. Because it is open source, it is frequently updated and improved, so some of the screen shots of the program in this chapter may be slightly out of date in the future.

Which brings up an important point. This chapter is a "bare bones" introduction to using *JASP*. Jasp-stats.org has an excellent on-line tutorial for using *JASP* to do data analyses. You should make extensive use of it as the course progresses—starting today! Google "Statistical Analysis in *JASP*" and you will get the following:

Statistical Analysis in JASP:
A Guide for Students
By Mark A. Gross-Sampson
2020

As an alternative to using *JASP* for data analysis, if your college or university has a license for the *Statistical Package for the Social Sciences* (*SPSS*) or you have purchased a copy, you can elect to use that in place of *JASP*. If you get the logic for the procedures introduced in this textbook, *SPSS* is sufficiently intuitive that you can make the substitution with a bit of effort.

Creating Data Files

All statistical packages including *JASP* and *SPSS* assume that the data to be analyzed exists in a separate data file which is "loaded into" the statistical package for analysis. We'll begin by describing how data files are structured.

Entering Data Using a Separate Spreadsheet

In *JASP* there are two options for entering and saving data to be analyzed. **The first option** is to use a separate spreadsheet program to enter and format the data. This is handy if you already are conversant with spreadsheets. If you have a copy of *Microsoft Office* or *Microsoft 365*, then you are all set because both have a spreadsheet program called *Excel*. If you don't have the Microsoft software, you're in luck because there is a free office suite program called *Apache OpenOffice* that includes a spreadsheet called *Calc*. *Apache OpenOffice* can be downloaded for Windows and Macintosh from

https://www.openoffice.org/

Excel and *Calc* work exactly the same way, and the comments about instructional YouTube videos for *JASP* applies as well to *Excel* and *Calc*. Any issues about entering or formatting data into either spreadsheet program can be solved with a simple Google search.

The advantage in using a separate spreadsheet program to create *JASP* data files is that spreadsheets are extremely feature rich and permit a great deal of flexibility in how data are entered and formatted. In addition, spreadsheets allow you to manipulate and combine entered data to create new variables.

The Structure of a Data File

All major data analysis packages including *JASP* assume that the data are formatted in a particular manner. This format is so standard that we'll call it the

"Iron Law of Data Files." We'll discuss the format using the Excel spreadsheet as our example. Basically, a spreadsheet is a grid made up of rows and columns. Here is what the upper left-hand corner of the Excel spreadsheet looks like: looks like

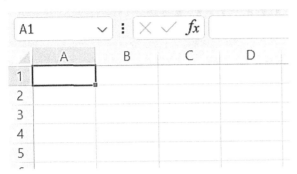

In a data file, the **columns are variables** and the **rows are the specific cases**. Do not violate this rule.

The first row of a spreadsheet contains the names that you have given to the variables. In the below screen shot the spreadsheet now contains the names of four variables: "*CaseNum*" (the code number for the participant), "*Gender*" (the case's gender), "*Condit*" (the condition of the experiment in which the case participated), and "*Score*" (the value of the dependent variable obtained by the case).

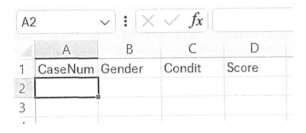

Let's enter hypothetical data for the first five cases. In most studies, *CaseNum* would be a 3-digit code number; *Gender* we'll code 1 = Female, 2 = Male, and 3 = Other; *Condit* is 1 = Control Condition and 2 = Experimental Condition; and *Score* is just the value of the dependent variable.

	A	B	C	D
1	CaseNum	Gender	Condit	Score
2	1	1	1	26
3	2	2	1	37
4	3	2	2	45
5	4	1	2	50
6	5	2	1	33
7				

The numerical coding for Gender and Condit makes it much easier to enter the data (which can be a laborious process) because typing a "1" is a lot quicker than typing "Female." Generally, you enter all the data for each case using the Tab key to move to the right. At the end of the row, hitting Enter moves you down a row and back to the first column.

The only problem with numerical coding is that you have to remember what the 1's and 2's stand for. Once all the data have been entered, you can use a "Find and Replace function" for each variable to change all the numbers to what they stand for. In *Excel*, the replace function is activated by Ctrl-H; in *OpenOffice Calc*, it's Ctrl-F. Here's what the search and replace window looks like in *Excel*:

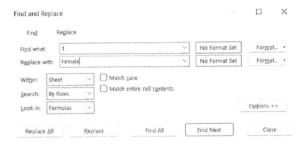

To apply the Find and Replace function, highlight a column (variable) by clicking on the column header indicated by a capital letter. We'll do that for the *Gender* Variable column B:

	A	B	C	D
1	CaseNum	Gender	Condit	Score
2	1	1	1	26
3	2	2	1	37
4	3	2	2	45
5	4	1	2	50
6	5	2	1	33

Now use the replace function to change the 1's to Female, the 2's to Male, and the 3's to Other, yielding:

	A	B	C	D
1	CaseNum	Gender	Condit	Score
2	1	Female	1	26
3	2	Male	1	37
4	3	Male	2	45
5	4	Female	2	50
6	5	Male	1	33

Repeat the process for the *Condit* variable and you have the final form of the data set ready to be saved:

	A	B	C	D
1	CaseNum	Gender	Condit	Score
2	1	Female	Control	26
3	2	Male	Control	37
4	3	Male	Experi	45
5	4	Female	Experi	50
6	5	Male	Control	33

We'll use the forgoing process of entering data into the *Excel* spreadsheet program using a simple set of data from a hypothetical study called "Brain Waves" which is displayed at the bottom of this page. **Read carefully through the description of the study before continuing.**

Figure 1 is a screen capture of the Excel spreadsheet with the data from the Brain Waves study already entered (shown for the first 10 cases). The variable names are "*Particp*", "*BWC*", "*IQ*", and "*GPA*." The *Particp* variable is simply the initials for each participant in the study and will not be analyzed. The remaining 15 rows contain the data for the 15 student participants. The student R.R. had quite a high *BWC* score (6), a modestly impressive *IQ* (116), and a very strong freshman *GPA* (3.21).

Figure 1

Screen Capture of Excel Spreadsheet After the Brain Waves Data Have Been Entered.

	A	B	C	D
1	Particp	BWC	IQ	GPA
2	R.R.	6	116	3.21
3	S.M.	3	121	2.05
4	W.Y.	3	110	3.00
5	L.A.	2	113	4.00
6	D.B.	3	121	3.95
7	J.H.	2	105	3.60
8	C.D.	2	117	3.33
9	B.M.	4	126	3.33
10	A.J.	3	112	2.75

After the data have been entered and checked for accuracy, the spreadsheet is saved as a file, which JASP

Brain Waves

Most neurobiologists and neuropsychologists believe that individual differences in intelligence are, in part at least, due to differences in brain functioning. Putting it crudely, some people have more "efficient" brains than others. In a preliminary study in this area, Drs. Watt and Probe discovered that two of their most (supposedly) intelligent graduate students showed unusually complex electroencephalographic responses to a patterned light display that they were testing. By contrast, an undergraduate student who wandered into the laboratory by mistake and was induced to cooperate by virtue of a $20.00 fee, showed a rather simpler brain-wave pattern. Encouraged, the two investigators conducted a larger study involving a sample of 15 sophomores enrolled in a required general biology course. Each student was tested for the complexity of his/her brain-wave pattern, and subsequently given a standardized IQ test (*IQ*) by the Director of the Counseling Center. Below are the data. The complexity of the brain-wave pattern was rated on a scale of 1-7 (*BWC*) by the investigators before they learned of the students' IQ scores. As a measure of academic performance, they also obtained the student's gpa (*GPA*) from his/her freshman year.

Particp	*BWC*	*IQ*	*GPA*	*Particp*	*BWC*	*IQ*	*GPA*
R.R.	6	116	3.21	A.J.	3	112	2.75
S.M.	3	121	2.05	J.M.	2	109	3.15
W.Y.	3	110	3.00	J.P.	7	135	4.00
L.A.	2	113	4.00	F.H.	5	128	2.00
D.B.	3	121	3.95	L.M.	3	119	2.15
J.H.	2	105	3.60	M.F.	4	120	2.90
C.D.	2	117	1.75	R.S.	6	142	4.00
B.M.	4	126	3.33				

will access for the purposes of statistical analysis. It's a good idea to name the file something that indicates the nature of the study. In the example, the file was saved as "BWC Data 2023.csv." **You need to use the "Save As" command to save it as a file with the extension "csv." Currently the default file type for *Excel* is not directly readable by *JASP*.**

When you activate *JASP*, you are presented with a friendly greeting screen that proclaims that the program is free, friendly, and flexible—and, indeed, it is! **Figure 2** below Is a capture of the opening screen. You will load your newly created data file by clicking on the three horizontal bars in the upper left-hand corner. That will bring up the main menu on the left-hand side of the screen.

Click on "Open" to view saved data files. *JASP* has many saved example files to practice with and keeps track of which files have been recently opened. First time around, you will have to click on the option "Computer", followed by "Browse" to locate where you have saved the file BWC Data 2023.csv. When you locate the file, double click on it and the data will appear in a screen with the analysis routines across the top of the display. **Figure 3** shows a screen capture of this screen. Note that spreadsheets and *JASP* drop trailing 0's in their displays.

Figure 3

Initial Display Screen in JASP After Brain Wave Data Have Been Retrieved.

Particp	BWC	IQ	GPA
1 R.R.	6	116	3.21
2 S.M.	3	121	2.05
3 W.Y.	3	110	3
4 L.A.	2	113	4
5 D.B.	3	121	3.95
6 J.H.	2	105	3.6
7 C.D.	2	117	3.33
8 B.M.	4	126	3.33
9 A.J.	3	112	2.75

Entering Data with *JASP*'s Data Editor

Your second option for entering data into *JASP* uses *JASP*'s own Data Editor. If you are not already familiar with spreadsheets, you may wish to use this option. The following discussion shows you how to use the Data Editor built into *JASP* to create data files. Click on the "New Data" option shown In **Figure 2**. This will bring up

Figure 2

JASP Home Screen with Two Options Indicated for Beginning Your Data Editing or Analysis Session.

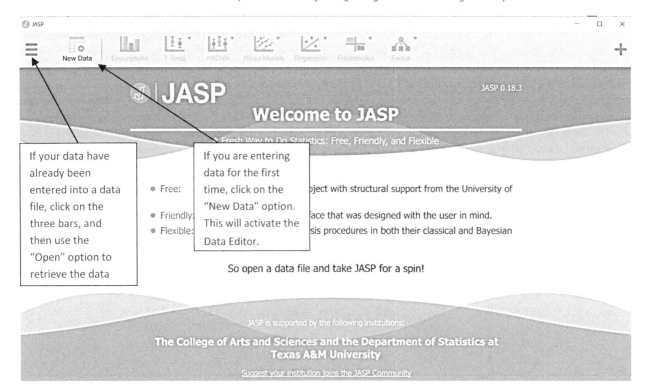

the Data Editor spreadsheet. **Figure 4** shows what the Data Editor looks like before any data have been entered.

Figure 4

Screen Shot of the JASP Data Editor Before Any Data Have Been Entered.

Double clicking on the column header labeled "Column 1" will open a window used for naming and formatting the variables. This is shown in **Figure 5**.

Figure 5

Entering the Variable Name in the JASP Spreadsheet Using the Name Field.

The Name field currently shows the name of the variable to be "Column 1." You will replace that with "Particp" and change the Column type field to "Nominal", i.e., a categorical variable. While still in the variable definition window, click on the second column header and name the variable "BWC" and leave the Column type as "Scale," i.e., a numerical variable. Repeat this process to label the third and fourth columns. When you have labeled the fourth column "GPA", exit the window by clicking on the red circle with an X in it on the far right portion of the window. That will take you back to the spreadsheet, which now looks like this (**Figure 6**).

Figure 6

JASP Data Editor After All Four Variables Have Been Named and their Type Defined.

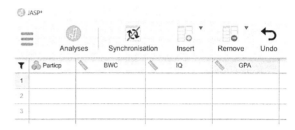

Begin entering data for participant R.R. Enter his initials in the first cell. Use the TAB key to move to the next cell and enter his *BWC* value, and then the rest of R.R.'s data in the cells in columns three and four. The ENTER key moves you down one row and the mouse cursor control will move you to any desired cell.

Figure 7 shows what the Data Editor looks like after all of the variable names have been changed and the data entered for the 15 cases.

Figure 7

JASP Data Editor Spreadsheet Showing the First Six Cases After All the Data Have Been Entered.

	Particp	BWC	IQ	GPA
1	R.R.	6	116	3.21
2	S.M.	3	121	2.05
3	W.Y.	3	110	3
4	L.A.	2	113	4
5	D.B.	3	121	3.95
6	J.H.	2	105	3.6

Notice that the trailing zeros for the data have been dropped. After checking the data for accuracy, if you click on the three bars you will have the option to save the data file using the "Save As" command. As noted above, it's a good idea to name the file something that indicates the nature of the study. Recall that the file created by the *Excel* spreadsheet was saved as "BWC Data 2023.csv."

Before beginning our discussion of data analysis with *JASP* it's important to discuss the need for organization of your work. Mastering statistics is enough of a challenge that you don't want to be hindered by computer disorganization. Organize a file system to keep track of data files and analysis outputs.

Once the data have been saved, click on the "Analyses" option and that will return you to the main data analysis screen.

Using *JASP* to Analyze the Data

If you have used a separate spreadsheet program to enter, format, and save your data, at this point you will click on the three bars at the home screen and use "Open" – "Computer" – "Browse" to locate your data file. Double click on the file and it will load into *JASP*'s initial data analysis screen.

The first analysis that we'll run uses the DESCRIPTIVES routines. Click on DESCRIPTIVES and a screen will open that contains the two large Display Panels shown in **Figure 8**. The panel on the left is labeled "Descriptive Statistics" and contains the variables and all the available statistical routines available. Call that the **Statistics Panel**. The right-hand Panel is labeled "Results" and contains a scrollable window where all the results of the analyses will appear. Call that the **Results Panel**.

To select variables for analysis, a variable appearing in the "list-of-variables" window in the left-hand side of the Statistics Panel is moved over into the window labeled "Variables" by highlighting the variable and clicking on the ▶ symbol. This has already been done with the *BWC, IQ, and GPA* variables. As many variables as desired can be moved over to the Variables window. As soon as a variable appears in the Variables window, a basic descriptive analysis table appears in the Results window to the right. *JASP* carries out analyses as soon as they are selected—which can fill up the Results window very fast indeed. **Figure 8** shows the situation. Notice that the mean, standard deviation, minimum, and maximum have already been computed and displayed in the Results window.

In the bottom of the *JASP* Statistics Panel in **Figure 8** are four categories of analyses that can be carried out—Statistics, Basic plots, Customizable plots, and Tables. We'll start by clicking on Statistics. That reveals an enormous number of possible statistics that could be computed and displayed for the *BWC* data. This is shown in **Figure 9**.

Figure 8

JASP Display Once the DESCRIPTIVES Option Has Been Selected and the BWC, IQ, and GPA Variables Have been Moved into the Variables Window.

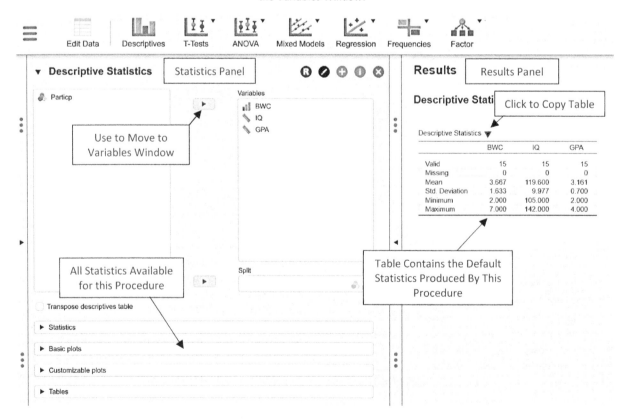

Figure 9

Clicking On the Statistics Option Shows All of the Available Statics That JASP Can Compute.

Figure 10

Screen Shot of JASP Results Window Showing the Expanded Results Table and the Histogram Plot of the BWC Data.

Results ▾

Descriptive Statistics

Descriptive Statistics

	BWC	IQ	GPA
Valid	15	15	15
Missing	0	0	0
Median	3.000	119.000	3.210
Mean	3.667	119.600	3.161
Std. Deviation	1.633	9.977	0.700
IQR	2.000	11.000	0.950
Minimum	2.000	105.000	2.000
Maximum	7.000	142.000	4.000

Distribution Plots

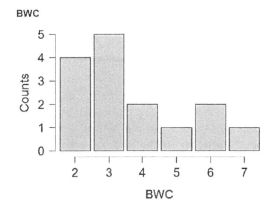

Under the category Central Tendency, the mean is selected by default. Under Dispersion, the standard deviation, minimum, and maximum are default. We'll expand our analysis by selecting Median under Central Tendency and Interquartile Range (IQR) under Dispersion. As each is selected, the Descriptive Statistics table in the Results window expands with the new statistics.

Clicking on the Basic Plots Option and selecting Distribution Plots will generate a histogram plot of the variables. This is displayed in **Figure 10** where the expanded statistics table and the histogram for *BWC* is seen. Scrolling down would show the plots for the other variables.

Brief inspection of the statistics indicates that the *BWC* data are highly positively skewed and extremely variable ($\bar{X}_{BWC} = 3.67, s_{BWC} = 1.63$). The average *IQ* score of 119 is over one standard deviation above the population mean and also quite variable. The freshman *GPA* is a solid "B" and also quite variable. Inspection of the frequency distributions for *IQ* and *GPA* (not shown) indicate that *IQ* is also modestly positively skewed while that for *GPA* is almost flat with a slight negative skewing.

We have one more analysis to perform. Recall that the purpose of the "Brain Waves" study was to determine whether measures of brain wave complexity were related to intelligence. The statistic to measure such a relationship is called **Pearson Correlation** which is symbolized *r* and will be explained in Chapter 5. For the moment, we'll simply carry out the analyses and briefly discuss the result.

The Pearson Correlation analyses are located under the REGRESSION procedures at the top of the *JASP* screen. Click on REGRESSION and select Correlation. When you do so, the *JASP* screen will reset to allow selection of various statistics associated with correlational analyses.

When a new procedure is selected, all of the variables are reset back into the list-of-variables window, and the Variables window is empty. We're going to move all three variables into the Variables window using the ▶ button, with BWC moved over first as the "predictor" variable. When all three variables are moved over, *JASP* automatically computes the value of *r* and displays it in a correlation table in the Results window. The statistics Pearson's *r* and Report significance are already checked.

Figure 11

JASP Correlation Procedure After All Three Variables Were Moved to the Variables Window and the Display Pairwise and Scatter Plots Options Were Checked.

To improve the layout, we'll also check Pairwise display. Under the Plots subhead, we'll check Scatter plots. When that's done, *JASP* will fill in the Results window with a modified correlation table and a graph called a scatterplot. That's shown in **Figure 11**.

Inspection of the correlation table shows that *BWC* is a strong predictor of *IQ* scores (.772) but is unrelated to *GPA* (.135). The scatterplot clearly shows that the relationship between *BWC* and *IQ* is linear. All of this is made clearer in the next chapter on Pearson correlation.

Formatting the Statistical Analyses

The last step in analyzing your data is to print it out and save it for further use. To save tables and figures, *JASP* provides a simple way to export and paste them into a word processor document. That's what we're going to do.

We'll begin by starting Microsoft *Word* and selecting a blank document. In the *JASP* Results window each output table or graph has a **solid down pointing triangle next to the title for the table or graph** (refer back to Figure 9). If you click on the triangle, a small dialog box opens. For tables and figures, we'll select the option "Copy." When you click on Copy, *JASP* will make a copy of the table or graph and paste it to the Windows Clipboard.

The usual shortcut for pasting a clipboard object into your word processor is Ctrl - V. You can copy and paste as many tables and graphs as desired.

Tips for Editing Saved Tables

JASP defaults to formatting tables in what is called APA Style. APA stands for the American Psychological Association and their *Style Manual* is the bible for manuscript preparation for many of the social sciences. When you paste a table into your word processor, it may appear different than your usual document formatting.

Editing parts of the table is done by highlighting whatever parts of the table you want to modify either by using Hold-Shift-Mouse or by right-clicking on the box in the upper left-hand part of the table. For example, you may wish to change the font style and/or point size to be compatible with text you have added. The size of the table can be changed by dragging the box on the lower right-hand part of the table.

Let's see what kinds of alterations you can make in the Descriptive Statistics table shown in **Figure 10** from the original when it's copied and pasted. **Figure 12** shows the table after it was pasted into a *Word* document.

The table is in *APA Style Manual* format and is a bit cramped. **Figure 13** immediately below it is what the table

Figure 12

Default Descriptives Statistics Table When Pasted Into a Word Document.

Descriptive Statistics

	BWC	IQ	GPA
Valid	15	15	15
Missing	0	0	0
Median	3.000	119.000	3.210
Mean	3.667	119.600	3.161
Std. Deviation	1.633	9.977	0.700
IQR	2.000	11.000	0.950
Minimum	2.000	105.000	2.000
Maximum	7.000	142.000	4.000

Figure 13

Descriptives Statistics Table When Pasted Into a Word Document and Edited.

Descriptive Statistics for BWC Study

	BWC	IQ	GPA
Valid	15	15	15
Missing	0	0	0
Median	3.000	119.000	3.210
Mean	3.667	119.600	3.161
Std. Deviation	1.633	9.977	0.700
IQR	2.000	11.000	0.950
Minimum	2.000	105.000	2.000
Maximum	7.000	142.000	4.000

looks like after it has been resized slightly and the boldfacing for the title and headers removed. I even added some text to the table title to make it clearer that it referred to the BWC study.

Best advice, take 20 minutes and play around with the possibilities until you can modify a table in any way desired.

Tips for Editing Figures

Figures can be edited within *JASP*, either changing the size or shape of the figure or modifying the axes labels and scaling. Changing the size of the figure is done by clicking on the triangle shaped handle on the lower right-hand part of the figure with your mouse and dragging it in or out.

Modification of the labeling and scaling of the figure axes is done by clicking on the small triangle next to the title of the figure in the *JASP* output window and selecting "Edit Image." This option is particularly useful when you want to make the axes labeling more precise, or you want to rescale the *Y* axis to make several figures more directly comparable.

Once pasted into your document, you can change the size by clicking on the figure and dragging the lower right-hand figure dot or, if you're especially adventurous, you can click on the figure after pasting it and use the Picture Format function in *Word* to sharpen and darken the elements comprising the figure.

Let's see how we could enhance the appearance of the scatterplot shown in **Figure 11**. The default format for a scatterplot produced by the Correlation procedure is shown below in **Figure 14**.

Figure 14

Default Scatterplot When Pasted Into a Word Document.

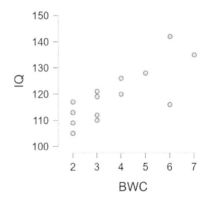

Figure 15 shows what the scatterplot looks like after it was edited in *JASP* and resized after being pasted into a *Word* document. The Picture Format function of *Word* was used to make it a bit sharper.

Figure 15

Scatterplot After Editing in JASP and Resized After Pasting Into Word.

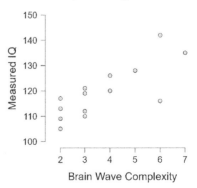

Selecting Data Within a Data File

Frequently in more complex data sets you might want to carry out an analysis on only a portion of the cases in the sample. For example, you might want to analyze the data separately for Males and Females or for only those cases that meet some numerical criterion. *JASP* provides a simple way to do that.

Lets imagine you have access to a large social science survey data set that contains the responses to the following questions for a representative sample of American adults ($n = 1200$):

1. Code Number (001 – 1200)
2. *Gender* (F = Female; M = Male)
3. *Age* in years
4. *Religious Preference* (Pros, Cath, Jewish, Other, None)

Those are followed by four contemporary social issue questions with scale responses that range from 1 = Extremely Liberal to 7 = Extremely Conservative.

5. *Sponsored Childcare* (*ChildCare*)
6. *Abortion Access* (*Abortion*)
7. *Immigration Policies* (*Immigrat*)
8. *LGTBQ+ Issues* (*LGBTQ+*)

Figure 16 shows a screen shot of the *JASP* Data Editor with the data for the first 10 cases of the data set.

Selecting Cases Based on Categorical Variables

Without getting into the details of the types of analyses that you will want to run (save that for Chapters 5–16), let's see how you would select the data for just the females in the sample.

Double click on the column header labeled "Gender" and the following window will open. It's the one that you used for the BWC study to define the variables. In this case the variables have already been defined and *JASP* has categorized *Gender* as a categorial (Nominal) variable.

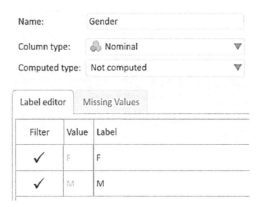

The checkmarks under the column labeled "Filter" indicate that both values are active in the data set.

Figure 16

Screen Capture of JASP Data Editor Showing the First 10 Cases From the Large Social Science Survey of Social Attitudes.

	CodeNum	Gender	Age	Rel Pref	ChildCare	Abortion	Immigrat	LGBTQ+
1	1	F	32	None	2	2	4	2
2	2	F	67	Cath	6	4	5	6
3	3	M	44	Cath	7	6	7	3
4	4	F	23	Pros	4	1	1	1
5	5	M	69	Cath	6	7	7	7
6	6	M	51	Pros	2	6	6	6
7	7	M	33	Jewish	7	7	5	5
8	8	F	75	Other	1	3	3	5
9	9	F	65	Pros	2	2	5	7
10	10	F	47	Cath	4	7	2	3

We're going to do two things here. First, let's change the labels for the values of *Gender*. As the data were entered, the values are "F" for female and "M" for male. Under the field "Label", change the F to Female and the M to Male. The new values will immediately appear in the spreadsheet.

Next we'll deactivate the data for the male participants. Click on the checkmark for M (Male) and see what happens. Volia! The checkmark is replaced by an X, indicating that the value is no longer active, i.e., has been "deselected" in the data set.

Label editor		Missing Values
Filter	Value	Label
✓	F	Female
✗	M	Male

This is what the data now look like in the Data Editor for the first six cases:

▼	CodeNum	Gender ▼	Age	Rel Pref
1	1	Female	32	None
2	2	Female	67	Cath
3	3	Male	44	Cath
4	4	Female	23	Pros
5	5	Male	69	Cath
6	6	Male	51	Pros

Any statistical analyses now run will include only the female respondents. Reversing the process will reselect the males for inclusion in the data set. The other categorical variable, *Religious Preference*, could be treated in the same way, e.g., selecting just Catholics by deselecting "Pros", "Jewish", "Other", and "None."

Selecting Cases Based on Numerical Variables

Often in analyzing numerical data, you may want to include only cases that meet some numerical criterion. In the present data set we might wish to study only respondents who are 50 years of age or older. To do so, we need to write a mathematical expression to identify them. Technically, it's called a **Boolean Expression** and it results in a "True-False" decision for each case to which it's applied—where True-False means include or not include. Begin by clicking on the funny black silhouette in the upper left-hand corner of the *JASP* Data Editor (I'm pretty sure it's a funnel-filter). Notice that the filter silhouette is also present on the Gender column indicating that this variable has been filtered.

▼	CodeNum	Gender ▼
1	1	Female
2	2	Female
3	3	Male

When you click on it, a new window opens that shows the blank numerical filter (**Figure 17** below).

Click on the variable *Age* on the left-hand side of the window. *JASP* knows that Age is a numerical variable and will allow all mathematical operations on it. We're going to write an expression to select all cases in which Age is equal to or greater than 50, i.e.

$$Age \geq 50$$

Figure 17

Screen Shot of JASP Data Selection Window After the Black Filter Silhouette Has Been Selected.

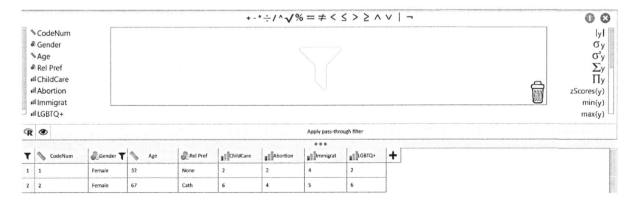

This is done in the Filter Window by clicking *Age* and the symbol "≥" and then typing in 50, i.e.,

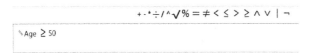

Clicking on the "Apply Pass Through Filter" at the bottom of the window will eliminate all cases less than 50 years of age. Exit the Filter Window by clicking on the Red-X in the upper right corner of the window. That closes the window and leaves the Data Editor looking like this:

▼	CodeNum	Gender	Age	▼	Rel Pref
1	1	Female	32		None
2	2	Female	67		Cath
3	3	Male	44		Cath
4	4	Female	23		Pros
5	5	Male	69		Cath
6	6	Male	51		Pros

To reverse the selection process, click on the filter symbol to return to the Filter Window. Double-click the "Trashcan" symbol on the right-hand side of the window. When the window is blank, click on the Red-X in the upper right corner of the window to exit.

The filter is capable of more complex Boolean Expressions. For example, to select only Males who are less than or equal to 60 years of age, you would click and type to enter the following expression:

The expression says, "Age less than or equal to 60 and Gender equal Male."

The following would select the Catholic females from the sample:

Finally, one can select cases based on statistics from the data itself. Here is the expression that would select females who are less than the mean age for the entire sample:

The statistic "mean ( )" is selected from the array of statistical values on the right-hand side of the Filter Window and the variable "Age" is clicked on to insert it into the parentheses.

Computing New Variables From Existing Variables

In the *JASP* Data Editor it's possible to combine existing variables to create a new variable. We'll do so by creating a new variable called "Tot-SI" which is the sum of the four social issue ratings.

Click on the "+" at the end of the columns and that will bring up the following dialog box:

Create Computed Column

Name: [   ]

[R] [👆]

[Scale] [Ordinal] [Nominal] [Text]

[ⓘ] [Create Column] [✕]

Fill in the variable name, "Tot-SI" and click on "Create Column." The column will then appear at the end of the data set, labeled "$\int_x$ Tot-SI". Double click on the column name and it will bring up a window to write an expression to generate the variable by combining already existing variables. Before you do that make sure that the variables *Childcare* through *LGBTQ+* are all defined as "scale" (numerical). ***JASP* can be prone to defining variables that take on only integer values as "nominal" (categorical) even it they are intended to be analyzed as numerical** (see below to change this). To change the variable type, click on the symbol next to the variable name and select "scale" from the small pop-up window.

Figure 18 on the next page shows what the computation window looks like. Each of the four SI

Figure 18

Screen Shot of Computed Column Window with the Four SI Variables Already Entered with + Signs Between Them.

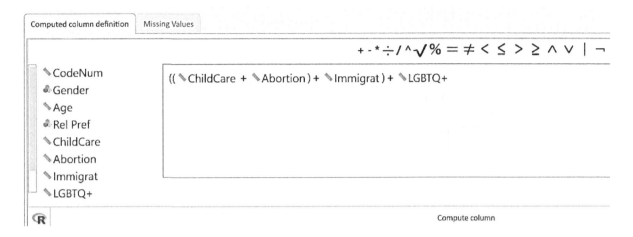

variables has already been moved into the window by clicking on each from the variable list to the left of the variable window. After each is moved over, the + sign is added by clicking on it. Once all four were entered, the "Compute column" tab at the bottom of the window was clicked and the computed values were added to the $\int_x$Tot-SI column. The window is exited by clicking on the Red-X button. Here is what the spreadsheet looks like with the new variable added:

ChildCare	Abortion	Immigrat	LGBTQ+	f_xTot-SI
2	2	4	2	10
6	4	5	6	21
7	6	7	3	23

Here is a qualifier regarding *JASP*'s capacity for computing new variables from existing ones. This function seems to be still under development and is incomplete in some ways. Moreover, it does not always work as anticipated. For example, adding the four SI variables together to produce a sum works correctly. But attempting to divide the sum by 4 to get an average sometimes generates an error message. Simply reentering the number generally solves the problem.

Customizing *JASP* for Your Needs

It is possible to change several settings in *JASP* to make the program work in ways that you choose. The menus for doing so are accessed by clicking on the three bars at the upper left corner of the main menus screen and selecting "Preferences."

The first submenu that appears is labeled "Data" shown in **Figure 19**. The one thing that you may wish to change is the "Import Threshold" setting. Its default is 10, which means that when a data file is first scanned, if there are fewer than 10 different integer values the variable will be categorized as (a) categorical if only two values occur or (b) ordinal if between 3–9 values occur. Only when 10 or more values are found will the variable be categorized as "scale" (interval or ratio). I'd suggest clicking on this option and changing the threshold to 3 or 4. If you have a categorical variable having more than two numerical coding values, you'll have to change the variable type in the *JASP* Data Editor (see below).

Figure 19

Screen Shot of the Data Preferences Submenu Menu.

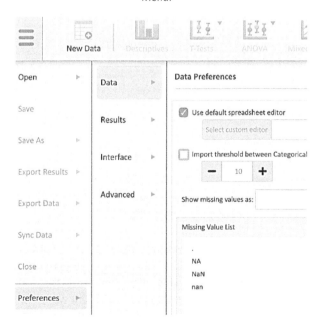

Changing the variable type is done by clicking on the type symbol in the variable name and selecting the desired type from the pop-up window, i.e.

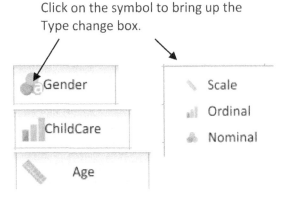

Next, click on the submenu "Results." **Figure 20** shows the options available. Here you may wish to change the number of decimals displayed in *JASP* results tables to something other than the default of three. Another set of options allow you to change the characteristics of graphs produced by *JASP* to increase their resolution and alter their background from white to transparent. The latter might be appropriate if you are going to use the graphs in a *PowerPoint* presentation.

Figure 20

Screen Shot of the Results Preferences Submenu.

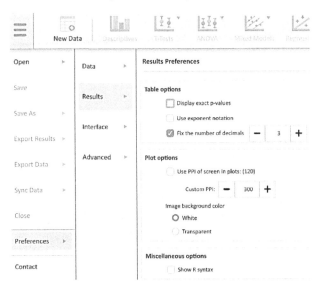

Finally, the "Interface" Preference option allows you to control the font used in the figures and tables that *JASP* produces (**Figure 21**). This allows you to match *JASP* outputs to whatever word processing font you use—which can save you a good deal of reformatting effort. You can also change the on-screen size of the text and buttons for all *JASP* displays. And you can change the

Figure 21

Screen Shot of the Interface Preferences Submenu.

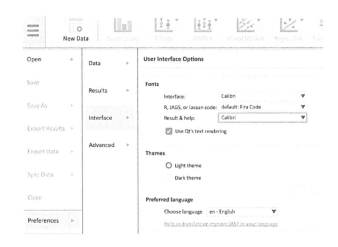

language for the *JASP* interface and output tables and figures to any one of 13 languages! (Pretty amazing)

Final Comments

As I mentioned at the beginning of this chapter, it is a relatively modest introduction to creating data files and using *JASP* to analyze data. Fortunately, *JASP* is very easy to use and you can quickly improve your data analysis skills by doing some or all of the data analysis problems at the end of the chapter. With a bit of effort on your part, you will soon be offering to help other students—and even professors—with their data analysis needs!!

Chapter 4, Getting Started with *JASP*
Chapter Exercises

Some questions are purely factual and have only one correct answer. Others require greater thought on your part and may have more than one possible answer.

1. Fill in the missing term(s) that makes the statement true.

    a. *JASP* stands for _____.
    b. Harold Jeffreys was a pioneer in _____ statistics.
    c. An alternative to using *JASP* is a program called _____.
    d. When uncertain how to do something in *JASP*, try _____.
    e. You can create a data file either using a separate _____ or *JASP*'s own _____.
    f. *Excel* and *Calc* are _____.
    g. In a data file, the rows are _____ and the columns are _____.
    h. The first row of a spreadsheet always contains the _____.
    i. For ease of data entry, use _____ which can later be changed to variable names using _____.
    j. To change "1" to "Female" for the variable *Gender* in *Excel*, use _____ to open the _____ window.
    k. Spreadsheets created by *Excel* must be saved using the "saved as" option using the _____ format.
    l. It's strongly suggested that you _____ your work using a consistent file directory system.
    m. The "Brain Waves" Study is not an experiment, but a _____ study.
    n. When using *JASP*'s built-in Data Editor, your first task is to name and define the _____.
    o. After you have entered and checked the data in the *JASP* Data Editor, save the file using the _____.
    p. Once a statistical procedure is selected in *JASP*, the screen shows the _____ Panel and the _____ Panel.
    q. Within a procedure, variables to be analyzed are moved from the _____ window to the _____ window.
    r. Statistical computations and graphs will appear in the _____ window.
    s. At the bottom of the Statistics Panel are all of the _____ available for the procedure.
    t. To print out tables and graphs, they need to be _____.
    u. A statistic that quantifies the correlation between variables is called _____.
    v. *JASP* formats tables in the style of _____.
    w. When copied to a word processing document, tables can be modified in what ways _____?
    x. Figures (Graphs) can be edited in *JASP* to change the _____ and the _____.
    y. Within a data set, a subgroup of cases can be selected for analysis by _____ one or more variables.
    z. In the hypothetical "Social Attitudes" data set, you could select Females less than 30 years of age by entering the expression _____ into the _____.
    aa. The expression you would use is technically called a _____.
    bb. *JASP* permits the creation of _____ from combinations of _____.

2. At the beginning of the semester, Professor Zandar has the students in her extremely popular *Introduction to Woman's Studies* class complete an anonymous questionnaire regarding a variety of issues of particular importance to young adults. One question asks, "Assuming you have the financial means, how many children would you like to have?" Here are the responses of the 32 women and the 12 men in the class.

Women					Men				
5	3	0	2	2	3	0	1	2	3
0	1	0	0	1	2	3	4	3	2
4	6	0	1	3	3	2			
2	1	2	3	0					
0	4	2	4	0					
2	2	5	2	3					
0	1								

Create a data file that contains the student's gender (*Gender*) and their desired number of children (*Children*). Omit the code number. Then use the DESCRIPTIVES procedure to generate a Descriptive Statistics table of relevant statistics for both the women and men. That's done by moving *Children* to the Variables window and *Gender* to the Split window, which is located below the Variables window. Under the Basic plots option select Distribution plot and under Customizable plots select Box plot. Using the *JASP* figure Edit function clean up the two plots by rescaling the Y axis as needed and providing more complete axes titles. Print out the table and figures and edit them further as needed.

3. Students in an introductory sociology class are interested in whether people who violate traffic laws have identifiable characteristics that distinguish them from those who obey traffic laws. One student believes that scofflaws are more likely to drive pick-up trucks, so the group decides to do a real-world observational study. The students select a busy intersection in a neighborhood school district. At randomly selected 30-minute periods throughout a given week, observers record the type of vehicles that come to the intersection and whether they stopped for the stop sign. For each observation period the proportion of pick-up trucks and cars that stopped are tabulated. Here are the data for the 15 observation periods:

Pick-up Trucks	Cars	Pick-up Trucks	Cars
.13	.35	.22	.23
.26	.33	.27	.19
.14	.19	.46	.67
.28	.41	.18	.12
.20	.27	.24	.29
.52	.46	.33	.46
.31	.29	.10	.09
.16	.11		

a. Enter these data into *JASP*'s Data Editor or a spreadsheet and then do whatever statistical analyses you think appropriate using *JASP*. What type of graph(s) might be appropriate?

b. There's something odd about the data. Can you figure out what's going on?

4. People vary a great deal in the extent and nature of their "on-line" presence. Students in an introductory communications course were asked to fill out a brief questionnaire about their use of the internet. Students reported their gender (Male, Female, Other), class rank (F, S, J, S), and their rated usage of *Facebook* (*FB*), *X* (formerly *Twitter*), and Googling topics (*G*). For each, they used a 5-point rating scale ranging from 1 (Not at All) to 5 (All the Time). Here are the data:

Gender	Rank	FB	X	G	Gender	Rank	FB	X	G
F	F	4	2	2	M	F	2	4	5
M	S	1	3	3	M	F	1	2	2
F	F	3	3	3	O	S	5	1	2
F	J	5	1	5	F	F	3	4	4
O	S	3	4	3	M	F	1	2	1
M	F	2	4	2	F	F	3	1	2
M	J	2	5	5	O	S	4	3	3
F	F	4	1	3	F	S	1	2	5
M	F	3	3	3	M	F	2	4	5
F	S	5	3	5	O	S	1	2	5
M	F	3	4	3	F	J	4	4	4
O	S	5	1	2	F	S	2	3	2
F	F	2	3	4					

One point. Given how few juniors and seniors there are, the Juniors and Seniors could be recoded as "Upper" for upper-class persons. Analyze these data. Use the opportunity to "play around" with different options, statistics, and plots. Practice copying and pasting tables and plots into a word processor document.

5. As part of a senior thesis in anthropology, an undergraduate student conducted an observational study of aggression among preschool children. The study was carried out at a private day care center that catered largely to upper middle-class families. The average daily attendance was 23 children, 11 Boys and 12 Girls. Two observers completed 30 45-minute observation sessions while the children were playing in a large fenced-in playground. During each session, one observer recorded all behaviors of three randomly selected target children, each for 15 minutes. The other observer carried out "area scanning," looking for instances of verbal and physical aggression. For each instance, the observer recorded the gender of the children involved (G = All Girls, B = All Boys, M = Mixed Girls and Boys), the circumstances of the aggression, the outcome, and the duration in seconds. Below are the duration data in seconds for the 61 aggressive encounters observed during the course of the study, indicating the group composition.

B, 41	B, 113	M, 33	G, 9	B, 21	B, 106	M, 51
G, 13	B, 39	B, 96	M, 94	G, 33	B, 83	B, 102
M, 18	B, 99	M, 76	G, 12	G, 9	M, 15	B, 98
M, 5	G, 13	B, 85	G, 32	B, 75	G, 35	M, 31
G, 22	B, 26	M, 33	G, 11	M, 23	B, 12	B, 13
M, 48	B, 108	G, 3	G, 15	B, 14	B, 27	M, 19
B, 17	B, 23	G, 16	B, 25	B, 28	B, 83	G, 26
B, 22	G, 25	M, 27	B, 22	G, 4	B, 82	G, 16
B, 89	M, 16	G, 9	B, 92	M, 11		

Use *JASP* to analyze these data as completely as possible, generating whatever tables and graphs seem appropriate.

Chapter 5
Pearson Correlation

What's In This Chapter?

Correlation is the tendency for values of one variable to be systematically related to values of another variable. When two variables are numerical in character, their relationship may be visualized in a type of graph known as a **scatterplot**. When the relationship between two numerical variables is **linear** (a straight line), the correlation may be quantified using a statistic called **Pearson r**. Pearson r is a bounded statistic whose value can range from -1.00 through 0 to +1.00. The sign of the statistic indicates whether the correlation is **positive** (**direct**) or **negative** (**inverse**). The absolute value of r quantifies the strength of the relationship. Because r is a measure of the linear relationship between two numerical variables, it will not adequately describe the correlation that exists when two variables are related in a non-linear manner.

Several factors besides nonlinearity may influence the interpretation of a particular value of r. First, Pearson r is an unstable statistic, so values of r obtained on small samples may depart markedly from the correlation that actually exists in the population of interest. Second, the fact that two variables are correlated does not mean that one is the cause of the other. Two variables may be correlated because both are causally influenced by a third variable. This is called a **spurious correlation**. Third, r is sensitive to **range restrictions** in the values of one or both variables. Restriction in the range of either variable can artificially reduce the sample value of r. Fourth, pooling together **discontinuous subgroups** of cases can markedly alter the resulting value of r, compared with the values that would be obtained for the subgroups taken separately.

The **coefficient of determination** (r^2) is a better measure than r of the strength of the predictive relationship between two variables. This statistic quantifies the proportion of the variability in one variable that is associated with (predicted by) changes in the other variable. The coefficient of determination should always be used to interpret the relative magnitudes of two or more values of Pearson r.

In actual practice, correlational studies measure more than two variables. This is because most events in the world are multiply determined by complex patterns of **direct** and **indirect causality**. When there are several **predictor variables** for a single **criterion variable**, researchers compute all possible correlations among the variables and organize them in a **correlation matrix**.

Spearman's rho and the **Point biserial correlation** and are two variants of Pearson r. The **correlation ratio eta** (η) is a measure of the relationship between a categorical independent variable and a numerical dependent variable. η^2 is used as a measure of **effect size** in a great deal of research.

Introduction

In Chapter 2 you learned a variety of techniques for describing important features of individual variables, i.e., the central tendency and variability of the scores, and the pattern of the distribution of the scores. In this chapter, you will learn some important techniques in bivariate descriptive analysis, specifically, procedures for describing the **correlation** between pairs of variables.

Correlation is **the tendency for changes in the value of one variable to be systematically related to changes in the value of another variable**. This definition is intentionally broad because there are many different ways in which the correlation between variables can manifest itself and many different statistics for measuring correlation.

This chapter focuses on the most widely used correlational statistics, the **Pearson product-moment correlation coefficient**. This statistic is named for Karl Pearson, the statistician who developed it in its present form in 1896, building upon the seminal work of Sir Francis Galton. The coefficient is symbolized *r* when it is a statistic computed on sample data and ρ (**Greek lower-case rho**) when it is a parameter of a population. Because nobody wants to keep saying "Pearson product-moment correlation coefficient," the statistic is usually referred to as "**Pearson r**" or just "*r*."

The last section of this chapter introduces three other correlation statistics, two of them based on Pearson r and one that will find a good deal of use in analyzing the results of experiments.

Scatterplots

We will begin our discussion of Pearson correlation by considering the type of graph used to display correlational relationships. This graph is variously called a **scatterplot**, a **scattergram**, or a **scatter diagram**. We will use the term "scatterplot" in this text.

Table 1 contains the data for 10 students in an introductory statistics course. On the first day of class, they were asked to rate how confident they were in their ability to "do well in statistics." They used a 7-point scale that ranged from 1 (*Not at All Confident*) to 7 (*Extremely Confident*). These confidence ratings were used to predict the students' performance in the course, and technically constitute the **predictor variable** of the study. Traditionally, the predictor variable is abbreviated "*X*." The students' performance in the course was measured by computing the average of all test scores expressed as a percentage of the total possible points. This measure constituted the **criterion variable** of the study. Traditionally, the criterion variable is abbreviated "*Y*."

Table 1

Confidence Ratings and Average Percentage Grades for 10 Students Enrolled in an Introductory Statistics Course.

Student	Confidence Rating (*X*)	Final %-Age Grade (*Y*)
D. K.	2	76
M. L.	3	77
J. M.	5	89
R. B.	6	75
S. C.	6	93
R. H.	7	97
M. D.	3	81
K. K.	4	80
W. L.	4	85
V. A.	5	87
$\bar{X}$ =	4.5	84.0
S_X =	1.5	7.1

Figure 1 shows a scatterplot of the students' data. The *X* axis is scaled for values of the predictor variable (*Confidence Rating*). The *Y* axis is scaled for values of the criterion variable (*Final Average*). The two values for each case (student) are represented by a point or dot positioned at the *intersection* of an imaginary vertical line corresponding to the value of the predictor variable and an imaginary horizontal line corresponding to the value of the criterion variable. The line imposed by *JASP* is called the

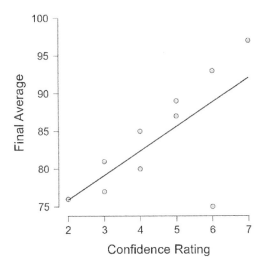

Figure 1

JASP Scatterplot of the Relationship Between Confidence Ratings and Final Average for 10 Students in Elementary Statistics.

"best fit straight line" or the "regression line" and is discussed in the next chapter.

For example, in the lower left-hand portion of the graph is a point that represents the data for the first student, D.K. D.K.'s two scores of 2 and 76 are represented by the point that is located above the value of 2 on the *X* axis and to the right of the value of 76 on the *Y* axis. The point in the extreme upper right-hand portion of the graph is the data point for R.H. (7 and 97), who was correctly confident about her aptitude for introductory statistics.

How do you "read" the scatterplot? Basically, one looks at the *general shape* of the "cloud" of points. In this example, there is a clear general tendency for low confidence ratings to be associated with low final grades and high confidence ratings with high final grades. The "cloud" is slanted upward, from lower left to upper right. We would say that there is a **positive** or **direct relationship** between the two variables. A positive relationship is one in which higher values of the predictor (*X*) variable are associated with higher values of the criterion (*Y*) variable.

The opposite of a positive relationship between two variables is called a **negative** or **inverse relationship.** In a negative relationship, increasing values of the predictor (*X*) variable are associated with decreasing values of the criterion (*Y*) variable.

Figure 2 shows what a negative correlation looks like. Our example here is the ominous relationship between the number of days absent from an introductory statistics class and the final grade for the class for a sample of 40 students. Notice that the cloud now slants downward from upper left

Figure 2

JASP Scatterplot of the Relationship Between Number of Absences and Final Average for 40 Students in Elementary Statistics.

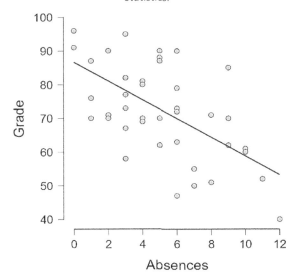

to lower right. In this example, the cloud is rounder or "fatter," indicating a less clear-cut tendency for higher values of X to be associated with lower values of Y (the moral should, however, be clear). We would say that the correlation between the two variables is not as "strong" as in the case for the data displayed in **Figure 1**.

If two variables are not correlated at all, the scatterplot will show a cloud of points that has no systematic tendency to "tilt" either up or down. **Figure 3** shows the scatterplot when we attempt to predict grades in an introductory statistics class from students' shoe sizes.

Figure 3

JASP Scatterplot of the Relationship Between Shoe Size and Final Average for 50 Students in Elementary Statistics.

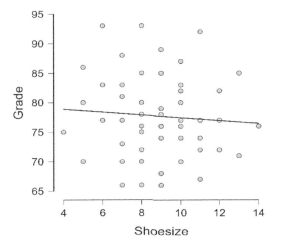

Pearson *r* Correlation Coefficient

Pearson *r* is appropriate for describing the correlation between two variables when the following conditions are met:

1. **Both variables are numerical (interval or ratio).**

2. **The relationship between the two variables is linear in the population.**

Pearson *r* is defined for sample data as follows:

$$r = \frac{\sum(z_X z_Y)}{n}$$

where: z_X = X score expressed as a z score,
z_Y = Y score expressed as a z score, and
n = number of pairs.

In words, Pearson *r* is the mean product of *z* scores. To get some feel for the meaning of this statistic, we will walk through its computation using the above definitional formula. The necessary computations are presented in **Table 2**, where the paired data from **Table 1** have been transformed into *z* scores.

	Table 2		
*Students Confidence Ratings and Average Percentage Grades from **Table 1** Expressed as z Scores, Plus the Products of the z Scores.*			
Student	Confidence Rating (z_X)	Average %-Age Grade (z_Y)	$z_X z_Y$
D. K.	− 1.67	− 1.13	+ 1.89
M. L.	− 1.00	− 0.99	+ 0.99
J. M.	+ .333	+ 0.70	+ 0.23
R. B.	+ 1.00	− 1.27	− 1.27
S. C.	+ 1.00	+ 1.27	+ 1.27
R. H.	+ 1.67	+ 1.83	+ 3.05
M. D.	− 1.00	− 0.42	+ 0.42
K. K.	− .333	− 0.56	+ 0.19
W. L.	− .333	+ 0.14	− 0.05
V. A.	+ .333	+ 0.42	+ 0.14
		$\sum(z_X z_Y)$ =	+ 6.86
	$r =$	$\sum(z_X z_Y)/n$ =	+ .69

Look at the table at D.K.'s scores. His confidence rating of 2 has a *z* score value of -1.67. It was calculated as (Score – Sample Mean) / Sample Standard Deviation, which is (2 – 4.5) / 1.50. D.K.'s final class grade of 76% yields a *z* score of

-1.13, which is computed as (76% −84.0%) / 7.10%. The last column in the table is the product of the paired z scores for each student. For D.K., the product is -1.67 times -1.13 which equals 1.89. D.K. didn't have much confidence in his ability to do well in statistics and he was basically correct. The z score equivalent for his confidence rating (-1.67) was substantially below the sample mean. The same was true of the z score for his final grade (-1.13). The product of the two negative z scores is, however, a positive number ($z_x z_y$ = +1.88), because the product of two negative numbers is positive.

Take another case. R.H. was very confident in her abilities (z = +1.67) and, in fact, did very well in the course (z = +1.83). The product of her z scores also is a positive value (+3.06). Clearly, then, when the predictor and criterion variables correspond in sign--either both negative or both positive--the resulting product of z scores will be a positive value. Consistent positive values for the z-score products means that the relationship between predictor and criterion variables is positive or direct. When these are summed for the entire sample and divided by n, they result in a large positive value for Pearson r.

Only when the two z scores differ in sign will the z-score product be negative. R.B. was quite confident in her ability to do well in the course (z = +1.00) but finished with a modest 75% in the course (z = -1.27). The product of her two z scores is −1.27. This negative product indicates inconsistency in the data (exceptions to a positive relationship) and reduces the sum of the z-score products for the sample, thereby reducing the magnitude of the correlation.

For the entire sample of 10 students, the sum of the products of the z scores is +6.86. The mean of the products is +.69 (rounded to 2 decimals), which is the value of Pearson r. The "+" means that the overall relationship is positive or direct, which we could tell from the scatterplot, and as we discuss below, the absolute value of .69 means that the relationship is quite strong, although still far from perfect.

What happens in the formula when an inverse relationship exists between two variables? Here, high scores in the predictor variable are systematically associated with low scores in the criterion variable and vice versa. As a result, when the scores are translated into z scores, most of the pairs of scores will have opposite signs, and the z-score products will be negative. In that case, the sum of the z-score products will be negative, making r a negative value.

Finally, if there is no systematic relationship between the predictor and criterion variables, positive and negative z-score products will occur about equally often. The positive and negative products will tend to cancel each other out, making the sum of products a value around zero.

Because Pearson r is based on z scores, it can range in value only between -1.00 to +1.00. As such, it is referred to as a **bounded statistic**. The algebraic sign of the statistic indicates the **direction of the relationship**. A plus sign means that the relationship is direct or positive, as is the case in our example. A negative sign in front of the statistic means that the relationship between the two variables is inverse or negative.

The absolute size of r indicates the **strength of the relationship**. A value of 0 means that there is no *linear relationship* between the two variables. The larger the size of Pearson r, the stronger is the linear relationship between the two variables. Values of +1.00 and -1.00 indicate a **perfect linear relationship** between two variables.

Figure 4 shows the scatterplot for a hypothetical data set in which there is a perfect linear relationship between the students' confidence ratings and their final scores.

Figure 4

JASP Scatterplot of a Perfect Positive Relationship Between Confidence Ratings and Final Average for Seven Students in Elementary Statistics.

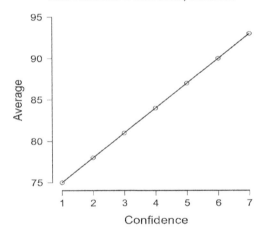

Many students are not quite sure how to interpret and to put into words the strength of the relationship indicated by different values of r. For our purposes, assume the values of r less than ± .20 are either meaningless (due to sampling fluctuation) or trivial (of no practical value). Values between ±.20 and ±.35 indicate "weak" relationships between variables. Values between ± .35 and ±.50 can be called "moderate" in magnitude, values between ± .50 and ± .65 are "strong," and those above ±.65 are "very strong."

Cautions in Using Pearson r

Nonlinearity

Pearson r is a remarkably useful statistic, but it has some important limitations. The first is that Pearson r is appropriate only for describing linear relationships between variables. Two variables can be related in a **nonlinear** fashion and Pearson r will not accurately reflect the relationship. An example of this is the well-known relationship between psychological "arousal" and performance on numerous cognitive ("thinking") tasks. For tasks of substantial difficulty, either very low or very high levels of "arousal" (anxiety, competitive spirit, etc.) tend to hurt performance. Optimal performance is normally achieved with intermediate levels of arousal. This type of relationship is often referred to as an "inverted U relationship" and is seen with some frequency in the social and physical sciences. **Figure 5** depicts what the scatterplot might look like for the correlation between psychological arousal and task performance.

Figure 5

JASP Scatterplot of the Relationship Between Psychological Arousal and Performance.

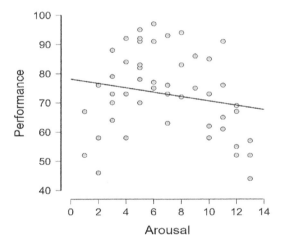

The point to be made here is that a nonlinear relationship is a perfectly legitimate way for two variables to be correlated, but Pearson r is *not* the way to describe the relationship. **If the relationship is nonlinear, r always will understate the strength of the association between the variables compared with a more appropriate curvilinear correlational statistic.**

Sample Size

Another important limitation of Pearson r is its **sampling instability**. Recall from Chapter 1 that statistics vary in their sampling stability and, therefore, the accuracy with which one can predict parameter values. The mean, we learned, tends not to vary markedly in value across repeated random samples even when sample size is fairly small. As a result, the mean of any given sample is likely to give you a fairly accurate estimate of the population mean. The same cannot be said of Pearson r. With small samples, r tends to be quite unstable.

Consider our hypothetical study of the ability of students to estimate the grade that they will receive in introductory psychology. We used 10 cases for the example just to keep the computations easy to follow. If we were really interested in knowing how accurately students "in general" could estimate their grades, we would need to use a much larger sample. An estimate based on only 10 cases could be wildly in error. For example, if the actual value of the population correlation (ρ) is +.50, about 15% of the random samples would yield a sample correlation of +.70 or greater. It's even worse when ρ is equal to 0. When there is no correlation at all between two variables in a population, random samples of size $n = 10$ would yield values more than ± .25 about 50% of the time!

So how big should the sample be to make use of Pearson r? Naturally there is no single, definitive answer to the question (this is statistics, remember). Sometimes small samples are the best you can do. If you are looking at the correlation between variables in an endangered species such as the manatee, you are not going to have very large samples. Here, researchers would be happy to have 30–40 cases. Social science researchers, on the other hand, often have thousands of cases drawn from large questionnaire studies. If your sample is a true random sample from the population of interest or if you have good reason to believe that it is at least representative of the population, you can probably draw meaningful conclusions with as few as 35 cases.

Inferring Causality

As we have already noted in Chapter 2, the simple fact that two variables are correlated does not establish that one is the cause of the other. There is, for example, a substantial negative (inverse) correlation between the amount of television viewing in children and grades in school. The more television that children watch, the poorer on average they do in school. It may be the case that excessive television viewing is one of the causes of poorer academic performance, but the correlation by itself does not establish the causal relationship. Some alternative explanations that may account for the observed correlation include the following:

1. Less intelligent children are more prone to watch television and because they are less intelligent, they tend to do poorer in school.

2. Children who are poorly supervised at home because of inadequate parenting tend to neglect their homework and also spend more of their time in front of the tube.

3. Culturally disadvantaged families have fewer resources to spend on recreational activities for children, who, therefore, spend more leisure time in front of the television. Such families also tend not to emphasize the importance of doing well in school, so the children's classroom performance is poorer.

Here is another classical example. Teenagers who drop out of school before receiving a high school diploma are much more likely to wind up in low paying dead-end jobs and/or drifting into criminal activities. A common response among educators is to urge the establishment of "dropout prevention programs" to keep these kids in school (thereby presumably saving them from a life of poverty and crime). The question, of course, is whether dropping out of school, *per se*, is a major cause of these sad outcomes? Can you come up with plausible alternative explanations?

The bottom line is that the relationship between correlation and causation is complex. This is illustrated in **Figure 6**, which shows several different ways that two variables, X and Y, could correlate. The arrows represent actual causal connections between variables; the direction of the causation is indicated by the direction that the arrow points.

Figure 6

Six Different Causal Mechanisms that Could Account for a Positive Correlation Between the Variables X and Y.

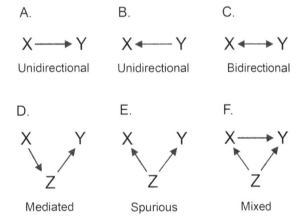

In Part A of the Figure, X is a cause of Y. This is the simplest interpretation of a correlation between two variables.

In Part B, Y surprisingly turns out to be the cause of X, thereby producing the correlation. If, for example, poor performance in school caused children to devote their time to television viewing, the supposed criterion would, in fact, be the cause of the ostensible predictor. The situations diagrammed in A and B are examples of what is called **unidirectional causation**. In Part C, X and Y act reciprocally to cause one another. This type of **bidirectional causality** is very common in many areas of social science research. For example, there is probably a positive correlation among college students in reported interest in a class (X) and amount of participation in class activities (Y). Here, the causality is almost certainly bidirectional. Greater interest leads to greater involvement, which in turn heightens interest.

The causal modes displayed in A-C are the simplest examples of situations in which a correlation between two variables does imply direct causation. Part D shows a more complex causal relationship that will result in a correlation between X and Y. Here X acts causally on another variable Z, which in turn causes changes in Y. To the extent that television does interfere with performance in school, its effects are probably of this type. Perhaps excessive television viewing impairs children's interest and ability to read effectively. Deficient reading skills then act in many aspects of school to impair performance. This is called **mediated causality**.

Direct and mediated causal patterns represent legitimate causal inferences from observed correlations between variables. Part E diagrams a causal pattern that results in what is known as a **spurious correlation** between two variables. The "spurious" doesn't mean that the correlation isn't real, only that it doesn't imply a direct or mediated causal relationship between the X and Y variables. In a spurious correlation, both X and Y are correlated with one another simply because both are causal effects of some third variable. The alternative explanations given for the negative correlation between television viewing and grades in school (above) are examples of how a third variable can produce spurious correlations. In the first alternative explanation, the correlation arises because a third variable, lack of intelligence, causes increased television viewing and impaired academic performance. In interpreting correlational data, you need to be alert for the possibility that correlations may be spurious.

Finally, Part F of **Figure 6** represents a common situation. Here, there is a directional causal connection between X and Y, and a third variable is also causally associated with each. In this case, *part* of the correlation between X and Y represents a legitimate causal association and *part* of the correlation is spurious. In the television viewing example, you could look at the correlation between television viewing and grades in school for a sample of students who were all the same intelligence. Because intelligence is held constant, it cannot be a factor in producing a correlation between the two variables. If the correlation disappears or is markedly reduced in magnitude when intelligence is held constant, you have good evidence to believe that the correlation is spurious. If holding intelligence constant has no effect on the correlation between television viewing and grades, then this alternative explanation has been eliminated from consideration. **Often in social science research, a great deal of the effort in establishing causal relationships based on correlational data is expended in eliminating such alternative explanations**.

In Chapter 7 you will learn about a technique called **partial correlation**, which allows one to compute the correlation between two variables after the effects of a third, spurious variable have been mathematically removed. Conceptually, partial correlation is equivalent to holding the third variable constant.

Range Restrictions

Here is a puzzle to ponder. At Clearwater College, the correlation between students' total SAT score and grade point average (GPA) at the end of the freshman year is +.52, which is a moderate to strong correlation. At Massachusetts Institute of Technology (MIT), the same correlation is about +.20, which signifies almost no relationship between the two variables. From these statistics you might conclude that the SAT is useful for predicting scholastic performance only at more modest institutions. If you were really snobby, you might claim that the pedestrian multiple-choice questions comprising the SAT don't test the more "refined" intellectual demands of an MIT. Alas, you would be wrong. The SAT works just as well for the MITs and the Harvards as it does for the local college. Actually, it may work better at such prestigious institutions.

Here is the correct explanation for the differences in correlations. At Clearwater College, most students who apply are accepted for admission. The freshman class will, therefore, contain a fair number of students who have very low SATs and a smaller number with very high SATs. Most students, of course, will be somewhere in the middle. This situation is represented in a scatterplot of hypothetical data in **Figure 7**. Notice that the "cloud" of points shows a gradual slope from lower left to upper right—higher SATs are associated with higher GPAs.

At MIT it's a different story. Here, SAT scores are used as a criterion for admission. Unless Daddy made a very large donation to the university, you are not likely to be admitted if your SATs are much below 1300. This restricts

Figure 7

JASP Scatterplot of the Relationship Between Total SAT Scores and Freshman GPA at Clearwater College.

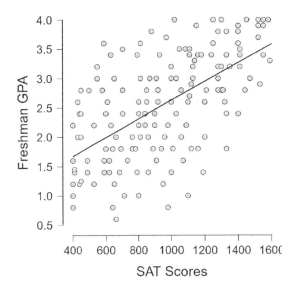

the range that the SAT variable can take and markedly reduces the correlation between SATs and freshman GPA.

The effects of using SATs as an admission criterion at Clearwater College is shown in **Figure 8**, where all points for SAT scores below 1300 have been eliminated. Notice that the remaining "cloud" of points doesn't suggest much of an association between the two variables. In general, when one or both of the variables in a correlational relationship are artificially restricted in value, the correlation will decrease in magnitude.

Figure 8

JASP Scatterplot of the Relationship Between Total SAT Scores and Freshman GPA at Clearwater College Only for those Students with SAT Scores Greater than 1300.

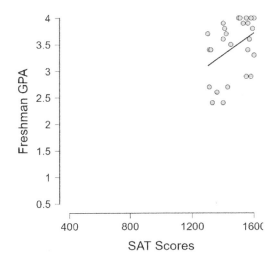

The moral to this little story is that the value of the correlation you compute is, in part, a function of how much spread or range you have in the scores. In doing or interpreting correlational research, you must be alert to the possibility that the sample on which the correlation is based may have been selected in such a way that the range of possible scores on one or both variables has been restricted.

Subgroup Discontinuities

Here is another story with a statistical moral. A good many years back, one of my classes in experimental psychology conducted a questionnaire study of the relationship between "hostility" and "friendship." There were several hypotheses tested in the study, but one of interest for our purposes concerned the relationship between a type of hostility called "expressive hostility" and the number of friends that a person has. Expressive hostility refers to the tendency to express one's anger and frustration by saying nasty things and being physically abusive. We reasoned that people who have a strong tendency toward expressive hostility would have fewer friends. After all, who wants to hang around with people who say and do nasty things.

A standard paper and pencil hostility test plus a class-constructed friendship questionnaire were distributed to the students in three introductory psychology classes. Complete data were available for 84 students. To test our hypothesis, we computed Pearson *r* between the students' expressive hostility scores (range 0–16) and the number of reported "close and casual friends" (0–40). The correlation was -.49. Naturally, we were pleased that our hypothesis had been confirmed. **Figure 9** shows what the scatterplot looked like for these data.

A correlation of -.49 is modest in magnitude, but in psychological research, researchers take what they can get. We then computed the correlation separately for male and female students to see if there was a difference between the sexes in the relationship. An odd thing happened. The correlation proved to be +.13 for male students and -.34 for females! Why did the relationship get weaker? The answer, strangely enough, is that it was really weak to begin with. The correlation computed on the combined male and female sample was an artifact of combining **discontinuous subgroups**.

A "discontinuity" is a sharp break or gap in something. In statistics, discontinuities occur in the values of variables. **Two groups are said to be discontinuous with respect to a variable if the two group means for the variable are substantially different in value.** This is what happened in our study. The female students had a markedly lower mean expressive hostility score than the males *and* a

Figure 9

JASP Scatterplot of the Relationship Between Hostility and Number of Friends for Male and Female College Students.

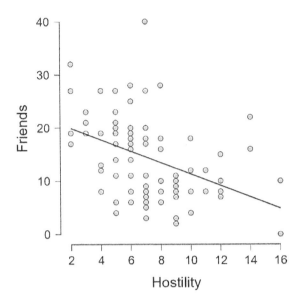

substantially higher mean friendship score than the males. The sexes were discontinuous with respect to both variables.

Separate scatterplots of the male and female data are depicted in **Figures 10** and **11**. Notice that for males, the cloud of points shows no relationship between the two variables.

Figure 10

JASP Scatterplot of the Relationship Between Hostility and Number of Friends for Just Male College Students.

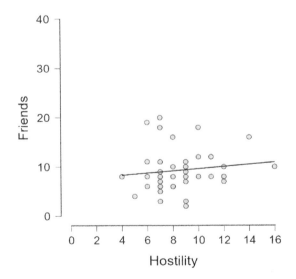

Figure 11

JASP Scatterplot of the Relationship Between Hostility and Number of Friends for Just Female College Students.

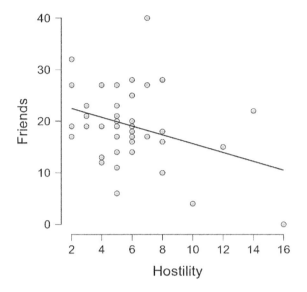

For women, there is still a negative correlation, but it's weak and is unduly influenced by one very hostile young lady who reported having no friends! But when gender is ignored and the two clouds combined (**Figure 9**), the resulting "super cloud" shows an inverse relationship between levels of expressive hostility and number of friends. We say that this combined relationship is artificial because the relationship is not clearly observed in either group taken separately. Said differently, the correlation between hostility and number of friends is mostly a spurious consequence of the third variable, gender.

Situations in which subgroup discontinuities in variables produce or obscure relationships is very common in social science research. Imagine all the possible subgroups—Asian Americans vs. Caucasians, males vs. females, Republicans vs. Democrats, Catholics vs. Protestants, workers vs. retirees, and so on. Pooling across multiple classes of subgroups may totally muddle numerous important relationships. This is one reason that researchers either work with relatively homogeneous populations (e.g., college sophomores) or analyze their data separately for obvious subgroups of subjects. As a beginning statistics student, you need to at least be aware of this problem in doing correlational research, and, when possible, compute scatterplots and correlations separately for obvious subgroups.

A Related Statistic - r^2

As you will learn in the next chapter, when two or more variables are correlated in a linear fashion, it is possible to derive a straight line equation for predicting values of the criterion variable Y from values of one or more predictor variables. This is called **linear regression** or **linear prediction**. When there is only one predictor variable in the equation, Pearson r is a measure of how accurately Y can be predicted from X. Unfortunately, r is not the *best* measure of the accuracy of prediction. The reason for this is that r is not related to accuracy of prediction in a simple, easy-to-comprehend fashion. **Figure 12** shows the nature of this relationship for positive values of r. The X axis contains values of r ranging from .00 to +1.00. The Y axis is the accuracy with which the linear equation will predict values of Y.

Notice that when $r = 0$, the accuracy of prediction is also 0. This is because an r value of 0 means that there is no linear relationship at all between X and Y. When $r = +1.00$, prediction is perfect. This makes sense because, as we have seen, when the correlation is perfect, all the data points in a scatterplot fall on a straight line. It's the shape of the function in between that is interesting. A value of r of .25 is associated with very low accuracy of prediction (about 6%). When r reaches .50 (a "moderate" to "strong" correlation), accuracy of prediction is only 25%. To reach just 50% accuracy of prediction, r must be .71.

Because of this complex relationship between values of r and accuracy of prediction, it is not easy to compare magnitudes of r. For example, recall that the correlation between students' confidence ratings and actual grades in introductory statistics was +.69. Suppose that the same test was carried out in an introductory literature class and the correlation was found to be +.46. How would we compare the magnitude of the two correlations? Clearly, the +.69 indicates a stronger association, but how much

Figure 12

Accuracy of Predicting Criterion Variable for Different Values of Pearson r.

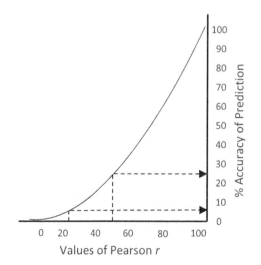

stronger? The value of the larger correlation is about 1.5 times bigger than the value of the smaller one. In terms of accuracy of prediction, however, inspection of **Figure 12** reveals that a correlation of +.69 represents about 48% accuracy of prediction, while a correlation of +.46 results in only 21% accuracy. In this respect, the larger correlation is over *twice* (2.3 times) as strong in predictive accuracy as the smaller one.

To help interpret the meaning of values of *r* and to facilitate comparisons between different values of *r*, statisticians make use of a related statistic called the **coefficient of determination**. This statistic is abbreviated r^2, because squaring *r* happens to be the simplest way to compute the statistic. The coefficient of determination (r^2) is a direct measure of accuracy of prediction. The fundamental formula definition of r^2 will be put off until we develop the necessary background knowledge in the next chapter. For the moment, we will use the following verbal definition. The coefficient of determination (r^2) is **that portion of the variability in the criterion variable, *Y*, that can be predicted from the predictor variable, *X*.**

In this definition, accuracy of prediction is expressed in terms of "predicted" variability in the *Y* scores. This is a new concept, but it is fairly easy to understand. Let's look at it by way of our example of guessing one's grades in introductory psychology. We had 10 criterion scores (final grades). These 10 scores varied considerably in magnitude—from R.B.'s low score of 75 to R.H.'s score of 97. Most of the differences in the final grades (i.e., variability of the scores) parallel differences in the students' confidence ratings. This is the predicted part of the variability. Some of the difference in final grades is not predicted from differences in confidence ratings. R.B., for example, got a relatively low final grade, adding to the total variability in the criterion scores, but his low grade was not predicted by his original rated confidence of 6. This lack of correspondence between anticipated and final grades is due to many factors, such as lack of insight into one's abilities, bad luck on exams, motivational changes during the semester, etc. Collectively, statisticians put these factors under the category of "error." The total variability of the criterion scores, then, consists of that portion which is predicted from the predictor variable, plus that portion which is not predicted and is due to error.

Note that r^2 is always a positive number and that it can range from 0 to 1.00. It is always a positive number because even if the correlation coefficient is negative, the square of a negative number is always positive. It ranges from 0 when *r* is 0 to 1.00 when *r* is ±1.00 . An r^2 value of .25 means that one quarter of the total variability among the criterion scores was predicted by differences in the predictor variable. This is the **predicted variability**. Three quarters of the variability among the criterion scores could not be predicted by the predictor variable. This portion is referred to as **error variability**. It is usual to convert the decimal fraction value of r^2 to a percentage, so we would say something like, "Twenty-five percent of the variability in the criterion variable was predicted by the predictor variable." Another way to say it is, "Twenty-five percent of the variability in the criterion variable was associated with variability in the predictor variable."

A good rule in correlational research is to compute both *r* and r^2. The former is the standard statistic for quantifying the degree of linear correlation between two variables; the latter is a useful statistic for helping interpret and compare different values of *r*.

Using *JASP* To Analyze A Correlational Study

What follows below is a strategy for analyzing correlational problems having more than two variables. We will develop the strategy in the context of the following example, called "The Right Stuff," which is presented on the next page. **Read through the description of the study before continuing**.

The first step is to create a data file of the Right Stuff data. This is shown in **Figure 13**. The file is labeled Right Stuff Data.csv.

Figure 13

Excel Spreadsheet Showing the Data for the First Five Cases of the Right Stuff Study.

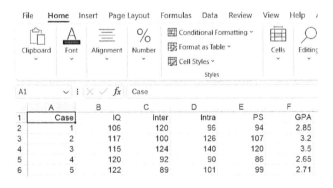

Univariate Analysis

Even though our primary interest in the problem lies in the relationships between variables, we still begin with a description of each variable—a univariate analysis. This is important for two reasons. First, by describing each of the variables, you build up a picture of the nature of the sample that you are dealing with. In our example, the values of the sample means will tell us the general ability level of the

The Right Stuff

Grad student Carol is interested in an area of research called "Practical Intelligence." The basic hypothesis underlying research in this area is that traditional intelligence tests (Stanford-Binet, Wechsler) measure only a portion of the cognitive abilities that we think of as intelligence. Probably everyone has encountered the academically gifted high school or college student who is so socially unskilled that he/she couldn't ask a member of the opposite sex for the time of day, much less for any more intimate form of social interaction. Such people seem deficient in what we might call "social" or "interpersonal" intelligence. Another type of practical intelligence is what is called "intrapersonal" intelligence. People high in this type of intelligence have an accurate understanding of their own strengths and weaknesses, know how to organize and regulate their lives in an efficient manner, and are free of debilitating beliefs and emotions. Remember Janie in high school, who was on the honor roll every semester, edited the school newspaper, did drama and debate, ran varsity track, and held a part-time job at McDonalds? This lady was high on intrapersonal intelligence. A third type of practical intelligence is called "problem solving" intelligence. People high in this domain have efficient strategies for defining and solving problems of all types and for acquiring new problem-solving skills. These are the "good ideas" people.

Carol and fellow student Joe conducted a correlational study to find out how these three aspects of practical intelligence were related to academic success in students at their institution, as measured by cumulative grade point average (*GPA*). For comparison purposes, they also included a standard test of academic intelligence. Below are the data for a sub-sample of 35 male students. The first four variables are the predictor variables: *Academic Intelligence (IQ)*, *Interpersonal Intelligence (Inter)*, *Intrapersonal Intelligence (Intra)*, and *Problem Solving Intelligence (PS)*. All variables are expressed in standard score form, with Mean = 100 and Standard Deviation = 15. The fifth variable was the criterion variable, College *GPA*.

Case	IQ	Inter	Intra	PS	GPA	Case	IQ	Inter	Intra	PS	GPA	Case	IQ	Inter	Intra	PS	GPA
01	106	120	96	94	2.85	13	132	85	100	126	3.50	25	120	102	100	111	2.50
02	117	100	126	107	3.20	14	115	103	100	84	1.75	26	120	121	113	89	2.80
03	115	124	140	120	3.50	15	126	115	131	100	4.00	27	117	107	121	107	3.15
04	120	92	90	86	2.65	16	110	86	120	92	2.46	28	111	94	96	90	2.90
05	122	89	101	99	2.71	17	120	106	105	122	2.50	29	122	105	100	96	2.90
06	105	100	85	108	2.93	18	106	102	72	78	2.00	30	112	75	106	100	2.85
07	110	116	138	102	4.00	19	106	94	96	110	2.15	31	139	130	126	130	3.50
08	129	108	132	107	3.75	20	108	121	124	93	3.25	32	104	106	132	92	3.00
09	109	127	110	92	3.00	21	114	111	118	102	3.00	33	116	118	87	117	2.95
10	110	111	86	107	2.00	22	126	100	100	129	3.10	34	122	104	106	109	3.20
11	106	140	103	104	2.85	23	128	81	111	135	3.65	35	107	92	102	100	2.75
12	120	122	120	129	3.75	24	114	93	90	117	1.90						

students with whom we are concerned, and the standard deviations give us an idea how homogeneous the students are with respect to ability levels. Second, careful analysis of each variable can alert you to potential problems in using Pearson *r* to describe the relationships among the variables. Small standard deviations or highly skewed distributions are tip-offs that a variable may show a restricted range in the sample you are studying, and a bimodal distribution almost always implies the presence of some type of subgroup discontinuity for a variable. The univariate statistics for the "Right Stuff" data produced by the *JASP* DESCRIPTIVES procedure are shown on the next page.

Figure 14 is a table of the descriptive statistics for the five variables of the study. The sample of students included show a high average *IQ* with only modest variation in their scores. *IQ*'s ranged from slightly above average to extremely high. In contrast to the *IQ* scores, the students were, on average, much less outstanding in the three areas of practical intelligence. The mean for each subtest was only modestly above the population average of 100. Also, unlike the *IQ* scores, students showed much greater variability in all three *Practical Intelligence* scores, with some students performing markedly below the population mean of 100.

The frequency distribution for all three subscales were irregular due to the relatively small sample size. **Figure 15** shows the distribution of the *IQ* scores. It's mode is at 105–110 and is fairly positively skewed (*Sk* = .631).

Figure 14

JASP Descriptive Statistics For the Sample of Students In the Right Stuff Study.

Descriptive Statistics for the "Right Stuff" Study

	IQ	Inter	Intra	PS	GPA
Valid	35	35	35	35	35
Median	115.000	105.000	105.000	104.000	2.930
Mean	116.114	105.714	108.086	105.257	2.941
Std. Deviation	8.584	14.847	16.610	14.364	0.570
IQR	11.500	23.000	22.500	20.500	0.545
Minimum	104.000	75.000	72.000	78.000	1.750
Maximum	139.000	140.000	140.000	135.000	4.000

The remaining distributions were either symmetrical about the sample mean (*Interpersonal* and *Problem Solving Intelligence*) or somewhat positively skewed (*Intrapersonal Intelligence*). The students in the sample were homogeneous with respect to *GPA*, with the mean being a B average. The distribution of *GPA* scores was quite flat and symmetrical about the mean.

Notice that the narrative interpretation of the tabled statistics tries to build up a "picture" of the sample of students—what they were like, compared with students as a whole. From such a picture, the reader can judge how generalizable the findings might be to other situations. This sample, for example, seems pretty "average" in practical intelligence compared with their *IQ* scores. Why might this be and what implications would it have for extending the results of this study to other groups of students?

Also notice that the students show relatively little variability in either their *IQ* scores or their *GPAs*. This has direct implications for the correlational analyses to be performed on the data. Restriction in the range of a variable reduces the correlation coefficient, so correlations involving *IQ* might be artificially low.

Correlation Matrix

Let's now move on to the analysis of the correlational information. We'll use the Correlation sub procedure under the REGRESSION procedure. When there are several variables in a correlational study, researchers begin by computing the correlation between *every possible pair of variables*. This may seem odd to compute all possible correlations, including those among the predictor variables. After all, wasn't the purpose of the study to find out whether measures of practical intelligence were related to the criterion of academic success? That objective would seem to imply that only four correlations be computed, i.e., each predictor with *GPA*. Not so. To make sense of the relationships, we need to know not only how the predictor variables relate to the criterion, but also how the predictor variables relate to one another.

The set of correlations among all possible variables is organized in what is called a **correlation matrix**. "Matrix" is just a fancy name for a table organized into rows and columns. In a correlation matrix, both the rows and columns of the matrix are labeled with the names of the variables in the same order. The *intersection* of any row and column contains the correlation between the two variables represented by the row and column. **Figure 16** shows the correlation matrix of the "Right Stuff" data produced by the *JASP* REGRESSION procedures, using the Correlation sub procedure.

In constructing the correlation matrix, it is a good idea to enter the variables so that the criterion variable is preceded by the predictor variables. If you do so, then the last row of the matrix contains the correlation between each predictor and the criterion.

Figure 15

Histogram Display of the IQ Scores for the Right Stuff Study.

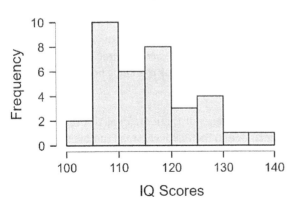

Figure 16

JASP Correlation Matrix for the Sample of Students in the Right Stuff Study.

Pearson's Correlations

Variable		IQ	Inter	Intra	PS	GPA
1. IQ	Pearson's r	—				
2. Inter	Pearson's r	-0.065	—			
3. Intra	Pearson's r	0.233	0.317	—		
4. PS	Pearson's r	0.557 ***	0.006	0.155	—	
5. GPA	Pearson's r	0.446 **	0.267	0.699 ***	0.392 *	—

* p < .05, ** p < .01, *** p < .001

If you inspect the row and column intersections, you will see the value of *r* displayed, followed by 0-3 asterisks. The asterisks indicate **the probability that one would obtain a sample value of *r* of the observed magnitude or greater by chance (random sampling), if the correlation in the population were equal to zero.** The key underneath the table indicates the ***p* value** associated with the number of asterisks. Because you are looking for real relationships, you want the *p* value to be small, indicating that the sample value represents a real correlation between variables rather than just sampling error. In practice, we normally require that the *p* value be less than .05 before we are confident that the sample value didn't occur through sampling error. We say that such correlations are **statistically significant**.

How do you interpret the information contained in the correlation matrix? Here is a useful approach. It involves four steps. We'll list the steps and use them to interpret the Right Stuff correlations displayed in **Figure 16**.

Step 1 Begin by examining the scatterplots of the relationship between each predictor variable and the criterion. If non-linearity is present, you need to be aware that the observed correlations may be understating the actual magnitudes of the relationships.

Figure 17 shows the four scatterplots for each of the four predictors and the criterion variable of GPA. There is considerable irregularity in the scatterplots, and it is difficult to be certain that all predictors relate to the criterion in a strictly linear manner. This is particularly true for the *PS* variable. The problem here is that the sample is sufficiently small that one or two discrepant scores can unduly influence the shape of the cloud of points. In the absence of clear-cut non-linearity, however, most investigators will simply proceed with the analysis.

Step 2 If the *p* value for a correlation is much greater than .05 or *r* < ± .20, assume that there is no linear

Figure 17

JASP Scatterplots for All Four Predictors and the Criterion Variable of GPA.

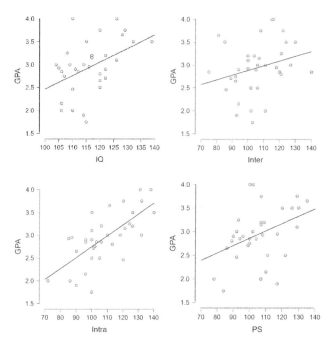

relationship between the variables. Eliminate these correlations from consideration. Inspection of **Figure 16** shows that there are five of them. *IQ* is not significantly correlated with either *Interpersonal* or *Intrapersonal Intelligence*. *Interpersonal Intelligence* isn't significantly correlated with anything, and *Intrapersonal* and *Problem Solving Intelligence* aren't significantly correlated.

Step 3 Examine the correlations between the predictor variables and the criterion variable. Identify which variables are significantly correlated with the criterion. Then, compare the relative magnitudes of the correlations between the predictors and the criterion to determine which are the stronger predictors. This is where r^2 can be very helpful.

Three of the predictors are significantly correlated with the criterion. Two of these, *IQ* and *Problem Solving Intelligence* are about the same magnitude ($r = +.45$ and $+.39$, respectively). These are moderate correlations. The third variable, *Intrapersonal Intelligence*, is strongly to very strongly correlated with *GPA* ($r = +.70$).

The r^2 values for *IQ* and *Problem Solving Intelligence* are .20 and .15, indicating that about 15% to 20% of the variability in *GPA* is associated with (predicted by) variability in each of these two variables. The r^2 value of *Intrapersonal Intelligence* is .49, which is 2.5 times larger than the values for the other two predictors. Almost 50% of the variability in *GPA* is predicted by variations in students' *Intrapersonal Intelligence*. We can conclude that *Intrapersonal Intelligence* is a *much* stronger predictor of *GPA* than either *IQ* or *Problem Solving Intelligence*.

Step 4 Examine the correlations among just those predictors that are significantly correlated with the criterion. **If two predictors are substantially correlated with one another, then their association with the criterion variable is mediated by some common process or mechanism**. If two predictors are not correlated substantially with one another, then their correlations with the criterion variable are via independent processes or mechanisms. In deciding whether two predictor variables are "substantially" correlated, use r^2 rather than r.

Notice first that *IQ* and *Problem Solving Intelligence* correlate $+.56$, which is a strong correlation. The r^2 value of .31 means that 31% of the variability in one of the variables is associated with variability in the other variable (in this case, neither variable is obviously the predictor or criterion). This means that the two predictor variables are correlated with *GPA*—at least in part—because of sharing something in common. One interpretation might be that scores on the subtest measuring *Problem Solving Intelligence* substantially reflect differences in plain old *IQ*, i.e., the two tests measure substantially the same thing.

How about *Intrapersonal Intelligence*? Here the situation is different. *Intrapersonal Intelligence* does not correlate significantly with either *IQ* or *Problem Solving Intelligence*. This means that its association with the criterion of *GPA* reflects the operation of some different process or mechanism. Clearly, regardless of one's *IQ* or *Problem Solving Intelligence*, people with higher levels of self-knowledge and self-management skills do better in college (but recall mean *IQ* was high with little variability).

To summarize the findings, here is a narrative description of the results of the correlational analysis:

Analysis of the correlations among the variables revealed that *GPA* was moderately predicted by *IQ* and *Problem Solving Intelligence* and strongly predicted by *Intrapersonal Intelligence*. *IQ* and *Problem Solving Intelligence* were strongly correlated, indicating that, at least in part, these variables were related to academic performance via some common mechanism. Probably the test of *Problem Solving Intelligence* is tapping basic *IQ* functions. To the extent that grades are influenced by academic intelligence (*IQ*), the test of *Problem Solving Intelligence* will also correlate with GPA.

Intrapersonal Intelligence was uncorrelated with the other two significant predictors, suggesting that self-management knowledge and skills may contribute to academic success independently of academic ability. Organized people who are not inhibited by neurotic problems and who know how to motivate themselves do better in college, regardless of level of academic intelligence.

Notice that this summary advances in a tentative way causal explanations for the observed correlations. These explanations may be wrong, of course, but it is important to remember that the purpose of doing correlational research of this type is to uncover cause-effect relationships.

Other Correlational Statistics

This chapter has introduced you to the most common correlational statistic, Pearson's r. It is by far the most widely used measure of correlation and is related to many other statistical procedures. But there are other correlation statistics that have various uses, either (a) for descriptive purposes when r isn't appropriate or (b) for quantifying how big an effect an independent variable has on a dependent variable. We'll consider three such correlational statistics—**Spearman's rho** (r_S), the **point-biserial correlation** (r_{pb}), and the **correlation ratio eta** (η).

Spearman's rho (r_S)

Spearman's rho is the Pearson correlation between two variables measured at the **ordinal level**. Recall from Chapter 1 that numerical data that are expressed as ranks are not appropriate for most statistical procedures in this text because ranking loses the actual distance between points in a variable. Spearman's rho is the exception.

Consider a study looking at the consistency of brands of wine over time. Cabernet Sauvignons from 10

California wineries are rank ordered for taste for the years 2010 and 2016. Here are the data.

Winery	2010	2016
A	3	2
B	8	6
C	5	7
D	2	1
E	7	4
F	1	3
G	4	5
H	6	8
I	9	10
J	10	9

The paired ranks data are entered into REGRESSION – Correlation procedure and Spearman's rho is selected as the statistic. *JASP* gives the value of rho as .818 with a *p* value of .007. We'll accept the conclusion that there is a very strong positive correlation between quality rankings over a six-year period for the wineries in question.

Interestingly, for these data the value for Pearson *r* and Spearman's rho are identical. That's not always or even usually the case. Pearson r measures the **linear relationship** between two variables while Spearman's rho measures the **monotonic relationship** between variables. Monotonic simply means that two variables systematically increase (positive relationship) or decrease (negative relationship) together. If, for example, a positive relationship between two variables is non-linear, Pearson r will be smaller than Spearman's rho computed on the same data when they are converted into ranks.

Figure 18 shows a simplified scatterplot of this situation. Notice that the best fit straight line doesn't precisely capture the shape of the points in the scatterplot, although it comes close. *JASP* computes Pearson r to be .929. When Spearman's rho is selected as the correlational statistic, the value is 1.00—a perfect monotonic relationship.

Spearman's rho is used purely for descriptive purposes. You cannot square rho to get the proportion of variation in the criterion variable predicted by the predictor variable and rho is not related to linear regression procedures discussed in the next chapter.

Point Biserial Correlation (r_{pb})

The point-biserial correlation (r_{pb}) is basically Person r in disguise. Recall that *r* requires that both variables are numerical and related in a linear fashion. Sometimes, however, one of the variables of interest is a categorical variable having just two values, such as political affiliation, traditional gender, or handedness. In that case, r_{pb} is the Pearson correlation between the numerical variable and the categorical variable coded 1 and 2.

For example, suppose 36 young adults identifying as male or female are asked to rate the importance of sex in a stable relationship using a 9–Point scale ranging from 1 "Not at All Important" to 9 "Critically Important." Here are the rating data for each gender:

Male (1)			Female (2)		
9	9	7	8	7	8	7
8	6	8	4	6	5
9	9	9	6	8	7
6	7	5	6	6	9
8	5	6	3	5	8
7	9		6	5	8

The data are entered into a spreadsheet with the value for gender paired with the rating. The data for the first case is 1,9, the second case 1,8, the third case, 1,9, and so on. You would use *JASP* REGRESSION – Correlation to compute the Pearson correlation between the paired values. This would be r_{pb}. *JASP* computes r_{pb} to be -.334, with an associated *p* value of .047. This means that there is a significant but very modest correlation between gender and rated importance of sex. The sign of r_{pb} is meaningless because it depends on how the variable gender was coded.

When squared, r_{pb} will show up in Chapter 11 on two-sample inference as a measure of what is called **effect size**, which is a way to quantify the practical importance of a statistically significant difference between two sample means.

Figure 18

JASP Scatterplot of a Curvilinear Relationship Between Variable X and Variable Y.

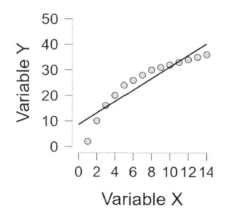

The Correlation Ratio eta (η)

Like r_{pb}, the correlation ratio (η) is a measure of the strength of the relationship between a categorical independent variable and a numerical (interval or ratio) dependent variable. When there are two categories of the independent variable (Chapter 11 on two-sample inference) r_{pb} and η are actually the same thing! It is rarely used for descriptive purposes, but when squared (η^2), it's an extremely useful measure of the effect size of an independent variable.

A better understanding of eta depends on the material in the next chapter on linear regression and the chapters on analysis of variance (Chapters 12–15), but a simplified example will get you started.

Imagine an experiment in which 30 student volunteers learn a list of 20 words and then are asked to recall them after two weeks. There are three conditions of the experiment depending upon the type of words making up the list, with 10 students randomly assigned per condition. In the first condition the words are unrelated. In the second condition the words are all descriptors of happy situations or emotions. In the third condition the words are descriptors of sad or scary situations.

Here are the number of words recalled for each of the three conditions:

Unrelated Words	Happy Words	Sad/Scary Words
10 8 8	12 10 13	9 16 8
6 4 9	9 11 14	10 7 8
4 7 2	7 10 12	11 9 11
5	13	17
$\bar{X}_U = 6.30$	$\bar{X}_H = 11.10$	$\bar{X}_S = 10.60$
$\hat{s}_U = 2.54$	$\hat{s}_H = 2.13$	$\hat{s}_S = 3.37$

Look at the 30 scores for the experiment. Obviously the scores vary a great deal—from a low of 2 words recalled in the Unrelated Condition to a high of 17 recalled in the Sad/Scary Condition. The variability of all 30 scores is called the **total variability**. The variability among the three condition means allows computation of what is called the **treatment variability**. Eta is equal to the square root of treatment variability divided by the total variability, i.e.

$$\eta = \sqrt{\frac{Treat\ Var}{Total\ Var}} = \sqrt{\frac{139.27}{340.67}} = .640$$

Eta is sensitive to any type of association between the independent variable and the dependent variable. It varies from 0 to 1.00 and the magnitude can roughly be interpreted in the same way as Pearson r.

As noted above, eta usually is not used for descriptive purposes, but instead it is used in the same way r^2 is used—to measure the proportion of the variability in the dependent variable accounted for by the independent variable. In this case, $\eta^2 = .410$, which means that the type of word accounts for 41% of the variability in the number of words recalled.

Final Comments

This is a fairly lengthy chapter that has introduced a good many new concepts and procedures. It is, however, an extremely important chapter because correlation and linear regression (next chapter) are probably the most widely used research tools in the social and biological sciences.

The four-step process for interpreting a correlation matrix described above is a very useful way to approach real-world correlational research—but it takes practice! Many students who believe that they completely understand the procedure turn out to have problems applying it when test time comes. To get some practice in applying the procedure, there are two realistic correlation problems at the end of the chapter.

Chapter 5, Pearson Correlation
Chapter Exercises

Some questions are purely factual and have only one correct answer. Others require greater thought on your part and may have more than one possible answer.

1. Fill in the missing term(s) that makes the statement true.

   a. _____ is the tendency for values of one variable to be systematically related to _____.
   b. Pearson correlation is abbreviated _____.
   c. ρ is the _____ in the _____.
   d. Pearson correlation is appropriate only when variables are _____.
   e. In correlation, X is usually the _____ and _____ is the criterion variable.
   f. _____ are graphs of correlational relationships.
   g. Negative values of Pearson correlation coefficients indicate _____.
   h. The straight line drawn in a _____ represents the _____.
   i. If the line in a scatterplot is horizontal it indicates that _____.
   j. r is defined as the _____ divided by _____.
   k. The closer the cloud of points hugs the best fit straight line, the _____.
   l. Because r can't go below -1.00 or above +1.00, it is called a _____.
   m. In an "inverted U" relationship between variables, r will always _____.
   n. Unlike $\bar{X}$, r is extremely _____ for small samples.
   o. A Pearson correlation of -.25 indicates a _____ relationship.
   p. If two variables are strongly correlated, we can conclude _____.
   q. In a _____ relationship, X and Y are not themselves _____ related.
   r. Computing the correlation between two variables with a third variable "controlled" is called _____.
   s. If X is a cause of Z, which in turn is a cause of Y, the relationship is said to be _____.
   t. If a variable shows a very small value for $\hat{s}$, we may be dealing with a restriction in _____.
   u. If two variables are uncorrelated in two subgroups, but highly correlated when the groups are combined, we are dealing with _____.
   v. SAT scores for students enrolled at Yale University are uncorrelated with grades because _____.
   w. If r = + .60, the coefficient of determination equals _____.
   x. An r^2 value of .40 indicates that the predictor variable accounts for _____ in the criterion variable.
   y. Analysis of the "Right Stuff" data begins with _____.
   z. A table of all possible correlations among variables is called _____.
   aa. W and X are strongly correlated with Y, but only weakly correlated with each other. From this we can conclude that _____.
   bb. $r_1 = .30$ and $r_2 = .60$. The second correlation is _____ times stronger than the first.

2. Drawing rough outlines of scatterplots is a useful way to make sense of how subgroup variable discontinuities can affect an overall Pearson r values. Here's an example. The correlation between X and Y for Group A = - .45 and for Group B = - .65. When the two groups are combined, r = - .20. What's going on? Draw a sketch of the situation, e.g.,

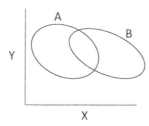

Here are three more problems:

   a. The correlation between X and Y for Group A = + .50 and for Group B it's - .50. When the two groups are combined, r = 0. What's going on?

b. The correlation between X and Y for Group A = + .05 and for Group B it's - .10. When the two groups are combined, r = + .80. What's going on?

c. The correlation between X and Y for Group A = + .40 and for Group B it's + .35. When the two groups are combined, r = - .75. What's going on?

3. Why do people "ghost" friends and acquaintances? The usual explanations include avoiding conflict, having hurt feelings, protection of self-esteem, and revenge for perceived slights. Proposed traits include such things as self-centeredness, avoidant personality, selfishness, etc. These seem pretty common sense, but they are really mostly descriptions of behavior. Are there stable personality characteristics that predispose people to ghost? As a class project in Psych 305, *Theory & Research in Personality*, one research team proposed to see whether any of the "Big Five" personality dimensions predicted ghosting behavior. The Big Five dimensions are:

Openness to Experience The extent to which a person is "open" to new ideas and experiences.
Conscientiousness Being organized, goal oriented, persistent.
Extraversion Engaged with the social, external world, enjoying people.
Agreeableness Willing to put other's needs and interests ahead of their own.
Neuroticism Sensitivity to stressors and proneness to negative emotions.

At the beginning of the course, all 25 students in the class had taken a version of the Big Five personality test and agreed to use the information for class projects. The test scores were coded for purposes of anonymity. The research team constructed a brief questionnaire to measure how often and persistently a person had ghosted others in the past two months. Class members completed the questionnaire and turned it in to the professor who scored the questionnaire and linked the results to the individual's personality test scores. Each student knew only his/her own scores, no one else's. Here are the data for the 25 class members. Each of the Big Five dimensions was scored using T scores (μ = 50 and σ = 10) and the Ghosting scores ranged from 0 to 15.

Student	O	C	E	A	N	G
1	52	64	60	45	50	0
2	40	60	50	52	40	5
3	55	50	55	46	42	6
4	60	65	44	45	35	1
5	58	48	44	55	60	7
6	65	55	35	48	55	4
7	53	40	44	35	38	10
8	50	58	44	50	52	3
9	35	54	50	40	45	9
10	40	42	40	40	38	15
11	57	52	50	45	47	7
12	60	46	42	45	56	8
13	52	64	42	50	30	0
14	65	60	58	45	35	1
15	50	54	46	60	40	2
16	50	55	42	58	52	4
17	56	50	43	52	60	6
18	42	44	35	60	55	0
19	62	48	44	50	54	5
20	45	32	35	44	62	9
21	50	50	55	40	42	5
22	55	55	44	40	45	3
23	45	54	45	45	42	7
24	68	66	40	65	50	1
25	73	46	44	50	45	6

Using your accumulated knowledge from Chapters 1–5, do a complete analysis of these data and write an interpretative summary of your findings.

4. Honda Motors is building a new electric vehicle (EV) assembly plant in Eastern Texas. Although labor is plentiful, virtually no potential workers have had experience building EVs. For this reason, management has developed an intensive three-week training program to teach new employees the skills and attitudes needed to assemble the new high tech cars. It is anticipated that it will cost $9000 to train each new employee, including those who do not pass the training program. In view of this cost, management wishes to develop a relatively short assessment battery to screen out individuals who are not likely to succeed in the training program. A consulting firm recommends that Honda employ a battery of three paper-and-pencil tests for selection purposes: (a) the "Conscientiousness" scale of a well-known personality test, (b) the Haversham-Jones *Test of Mechanical Aptitude*, and (c) a widely used multiple choice IQ test. The following study was carried out to evaluate the effectiveness of this test battery before employing it for actual worker selection.

Job candidates were randomly selected from the much larger applicant pool and invited to complete the normal application process, plus the new selection test battery. Regardless of their scores on the selection test battery, 50 persons who passed the usual Honda interview were then put through the training course at the company's assembly plant in Marysville, Ohio. At the end of the course, each prospective employee was given a composite performance score, which was the weighted average of supervisor ratings, instruction test scores, on-the-job speed and quality scores, and peer ratings. Below are the test battery scores and performance scores for the 50 workers (each set of four scores in a row represents the data from one prospective employee). Note: CS = Conscientiousness Scale (0–25), HJ = Haversham-Jones *Test of Mechanical Aptitude* (0–50), IQ = IQ test (75–150), and PS = final performance score (0–100).

CS	HJ	IQ	PS	CS	HJ	IQ	PS	CS	HJ	IQ	PS	CS	HJ	IQ	PS
23	19	107	92	21	42	102	83	21	28	110	90	15	26	119	91
17	29	118	84	19	37	120	96	22	21	131	55	9	19	113	59
18	18	91	65	14	26	114	80	17	29	118	82	14	29	97	51
19	21	100	78	11	32	106	75	21	43	102	68	18	47	108	71
19	29	112	88	20	18	113	87	10	36	126	64	18	28	113	84
12	33	128	98	22	29	110	91	19	27	93	46	15	38	147	62
15	18	111	83	15	26	134	51	25	31	112	96	12	34	107	73
23	26	106	89	8	11	98	66	14	35	106	76	20	41	103	78
26	29	127	71	19	31	113	93	10	38	105	67	8	38	111	61
21	30	112	79	13	17	126	76	10	37	100	51	17	26	101	77
5	25	133	43	17	19	112	85	15	26	100	73	16	31	110	72
20	28	126	71	17	29	86	51	13	17	99	63				
15	21	146	48	13	33	95	62	16	25	96	74				

Use *JASP* to do a complete analysis of the data and make your recommendations concerning which, if any, of the three tests should be used for employee selection.

Chapter 6
Linear Regression

What's In This Chapter?

Linear regression is a statistical procedure in which a linear equation is used to predict values of a criterion variable from knowledge of the values of one or more predictor variables. If there is only one predictor variable, the procedure is called **bivariate linear regression**. When there are two or more predictor variables, the procedure is called **multiple linear regression**. The equation that best describes the relationship between the predictor variable(s) and the criterion variable is derived using the **least squares criterion**. When there is only one predictor variable, the predictive equation generates a straight line. Two or more predictor variables generate predictive **planes**.

Analyzing data with regression analysis is a multi-step process in which one begins by including all predictor variables in the regression equation and then refining the equation by eliminating those variables that do not contribute significantly to predicting the criterion.

In the **raw score form** of the regression equation, the *B* **values** indicate the number of units that the criterion variable will change for each unit increase in the predictor variable. The raw score form of the regression equation is useful when we want to make predictions for members of the population who were not included in the sample on which the equation was derived. In the **standard score form** of the regression equation, the **beta (β) values** indicate the number of standard deviations that the criterion variable will change for each standard deviation increase in the predictor variable. Because the β values are "standardized," their magnitudes directly reflect the strength of the predictive relationship between each predictor variable and the criterion variable.

The **coefficient of multiple correlation (*R*)** is the Pearson correlation between the actual values of the criterion variable and the values predicted by the regression equation. The **coefficient of multiple determination (R^2)** is the proportion of the total variation in the criterion variable predicted by the regression equation. **Adjusted R^2** is the value of R^2 corrected for sampling errors in estimating the regression coefficients. The **standard error of estimate ($s_{Y.X}$)** is the standard deviation of the errors in predicting the values of the criterion variable from the regression equation. It is used primarily to establish **confidence intervals** for predictions made using the regression equation.

The coefficient of determination (r^2) introduced in Chapter 5 is defined in terms of the predicted and total variance components derived from the bivariate regression equation.

Introduction

In Chapter 5, you learned how to use r and r^2 to describe the direction and magnitude of the correlation between two numeric variables. The basic assumption for using these statistics was that the two variables were related in a linear or straight-line fashion. In this chapter, we will focus on actually finding the linear equation that best characterizes the relationship between one or more predictor variables and a criterion variable. We will then use this equation to accomplish two important functions: (a) **predicting the value of the criterion variable for different values of the predictor variable or variables**, and (b) **when there are two or more predictor variables, measuring the magnitude of each variable's *unique* association with the criterion variable**.

We are going to approach the topic of linear regression in a manner that is different from many other introductory statistics texts. The usual way to explain regression is to limit discussion to situations in which there are only two variables—the predictor and criterion variables. This is called **bivariate linear regression** or **bivariate linear prediction**. The reason for doing this is that bivariate regression is closely related to Pearson correlation and many authors want to emphasize the relationship between the two techniques.

Unfortunately, as you learned in Chapter 5 on Pearson correlation, contemporary correlational research rarely is concerned with just two variables. Almost always, one is interested in how several predictor variables relate to some criterion variable of interest. For this reason, we are going to concentrate on a more general conception of regression that permits one to examine the relationship between any number of predictor variables and the criterion variable. This approach is called multiple linear regression or **multiple linear prediction**. As you will see below, bivariate linear regression is simply a "special case" of this more general approach. We'll begin with a general review of linear equations.

Linear Equations

You learned in high school algebra that any straight line can be defined by an equation of the following form:

$$Y = A + BX$$

where: Y = value of the function,
X = value on the X axis,
A = Y intercept, and
B = slope of the line.

In this equation the variable Y is a function of the variable X. The precise way that Y changes as a function of X is determined by the values for the coefficients or constants of the equation—"A" and "B." The coefficient A is the **Y intercept** when $X = 0$, i.e., where the line crosses the Y axis. The coefficient B is the **slope** of the line, i.e., how much Y will increase or decrease for every unit increase in X.

By way of example, **Figure 1** shows the plots for three linear equations. Line A intercepts the Y axis at +12 and has a negative slope of –3.00, i.e., the line decreases 3 units for every unit increase in X. Line B has a very gradual slope of +.33 and a positive Y intercept of +2.00. Line C slopes upward (+.50) and intercepts the Y axis at –3.00.

Figure 1

Plots of Three Straight Lines Having Different Slopes and Y Intercepts.

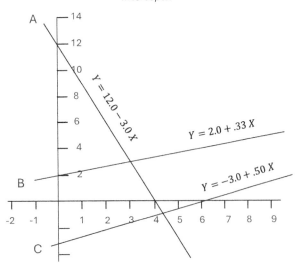

We can expand the concept of linear equations by making Y a function of more than one variable. Here is a linear equation in which Y is an additive function of two variables, X_1 and X_2:

$$Y = A + B_1X_1 + B_2X_2$$

In this equation, B_1 indicates how much Y changes for every unit increase in X_1 and B_2 indicates how much Y changes for every unit increase in X_2. The constants B_1 and B_2 are still slopes. The constant A is still the value of Y when both X_1 and X_2 equal 0. It is still the Y intercept.

When there is only a single predictor variable, graphing a linear equation produces a straight line that extends to infinity in each direction. What gets generated when Y is a linear function of two variables? The answer is that linear equations of this type generate linear *surfaces* called **planes**. To graph a function of this type, we need a three-dimensional axis system. **Figure 2** shows what the coordinate system would look like.

Figure 2

Three-Dimensional Coordinate System Needed to Graph an Equation in Which Y is a Linear Function of Two Variables.

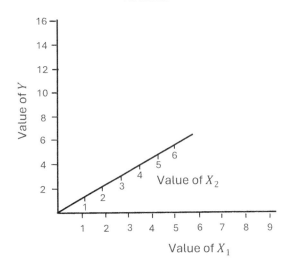

The Y axis is still the vertical axis, but X_1 and X_2 are axes at right angles to one another and to Y. We'll use this coordinate system to graph the following linear equation:

$$Y = 10.00 - 1.00X_1 + 1.75X_2$$

According to this equation, Y *decreases* 1.00 unit for every unit increase in X_1 and *increases* 1.75 units for every unit increase in X_2. When X_1 and X_2 are both equal to 0, Y is equal to +10.00. **Figure 3** shows a portion of the plane generated by this equation (the plane technically extends to infinity in all directions). Notice that the plane is tilted

Figure 3

Three-Dimensional Plane Generated by Linear Equation Having Two Predictor Variables.

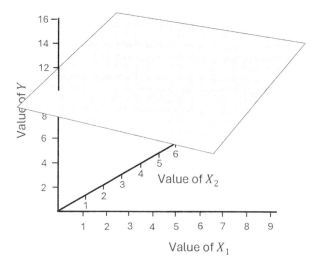

down in the X_1 axis, but in the X_2 axis direction, the plane tilts *up*.

There is no limit to the number of variables that you can have in a linear equation. Here is an equation in which Y is an additive function of three variables, X_1, X_2, and X_3.

$$Y = -7.25 + 1.50X_1 - 2.35X_2 + .75X_3$$

According to this equation, *Y increases* 1.50 units for every unit increase in the variable X_1, *decreases* 2.35 units for every unit increase in X_2, and *increases* .75 units for every increase in X_3. When X_1, X_2, and X_3 are all equal to 0, Y equals -7.25. We have three "slopes" and one Y intercept. The only problem with this equation is that it's not easy to draw a picture of the plane which it generates. To do so, we would need a four-dimensional axis coordinate system in which the axes X_1, X_2, and X_3 are at right angles to one another and to the Y axis. Unfortunately, our three-dimensional world doesn't lend itself to drawing four-dimensional axis systems. Happily, the fact that we can't easily draw the plane generated by this and other "higher order" linear equations doesn't diminish their usefulness. Mathematicians call such undrawable planes "hyperplanes" and are not at all disturbed that we can't draw pictures of them.

Linear Regression Analysis

Having introduced the concept of linear equations in a general way, we will now see how they are used in statistics. The basic assumptions of linear regression are the same as those for Pearson correlation:

1. All the variables involved in the regression analysis are numerical.

2. The predictor variables are related to the criterion variable and to one another in a linear fashion.

Let's begin with an explicit statement of the purpose of the statistical procedure known as linear regression. **Linear regression is a statistical procedure whose purpose is to write a linear equation that best allows us to predict values of a criterion variable from values of one or more predictor variables.** By statistical convention, the equation takes the general form:

$$Y = A + B_1X_1 + B_2X_2 + \cdots + B_iX_i$$

The "+...+" notation means that you can have as many predictors in the equation as desired. Note that bivariate linear regression is just a special case of this more general formulation of linear regression. In regression terminology, the Y intercept is simply called the **Constant**, and the slopes are called the ***B* Values**.

Linear regression works in the following general way. In some population of interest, we want to investigate the extent to which a criterion variable can be predicted by one or more predictor variables. We may be interested in the predictive relationship because we want to find a convenient and inexpensive way to predict some future event, such as predicting success in college from performance in high school, or we are seeking a theoretical understanding of the causal connections between variables. In either case, what we will do is to select a random or representative sample of cases from the population and find the equation that permits the most accurate prediction of the criterion variable from the one or more predictor variables.

Our ultimate interest, of course, is not in predicting values in the sample data. We already know those values. The interest, as usual, lies in using the sample data to estimate what the relationships are like in the population. More specifically, **the *A* and *B* values obtained from the sample are only estimates of their corresponding population parameters.** To the extent that the sample is not representative of the population, the *A* and *B* values of the sample equation may be substantially in error and will generate erroneous predictions.

To help us develop our understanding of linear regression analysis, we will conduct an analysis of the data from the "Right Stuff" study described in Chapter 5. Recall that in that study, two students were interested in how different types of "practical intelligence" are related to success in college. The predictor variables used in the study were scores on a traditional IQ test (*IQ*) and tests of "Interpersonal Intelligence" (*Inter*), "Intrapersonal Intelligence" (*Intra*), and "Problem Solving Intelligence" (*PS*). The criterion variable was cumulative grade point average (*GPA*). Our focus then was on describing the magnitude of the linear relationships between the predictor variables and the criterion using Pearson correlation. In a regression analysis of these data, our purpose will be to generate the optimal linear equation for predicting values of the criterion variable (*GPA*) from some combination of the predictor variables.

We will begin our explanation of regression analysis by focusing on the bivariate relationship between *Intra* and *GPA*. **Figure 4** shows the scatterplot of these data.

Figure 4

JASP Scatterplot of the Relationship Between Intrapersonal Intelligence and Freshman GPA.

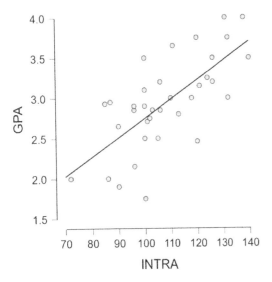

Looking at the scatterplot, it is obvious that there are an infinite number of different straight lines that could be drawn through the cloud of points making up the scatterplot, each defined by a different pair of values for *A* and *B*. Why is the straight line imposed by *JASP* the one that leads to the most accurate prediction of *Y* for the different values of *X*? In situations like this, statisticians use what is called the **least squares criterion** for selecting the values of *A* and *B* to maximize accuracy of prediction. **The least squares criterion says to choose values for *A* and *B* that make the expression $\sum(Y - Y')^2$ as small as it can be for the set of data.**

That is, we choose values for *A* and *B* such that

$$\sum(Y - Y')^2 \to Min$$

In the expression, *Y* stands for the actual value of the *Y* score for a given value of *X*, and *Y'* is the predicted value of *Y* for that value of *X*. Thus, (*Y* − *Y'*) is the discrepancy between the predicted and actual value of *Y*. This discrepancy is called the **residual**, or sometimes the **residual error of prediction** or **residual error**. The discrepancies are squared to eliminate negative values so that the sum of the values will always be a positive number. The least squares criterion has the advantage over other possible criteria in that it always yields a unique pair of values for *A* and *B*, i.e., only one set of values satisfies the criterion.

Here is an important point. Although formulas derived from the least squares criterion are available for computing the values of *A* and *B* "by hand" in the bivariate situation, it is not feasible to do so when there is more than one predictor variable. As a result, we will let the REGRESSION – Linear Regression procedure in *JASP* do all the computations and we will focus purely on developing a conceptual understanding of the results of those computations.

For these data shown in the scatterplot in **Figure 4**, the Linear Regression procedure computed the value of the *Y* intercept (*A*) and slope (*B*) for the best fitting straight line to be:

$$Y' = .351 + .024X$$

Or substituting the actual name of the predictor variable in the equation:

$$Y' = .351 + .024 \, (Intra)$$

Figure 5 shows the scatterplot of the *Intra* and *GPA* variables. For three of the cases, the (*Y* − *Y'*) discrepancies have been indicated by vertical lines, along with the actual magnitude of the discrepancy. If these values, along with the discrepancies for the other 32 data points were squared and then added up, the resulting sum would be as small as it could possibly be for this set of data, $\sum(Y - Y')^2$ would be minimized. **If any other line were used, the accuracy of prediction would be reduced.**

One consequence of using the least squares criterion for choosing the best-fit line is that the line will always pass through the point defined by the intersection of a horizontal line drawn at the mean of *Y* values and a vertical line drawn at the mean of the *X* values. Because of this, when the value of *X* equals $\bar{X}$, the predicted value of *Y* equals $\bar{Y}$.

What might we do with this equation? In a practical sense, we could use it to predict how well other students

Figure 5

JASP Scatterplot of the Relationship Between Intrapersonal Intelligence and Freshman GPA with Discrepancies for Three Cases Identified.

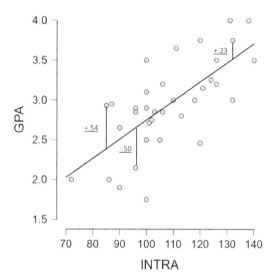

would do in school, knowing only their scores on the Intrapersonal Intelligence test. We might, for example, give the test to incoming first-year students and get each one's predicted *GPA*. If we believe that self-management skills are causally related to performance in college, we might identify those students whose predicted *GPA* is unacceptably low and give them special attention aimed at improving their self-management skills.

For example, Joe College takes the Intrapersonal Intelligence test during freshman orientation and receives a score of 78. What is his predicted freshman GPA? We would use the prediction equation derived from our sample and solve the equation for Joe's score of 78:

$$Y' = .351 + .024 \ (Intra)$$
$$= .351 + .024 \ (78)$$
$$= .351 + 1.87$$
$$= 2.22$$

Based on his limited Intrapersonal Intelligence, the equation predicts that Joe is going to wind up with a C– average. Of course, given that the correlation between Intrapersonal Intelligence and freshman *GPA* is only .70, we would expect considerable error in our prediction. The r^2 value of .49 ($.70^2$) means that over half of the variability in *GPA* is not predicted by Intrapersonal Intelligence. Joe could do better than his predicted score or worse, but we would bet that his *GPA* would at least be in the vicinity of 2.22.

It is important to understand that **we would be justified in using the regression equation to make predictions for other students only if the values of *A* and *B* derived from the sample data were close to the values of *A* and *B* for the population as a whole**. In other words, the sample values of *A* and *B* must be accurate estimates of their corresponding population parameters. This will be true when the sample on which *A* and *B* are computed is *large* (35 is too small to obtain really adequate estimates) and *representative* (ideally, a random sample from the population of interest, i.e., several years' worth of incoming students).

Let's expand our example a bit. Recall that although *Intra* was the strongest predictor of *GPA*, traditional *IQ* and *PS* also were significantly correlated with *GPA*. Of the two, the correlation for *IQ* was a bit stronger. Recall also that *IQ* and *Intra* were not correlated with each other, indicating that each was related to *GPA* via independent mechanisms or pathways. Given that *IQ* was substantially correlated with *GPA*, we should be able to improve our ability to predict *GPA* by using both *IQ* and *Intra* in the prediction equation. We can visualize this situation by starting with the three-dimensional scatterplot of the data which is shown in **Figure 6**.

Figure 6

Three-Dimensional Scatterplot of the Relationship Between Intrapersonal Intelligence and IQ and Freshman GPA.

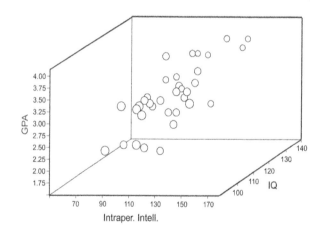

In **Figure 6**, the *Y* axis represents values of the criterion variable, *GPA*. The axis in the plane of the page (the usual *X* axis) is scaled for the predictor variable, *Intra*. The implied third dimension axis is scaled for *IQ*. The "balls" floating in our hypothetical three-dimensional space represent the location of each case with respect to values for the three variables. The size of the balls is used to imply depth into the third dimension—with larger balls being closer to the plane of the page. The "cloud" of balls is now a three-dimensional cloud, like the ones outside your window. Notice that the cloud rises along both of the horizontal axes becoming higher with increasing *Intra* scores and higher with increasing *IQ* scores.

Because the values of *GPA* now exist in a three-dimensional cloud, a straight line is not an effective way to predict the criterion variable. Instead, we need a three dimensional entity for making predictions. We need, in fact, a *plane* of predicted *Y* values to accurately approximate our cloud of actual *Y* values. And that is what our prediction equation will generate. The linear equation used for predicting *GPA* from *Intra* and *IQ* will take the following form:

$$Y' = A + B_1\,(Intra) + B_2\,(IQ)$$

Once again, the task is to select values of A, B_1, and B_2 to maximize accuracy of prediction. And the solution, once again, is the least squares criterion. The coefficients A, B_1, and B_2 are given numerical values that cause the sum of the squared predictive errors $\sum(Y - Y')^2$ to be as small as it can be for these data. As was the case for the situation involving one predictor variable, the least squares criterion generates a unique solution.

For the "Right Stuff" data, the actual equation generated by *JASP* is:

$$Y' = -1.693 + .022\,(Intra) + .020\,(IQ)$$

According to this equation, predicted *GPA* increases by .022 units for every unit increase in *Intra* and .020 units for every unit increase in *IQ*. On first glance, the *B* values for the two predictor variables seem pretty small. But think about it some more. An *IQ* point is pretty small itself. It stands to reason that you wouldn't get a huge increase in *GPA* for one point increase in *IQ*. But increasing 10 *IQ* points gets you a predicted *GPA* increase of .02 x (10) = .20, which is a meaningful change in *GPA*.

Figure 7 shows the predictive plane generated by this equation "inserted" into the three-dimensional cloud of the scatterplot. Notice that the plane tilts upward in both horizontal axes. Data points above the plane are represented by white balls, while those below the plane by black balls. Inspection of the scatterplot suggests that the plane representing the predicted values of *Y* for the values of the two predictor variables does a rather good job of approximating the actual value of *Y*. We will learn a couple of statistics in the next section that tell us just how well the equation can do in predicting the criterion variable.

Just as the regression line in bivariate regression passes through the intersection of the means of *X* and *Y*, the plane generated by two predictor variables "pivots"

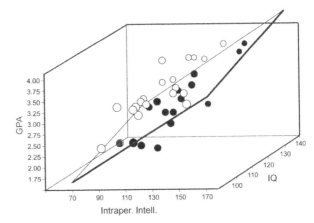

Figure 7

Three-Dimensional Scatterplot of the Relationship Between Intrapersonal Intelligence and IQ and Freshman GPA Showing the Best-Fitting Plane.

around the three-dimensional intersection of the means of X_1, X_2, and *Y*.

We could use this equation to make an even more accurate prediction regarding Joe College's academic success during his freshman year. Recall that Joe got a 78 on the test of Intrapersonal intelligence and based on that, his predicted *GPA* was 2.22. Suppose that Joe's *IQ* score is also available. He's a bright guy, having a measured *IQ* of 125, so we might expect that his academic intelligence could compensate somewhat for his weak Intrapersonal intelligence. Putting his two scores into the regression equation yields the following:

$$\begin{aligned}Y' &= -1.693 + .022\,(Intra) + .020\,(IQ) \\ &= -1.693 + .022\,(78) + .020\,(125) \\ &= -1.693 + 1.716 + 2.500 \\ &= 2.52\end{aligned}$$

Well, it helped a little bit—now we predict a solid "C" average. The fact that predicted *GPA* increased by only .30 points reflects the fact that *IQ* has a much weaker association with *GPA* than Intrapersonal intelligence (r = +.45 v .70).

Let's finish up our introduction to multiple linear regression by adding the *PS* variable to the regression equation. The least squares criterion yields the following "best fit" equation:

$$Y' = -1.662 + .021\,(Intra) + .013\,(IQ) + .007\,(PS)$$

We now have too many variables to draw the scatterplot of the data and associated plane generated by the equation. That would take four dimensions, and we have only three at hand. But no matter. The idea is exactly the same. We can say that predicted *GPA* increases by .021 units for every increase in *Intra*, by .013 units for every unit increase in *IQ*, and by .007 units for every increase in *PS*.

Joe got a 137 on the Problem Solving intelligence test. Putting in all three of his scores in this regression equation yields:

$$\begin{aligned} Y' &= -1.662 + .021\ (Intra) + .013\ (IQ) + .007\ (PS) \\ &= -1.662 + .021\ (78) + .013\ (125) + .007\ (137) \\ &= -1.662 + 1.638 + 1.625 + .959 \\ &= 2.56 \end{aligned}$$

If it doesn't seem like Joe is making much progress, it's because he isn't. As we'll discuss below, adding Problem Solving Intelligence to the equation does little to improve accuracy of prediction, because *IQ* and Problem Solving Intelligence are themselves fairly strongly correlated. Once one of them is entered into the equation, addition of the other has little effect.

We could, if we wished, expand the regression equation one more step by including Interpersonal Intelligence as a fourth predictor variable. We won't bother to do so, because by now you probably have caught onto the idea.

This ends our "conceptual" introduction to multiple linear regression. Before going on to the more detailed treatment of the topic in the next section, make sure that you understand the basic ideas involved. If you don't, you are likely to get bogged down in the statistical detail that follows. If you are uncertain about your understanding of the topic, go back and re-study the section on linear equations and this section on their application to making predictions.

Doing Regression Analyses

Initial regression Analysis

In our conceptual introduction to multiple regression, we deliberately skipped over most of the details associated with the procedure. What follows now is a more elaborated presentation of how you conduct a regression analysis of a set of data. As noted above, the appropriate *JASP* procedure is called REGRESSION – Linear Regression. **Figure 8** shows a screen capture of the *JASP* Statistics Panel after GPA has been entered into the Dependent Variable window and the four predictors entered into the Covariates (predictor variables) window.

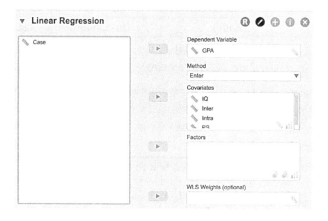

Figure 8

JASP Statistics Panel Showing the Linear Regression Procedure.

JASP provides four ways to proceed with the analysis under the Method option just below the Dependent Variable window. The default is "Enter," which means that all predictor variables are entered simultaneously. The other options are "Backward," "Forward," and "Stepwise" in which *JASP* automatically selects variables for inclusion or exclusion. For expository purposes, we'll stick with the default "Enter" option.

Entering all variables simultaneously gives you the best overview of what is going on. Based upon the initial examination of the equation and associated statistics, you can have the computer re-compute the equation after removing variables that do not contribute significantly to predicting the criterion variable.

Coefficients Table

We will begin with a display of the *JASP* Coefficients Table shown in **Figure 9**. It contains the regression equation statistics for the "Right Stuff" data. We will go through the entire table and explain each of the statistics in turn.

First, a bit of terminology. In regression analysis, *JASP* refers to the regression equation as the "Model," a term which comes from economics. The "Model" here is a linear equation having four predictor variables. The Null Hypothesis (H_0) model is virtually never of interest because of arbitrary scaling of variables in the social sciences.

Now, look at the components of the table. The rows in the table are the components of the regression equation, beginning with the H_1 Intercept (Constant). The first column of values contains the value of the H_1 Intercept (the Constant = -2.021), followed by the *B* Values (labeled "Unstandardized") for each of the predictor variables (.014, .004, .020, .007).

Figure 9

JASP Regression Coefficients Table for the Equation That Includes All Four Predictor Variables.

Coefficients Table For the "Right Stuff" Data

Model		Unstandardized	Standard Error	Standardized	t	p
H₀	(Intercept)	2.941	0.096		30.534	< .001
H₁	(Intercept)	-2.021	1.034		-1.954	0.060
	IQ	0.014	0.009	0.212	1.492	0.146
	Inter	0.004	0.005	0.092	0.750	0.459
	Intra	0.020	0.004	0.592	4.702	< .001
	PS	0.007	0.005	0.182	1.312	0.199

Note: I added text to the title of the table.

If you wanted to write out the prediction equation, you could copy the B values from the Coefficients Table. Here, with all four predictor variables in the equation, the B value for *IQ* rounded to three decimals is +.014, for *Inter*, +.004, for *Intra*, +.020, and for *PS*, +.007.

The equation then is:

$$Y' = -2.021 + .014\ (IQ) + .004\ (Inter) + .020\ (Intra) + .007\ (PS)$$

This is called the **raw score** form of the regression equation, and the B values are technically known as the **unstandardized partial regression coefficients**. "Unstandardized" means that the magnitude of the B value reflects not only the predictive strength of the predictor variable, but also the size of the units of measurement of the predictor and criterion variables. The term "partial" means that **each B value reflects the unique contribution of a variable, independent of the influence of the other predictor variables**. Said differently, the B value is the amount that the criterion variable will change for each one-point change in the predictor variable, **when all other predictor variables are held constant**.

Because the B values are "unstandardized," one *cannot* directly compare the size of the B values to see which predictor variables are most important in the equation in determining the value of the criterion variable. For example, the B value for *IQ* (.014) is about three-fourths the size of the B value for *Intra* (.020). Does that mean that *Intra* is only about a third more powerful in predicting *GPA* as *IQ*?

The answer is no, because one unit of change does not mean the same thing for the two variables. The standard deviation for *IQ* is only 9.3 units, so one unit of change is relatively large, i.e., about 1/9 of a standard deviation. By contrast, the standard deviation for *Intra* is 16.4 units, so one unit of change is only about 1/16 of a standard deviation. A smaller increment of change for *Intra* (1/16 of a standard deviation) produces a larger effect on *GPA* (.020) than does a much larger increment of change (1/9 of a standard deviation) for *IQ* (.014). Thus, *Intra* must have a much stronger predictive association with *GPA* than is implied by the magnitude of the B values.

If this idea isn't clear, reread the paragraph again and think about it. Although the B values cannot be used to compare the relative importance of the predictor variables, there is another statistic to be described below that can.

The next column of the table is labeled "Standard Error," which stands for the **standard error of the B value.** Right now, this statistic isn't going to mean much to you. Basically, the idea is this. The B values in the table are statistics obtained on sample data. They are estimates of the B values that exist in the population from which the sample was drawn, i.e., the parameter B values. You know that all statistics show sampling variability, and that the greater the sampling fluctuation, the greater the error there may be in estimating the corresponding population parameter. **The standard error of the B value is an estimate of the extent of sampling variability for each B value.** The larger the value of Standard Error for any given B value, the more error there is in estimating what the B value actually is in the population. As we will see below, the Standard Error is used to decide whether a predictor variable contributes "significantly" in the equation to predicting the criterion variable.

The next column labeled "Standardized" contains what are called the β **values** (Greek "**beta**") for the predictor variables. The technical name for the ß values is **standardized partial regression coefficients**. These are the values of the regression coefficients when both the predictor variables and the criterion variable are converted to z scores relative to their own means and standard deviations. Because all variables are measured in common units (z scores) the magnitude of the β value *directly* indicates the relative importance of each variable in predicting the criterion variable. This is important and bears repeating. **The size of the β value directly indicates the magnitude of the association between a predictor and the criterion variable when the effects of the other predictor variables are held constant.** Transformation of the scores to z scores makes the Constant $A = 0$, yielding the **standard score (z score)** form of the regression equation:

$$z_{Y'} = \beta_1 z_1 + \beta_2 z_2 + \cdots + \beta_i z_i$$

We can rewrite the regression equation for predicting *GPA* in standard score form by abstracting the β values from the table. It would look like this:

$$z_{Y'} = +.212 z_{IQ} + .092 z_{Inter} + .592 z_{Intra} + .182 z_{PS}$$

According to the standard score form of the regression equation, *GPA* will change .212 standard deviations for each standard deviation increase in *IQ*, .092 standard deviations for each standard deviation increase in *Inter*, .592 standard deviations for each standard deviation increase in *Intra*, and .182 standard deviations for each standard deviation of increase in *PS*. *Intra* is, then, about three times as important as *IQ* in predicting *GPA*. *PS* has about the same magnitude of predictive association with *GPA* as does *IQ*. *Inter* seems to have little predictive relationship to *GPA*.

The β and *B* values are related to one another according to the following formula:

$$\beta_i = B_i \left(\frac{S_{X_i}}{S_Y} \right)$$

where: β_i = β value for the ith predictor variable,
B_i = *B* value for the ith predictor variable,
S_{X_i} = standard deviation of the ith predictor variable, and
S_Y = standard deviation of the criterion variable.

The next column labeled "t" contains the values of an inferential statistic that is used to decide whether the *B* value for a predictor variable is "statistically significant." Recall in Chapter 5 that even if two variables were uncorrelated in the population (i.e., ρ = 0), Pearson *r* for any random sample is not likely to be exactly equal to zero. In fact, because *r* is such an unstable statistic, a rather large sample value of *r* could occur by "chance." Thus, it was necessary to test whether *r* was sufficiently large that we could conclude that ρ was some non-zero value in the population.

The same situation holds for *B* values. Even if the *B* value for a variable were equal to zero in the population, the sample *B* value is unlikely to exactly equal zero. In fact, *B* values are like correlation coefficients in being unstable statistics. For this reason, it is necessary to decide whether a sample value of *B* is large enough in magnitude that we will conclude the population value is something other than zero, i.e., that we are dealing with a "significant" *B* value. The actual test is carried out as

$$t = \frac{B \ value}{Std. \ Error}$$

The *t* in the equation is a test statistic which will be discussed in detail later in your text. For the moment, don't worry about the precise meaning of the equation.

The last column labeled "p" is used to interpret the *t*-test values. "p" stands for the *p* value of the test, just as was the case with correlation coefficients. The numbers in this column are the probabilities that, if the *B* value equals 0 in the population, you would **through sampling error** obtain a sample *B* value of this magnitude or larger. If the *p* value is $\leq .05$, you can confidently conclude that the population value of *B* is greater than 0. In such a case, we would say that the variable in question contributes "significantly" to predicting the criterion variable. **Variables in which the *p* value is appreciably larger than .05 should not be included in the *final* form of the regression equation.** Because the *B* values and β values are simple linear transformations of one another, if *B* is significant, β is also significant.

JASP also tests whether the Constant (*A*) is significantly greater than 0, but, because most psychological measures are arbitrarily scaled, we are rarely interested in this test.

Deciding Which Variables to Include in the Equation

Again, using **Figure 9**, let's consider the *p* values for each of the variables in turn. Notice that the *p* value for *IQ* is .146. This means that there is about a 15% chance that one would obtain a *B* value of ± .014 or larger if the *B* value in the population were actually zero. This probability is fairly low, although nowhere near .05. In the present form of the regression equation in which all four variables are included, *IQ*'s unique association with the criterion is not statistically significant.

The *p* value for *Inter* is very large (.459) indicating that the sample *B* value could have readily occurred by chance, even if the population value of *B* were 0. Thus, in the present form of the regression equation, *Inter* does not contribute significantly to predicting *GPA*.

The *p* value for *Intra* is .001. This means that there is almost no possibility that one would obtain a *B* value of ± .020 or larger by chance if the population *B* value were zero. We can be very confident in concluding that *Intra* is a real predictor of *GPA*.

Finally, the *p* value for *PS* is .199, which means that there is about a 20% chance that one could get a *B* value of ± .007 or greater through sampling fluctuation even if the population *B* value were zero. This tells us that the unique contribution of *PS* is not significant in predicting *GPA* when all four variables are included in the equation.

The above paragraph included the phrase "when all four variables are included in the equation." **This is a critically important qualifier**. The values of *B* and β for any predictor in an equation are influenced by the other predictor variables that are included in the equation. This is because the *B* and β values reflect the variable's *unique* association with the criterion, **relative to the other variables that are in the equation**. If two significant predictor variables are substantially correlated, when both are included in the equation, the *B* and β values for each will be smaller than would be the case if only one has been entered into the equation. The two variables have to "split up" their predictive power. As a result, the unique predictive power of one or both variables may be sufficiently small that it is no longer statistically significant.

This is what happened with *IQ* and *PS*. We learned in Chapter 5 that *IQ* and *PS* were both significantly correlated with *GPA* (Pearson *r*'s +.45 and +.39, respectively). But *IQ* and *PS* were themselves substantially correlated (*r* = +.56), i.e., they overlap a great deal in what they measure. When the regression equation "splits up" their *unique* association with *GPA*, the resulting *B* and β values are sufficiently reduced that neither is statistically significant. (Technically, what happens is that the Std. Error for the B values gets larger, resulting in a value of *t* that is no longer significant.)

This brings us to a general principle in doing correlational research. **When you are trying to predict a criterion variable, it is important to identify and measure relatively *independent* predictor variables**. There is little advantage to including two or more predictors that correlate substantially with one another, because so much of their predictive association with the criterion is shared. Once one of these variables is entered into the regression equation, inclusion of the other(s) will add little to the predictive power of the equation. **This condition in which two or more predictor variables are substantially correlated either in a bivariate or multivariate fashion is called multicollinearity. Multicollinearity is <u>always</u> undesirable.**

Ideally, in regression research, you should have predictors that intercorrelate ± .30 or less. Once the correlation between two predictor variables exceeds ± .50, there is usually little advantage to having both in the equation. **The moral here is that you should think long and hard about what you are going to measure before you undertake correlational research, and that once the data are collected, you should carefully study the correlation matrix of predictor and criterion variables before beginning the regression analysis of the data.**

What we're going to do to handle this problem is to re-compute the regression equation using only those variables that will contribute significantly to predicting the criterion. So what variables should we include? Clearly, *Intra* should be included because the *B* and β values were highly statistically significant. Just as clearly, *Inter* should not be included, because the *p* value for this variable did not even approach statistical significance. But, which of the other two variables—*IQ* and *PS*—should be included?

The first rule of thumb is to include the variable that has the stronger bivariate correlation with the criterion—in this case, *IQ*. The second rule of thumb is to include the variable that is theoretically the more fundamental or important—also, in this case, *IQ*. We say that *IQ* is more fundamental, based on the theoretical assumption that problem solving effectiveness derives from general intelligence rather than the other way around.

Figure 10 shows the Coefficients Table for the regression equation recomputed using only the two predictor variables of *IQ* and *Intra*. (Note: This is done by rerunning the analysis after removing *Inter* and *PS* from the list of Covariate (predictor) variables in the regression dialog box.) Notice that the *B* and β values for *Intra* are similar to what they were in the original 4-variable equation. This is because *Intra* doesn't correlate appreciably with any of the other predictor variables and, thus, never had to share any of its predictive power.

In contrast, the *B* and β values for *IQ* are now much larger (+.020 and +.299, respectively) and the *p* value of .017 for *IQ* indicates that the variable is statistically significant in its association with the criterion. **The β values indicate that *Intra* is over twice as strong a predictor as *IQ*.**

Significance Test of the Overall Regression Equation

Figure 11 shows the "ANOVA" table for the regression equation (Model). We won't get to this statistical procedure until Chapter 12, so, like the *t* test, it isn't going to mean much to you. Suffice it to say, the purpose of this inferential procedure is to determine whether the

Figure 10

JASP Regression Coefficients Table for the Regression Equation that Includes Just IQ and Intrapersonal Intelligence as Predictor Variables.

Coefficients Table For the "Right Stuff" Data

Model		Unstandardized	Standard Error	Standardized	t	p
H_0	(Intercept)	2.941	0.096		30.534	< .001
H_1	(Intercept)	-1.693	0.922		-1.837	0.076
	IQ	0.020	0.008	0.299	2.513	0.017
	Intra	0.022	0.004	0.629	5.289	< .001

Note: I added text to the title of the table.

Figure 11

JASP Test for the Significance of the Regression Equation.

ANOVA Table for the Two Variable Regression Solution of the "Right Stuff" Data

Model		Sum of Squares	df	Mean Square	F	p
H_1	Regression	6.321	2	3.160	21.414	< .001
	Residual	4.723	32	0.148		
	Total	11.043	34			

Note. The intercept model is omitted, as no meaningful information can be shown.

Note: I added text to the title of the table.

regression equation is able to significantly predict the criterion variable. The procedure uses the test statistic *F*, and the results of the test are displayed in the table whose headers are "Sum of Squares," "df," "Mean Square," "F," and "p." As was the case with the *t* test for the significance of each of the *B* values, if the *p* value for the computed *F* is ≤ .05, the equation as a whole significantly predicts the criterion variable. Generally, if at least one of the *B* values for variables in the equation is statistically significant, the *p* value for the test of the overall equation also will be significant.

Assessing Accuracy of Prediction

JASP computes four statistics which quantify how *accurately* the regression equation can predict the criterion variable. These are displayed in the *JASP* Model Summary Table, which is shown as **Figure 12**.

The first statistic is abbreviated "*R*," which stands for **coefficient of multiple correlation**. This statistic is the Pearson correlation coefficient computed between *Y* and *Y'* for all cases in the sample. If the regression equation is highly effective in predicting the criterion variable, then *Y* and *Y'* will be similar for most cases, and the Pearson correlation between them will be large. *R* ranges in magnitude from 0 to +1.00. It cannot be negative in value, because there will always be a positive relationship between the actual and predicted values of the criterion variable.

Although you could compute *R* by finding the predicted value of the criterion for each case and computing Pearson *r* on the pairs of *Y* and *Y'* scores, it turns out that *R* is computed much more conveniently by taking the square root of another statistic, R^2. The *R* for the "Right Stuff" data is .757, which means that there is a very strong

Figure 12

JASP Model Summary Table that Includes Statistics for Assessing the Accuracy with which the Equation Predicts the Criterion Variable.

Model Summary - GPA Predicted By *IQ* and *Intra*

Model	R	R^2	Adjusted R^2	RMSE
H_0	0.000	0.000	0.000	0.570
H_1	0.757	0.572	0.546	0.384

Note: I added text to the title of the table.

correlation between the actual values of the criterion variable and those predicted by the regression equation.

The next statistic is the **coefficient of multiple determination (R^2)**. R^2 is interpreted in the same way that r^2 is interpreted. It represents the proportion of the total variability in the criterion variable that is predicted by the regression equation. For the sample data R^2 is .572, which means that about 57% of the variability in GPA is predicted by the regression equation, i.e., by the linear combination of *IQ* and *Intra*.

We can use the value of R^2 to compare the effectiveness of different regression equations in predicting a criterion variable. Recall that *our* insightful student wondered about the predictive cost of not including *PS* in the regression equation. We said that there would be little loss of predictive power. Our answer was based on knowledge of the value R^2. If you do include *PS* in the equation along with *IQ* and *Intra*, R^2 becomes .60. This means that inclusion of the third variable would increase the amount of variability in *GPA* predicted by the equation by only about 3%. This is too little gain to justify the slippery practice of including variables in the equation for which the *B* values are not statistically significant.

The third statistic is called the **adjusted R^2 (adj R^2)**. This statistic gives a somewhat better estimate of the proportion of variance in the criterion that could actually be predicted if the regression equation is applied to another sample of data. It is needed because the regression coefficients of any set of data are "optimal," in the sense that they have been selected to maximize accuracy of prediction. If the same coefficients are used for a completely new sample of data, the value of R and R^2 would be lower—would "**shrink**."

The amount of shrinkage is a function of the **number of population regression coefficients to be estimated and how accurately the population regression coefficients were estimated in the original sample**. Large samples provide accurate estimates, leading to less shrinkage. To give an honest estimate of the extent to which the regression equation can predict the criterion, some researchers always report *adj R^2* rather than the simple R^2 itself.

The *adj R^2* value of .546 for the "Right Stuff" data reflects only a very modest "shrinkage" of R^2. In order for shrinkage to be severe, sample sizes must get quite small, or a very large number of *B* values must be estimated in the equation.

The last statistic is abbreviated *"RMSE"* which stands for Root Mean Square Error. This terminology is rather more general than our present application, so we'll use another term. Another name for this statistic is the **standard error of the estimate ($s_{Y.X}$)**. When the regression equation is used to predict the value of the criterion variable (Y') for a specific case, the amount of error in the prediction is quantified by $s_{Y.X}$. This statistic is the **standard deviation of the distribution of residuals for the entire sample**. The definitional formula for $s_{Y.X}$ is as follows:

$$s_{Y.X} = \sqrt{\frac{\sum(Y - Y')^2}{n}}$$

A related statistic not given by *JASP* is the estimated standard error of estimate (est. $s_{Y.X}$). This the estimated value of the standard error of estimate in the population and is defined as

$$s_{Y.X} = \sqrt{\frac{\sum(Y - Y')^2}{n - k - 1}}$$

where: n = sample size, and
k = number of predictor variables in the equation.

The two variants of the same statistic are generally close in value—particularly for large samples. As *JASP* gives us the former, we'll use it for the below computations.

$s_{Y.X}$ varies in value between zero when there is no error of prediction to approximately S_Y (standard deviation of the *Y* scores) when the predictors are all completely uncorrelated with the criterion. $s_{Y.X}$ for the "Right Stuff" data is .384, which is to say that the "average" error in prediction is about .38 *GPA* points.

Making Confidence Interval Estimates

$s_{Y.X}$ is not normally used for descriptive purposes, but it is used in applied regression situations to establish confidence interval estimates of predicted scores. Recall that when we predicted Joe College's freshman *GPA* from his Intrapersonal intelligence and *IQ* scores, we got a predicted value of 2.52. This value is technically a point estimate of Joe's *GPA* and as we discussed in Chapter 1, point estimates are almost certainly in error to some degree. Because $s_{Y.X}$ is a measure of the error of prediction, it can be used to quantify how much error we would expect in a point estimate, and from that establish a range of values that we are confident will include the actual value of the score.

Figure 13

Computation of the 90% Confidence Interval for Joe's Predicted GPA of 2.52.

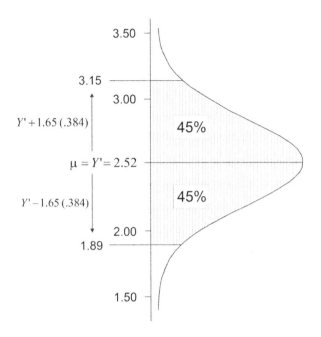

The *JASP* Linear Regression procedure has no provision for making interval predictions from a regression equation, but it is possible to do so manually without too much trouble. We begin by making the following assumptions regarding the errors of prediction:

1. **In the population from which our sample was drawn, $s_{Y.X}$ is an accurate estimate of the average errors of prediction for all possible values of Y'.**

2. **In the population, errors of prediction are approximately normally distributed for all values of Y'.** Inspection of the plot of residuals produced by *JASP* Plots option (not shown) suggests that this assumption is reasonable.

These are restrictive assumptions that are unlikely to be fully satisfied in most situations. To compensate for this, statisticians have developed quite sophisticated techniques to deal with the problem. The simplified formula below shows the logic of the procedure but is susceptible to a good deal of error if the assumptions are not met.

Let's say that we want to establish the **90% confidence interval estimate** of Joe's actual freshman *GPA* based upon the regression equation involving Intrapersonal intelligence and *IQ*. Confidence intervals are discussed fully in Chapter 9, but for immediate purposes, we'll define a confidence interval in regression as **a range established around a point estimate of a score that has a known probability of including the actual value of the score.** The 90% confidence interval estimate, for example, is a range that we are 90% confident includes the actual score.

Here is the logic of the procedure. We'll begin by considering Joe to be a randomly selected member of a sub-population of students, all of whom have a predicted *GPA* of 2.52. We will assume that in this sub-population, the actual *GPA*'s are normally distributed about the predicted value of 2.52, making the Y' of 2.52 the mean of this distribution. Finally, we'll assume that the standard deviation of this normal distribution is accurately estimated by $s_{Y.X}$. This situation is diagrammed in **Figure 13**.

We will establish the 90% confidence interval for Joe's *GPA* by finding two points symmetrically placed about the mean value of 2.52 that encompass the middle 90% of all actual *GPA* scores. Because Joe is assumed to be a randomly selected member of this sub-population, he will have a 90% chance of being one of those students included in this range. According to the normal curve **Table A**, 90% of the area under the normal curve lies between *z* scores of −1.65 and +1.65. Thus, the 90% confidence interval estimate would be given by the following equation:

$$\begin{aligned} 90\%\ CI &= Y' \pm 1.65\ (S_{Y.X}) \\ &= 2.52 \pm 1.65\ (.384) \\ &= 2.52 \pm .634 \\ &= 1.89 - 3.15 \end{aligned}$$

Given his Intrapersonal intelligence and *IQ* scores, we predict that Joe's freshman *GPA* will be somewhere between 1.89 and 3.15. The wide range of possible GPAs encompassed by the 90% confidence interval ("D+" to "B-") makes it clear that point estimates can be deceptively precise, and one needs to be very cautious in making practical decisions based on predicted scores.

Explained and Unexplained Variance

In Chapter 5 we defined the coefficient of determination (r^2) to be the proportion of the variability in the criterion variable associated with or predicted by the predictor variable. At that point, the idea of "predicted variability" was necessarily vague. Having studied regression, we are now in a better position to explain the idea of predicted variability. We'll do this with reference to the scatterplot shown in **Figure 14**.

The scatterplot is based on five cases, whose paired *X* and *Y* scores are (1,2) (3,3) (5,8) (7,5) (9,7). The correlation between the *X* and *Y* scores is +.74. The best-fitting straight

Figure 14

Scatterplot showing predicted and unpredicted parts of the discrepancies between Y scores and the mean of Y.

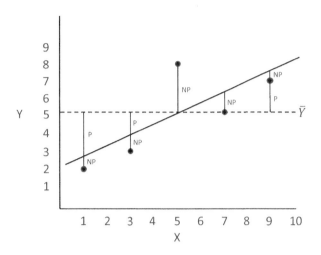

line has been drawn in on the plot. The equation for this best-fitting line is $Y' = 2.00 + .60X$. The mean of the five Y scores is 5.00 and this is indicated by the dashed line. Also, the distance between each score's Y value and the mean of Y has been indicated with a line.

The purpose of this figure is to show that the "distance" that each score's Y value lies from the mean of Y is made up of two components—a portion that is "predicted" by the regression equation and a portion that is "not predicted" by the regression equation. Consider the left-most data point (1, 2). The point is 3.00 units below the mean of Y. We can ask, rhetorically, "Why is it 3.00 units below the mean of Y?" Part of the answer is that there is a positive correlation between X and Y, and the regression equation predicts that X scores having a value of 1 will have a Y value of 2.60, i.e.,

$$Y' = 2.00 + .60X$$
$$= 2.00 + .60(1.00)$$
$$= 2.60$$

The predicted Y score of 2.60 is 2.40 units below the mean of Y, which is to say, the regression equation has predicted that portion of the distance between the score's Y value and the mean of Y. This predicted portion of the distance is indicated by the letter P. The remaining portion of the distance (.60 units) is not predicted by the regression equation, and in the figure is labeled "NP." For this data point, most of the distance between its Y value and the mean of Y has been predicted by the regression equation. In contrast, consider the third data point (5, 8). For an X value of 5.00, the regression equation predicts a Y value of 5.00 (the mean of Y), but the actual Y value is 8.00. The discrepancy from the mean of Y is 3.00 units, none of which is predicted by the equation.

Finally, look at the far right data point (9, 7). For an X value of 9.00, the regression equation predicts the Y score to be 7.40, which is very close to the actual Y value of 7.00. Here, the regression equation predicts almost the entire distance of the score from the mean of Y.

We can express the relationship between the total distance that a score lies from the mean of Y and the predicted and unpredicted components of that distance in a simple equation:

$$(Y - \bar{Y}) = (Y - Y') + (Y' - \bar{Y})$$

$$\underset{\text{Distance}}{\text{Total}} = \underset{\text{Distance}}{\text{Unpredicted}} + \underset{\text{Distance}}{\text{Predicted}}$$

What we are going to do next is some simple algebra to convert these distance components into components of Y variability, which we will quantify using a statistic called the **variance**. The sample variance is abbreviated S^2 because it is just the standard deviation squared. The variance of the Y scores is equal to

$$S^2 = \frac{\Sigma(Y - \bar{Y})^2}{n}$$

We are going to algebraically manipulate our distance component equation given above so that the **left-hand side of the equation equals the variance of the Y scores**. We'll see what happens to the right-hand side of the equation as we do so. We'll begin by squaring the distance that each score's Y value lies from the mean of Y, yielding:

$$(Y - \bar{Y})^2 = [(Y - Y') + (Y' - \bar{Y})]^2$$

which expands to:

$$(Y - \bar{Y})^2 = (Y - Y')^2$$
$$+ 2(Y - Y')(Y' - \bar{Y})$$
$$+ (Y' - \bar{Y})^2$$

Having squared the distance for each data point, we next need to sum the squared distances for all five data points. Algebraically, this is represented as:

$$\sum(Y - \bar{Y})^2 = \sum(Y - Y')^2$$
$$+ 2\sum(Y - Y')(Y' - \bar{Y})$$
$$+ \sum(Y' - \bar{Y})^2$$

The middle of the right-hand side of the equation is called the "cross-products" term. It can be shown that when summed across all the cases, that:

$$+2\sum(Y-Y')(Y'-\bar{Y}) = 0$$

With the cross product gone, the equation becomes:

$$\sum(Y-\bar{Y})^2 = \sum(Y-Y')^2 + \sum(Y'-\bar{Y})^2$$

We're just about done, but first a slight discursion. The three components above are technically called **sums of squares** (sums of squared deviations). Notice that they are additive. This idea of additive sums of squares is going to be extremely important in Chapters 12–15 on analysis of variance. Tuck that bit of information away. Now back to the problem at hand.

The last step in generating the variance of the Y scores on the left-hand side of the equation is to divide by the sample size (*n*). This yields:

$$\frac{\sum(Y-\bar{Y})^2}{n} = \frac{\sum(Y-Y')^2}{n} + \frac{\sum(Y'-\bar{Y})^2}{n}$$

This is where we want to be. The left-hand side of the equation is the total Y variance. Our algebraic manipulations show that it is equal to the sum of two separate variances. The first is the variance of the Y scores about their predicted values (i.e., about the regression line); the second is the variance of the predicted values of Y about the mean of Y. In words, the total Y variance is made up of two components—the **unpredicted variance** and the **predicted variance**. This relationship is usually stated mathematically in the following way:

$$S_Y^2 = S_{Y.X}^2 + S_{Y'}^2$$

$$\frac{Total}{Variance} = \frac{Unpredicted}{Variance} + \frac{Predicted}{Variance}$$

Because we are dealing with a very small sample, it's possible to manually compute each of these variance components from the raw data. **Table 1** contains the actual X and Y data, along with the computed values for Y', $Y-Y'$, and $Y'-\bar{Y}$. The last line of the table contains the computed variance for the Y scores and for the values in the last two columns. Substituting these computed variances into the total variance equation yields:

$$S_Y^2 = S_{Y.X}^2 + S_{Y'}^2$$
$$= 2.32 + 2.88$$
$$= 5.20$$

Table 1

Computation of Variance Components for Scatterplot Shown in Figure 12.

			Unpred. Distance	Predict. Distance
X	Y	Y'	Y - Y'	Y' - $\bar{Y}$
1	2	2.60	-.60	-2.40
3	3	3.80	-.80	-1.20
5	8	5.00	3.00	.00
7	5	6.20	-1.20	1.20
9	7	7.4	-.40	2.40
S^2's =	5.20		2.32	2.88

Recall that in Chapter 5 on Pearson Correlation, we defined the coefficient of determination (r^2) verbally as **that portion of the variability in the criterion variable, Y, that can be predicted from the predictor variable, X.** Based upon our new-found understanding of the components of the total Y variance, we can now offer a more complete and formal definition of the coefficient of determination (r^2):

$$r^2 = \frac{S_{Y'}^2}{S_Y^2} = \frac{Pred.Variance}{Total\ Variance}$$

Using this formula, we can compute the coefficient of determination for our small sample of data:

$$r^2 = \frac{S_{Y'}^2}{S_Y^2} = \frac{2.88}{5.20} = .554$$

According to our calculations, about 55% of the variability in the criterion variable is associated with variation in the predictor variable, i.e., is predicted by the regression equation. This was the hard way to compute r^2. The much easier way, as we know, is to square *r*, which yields $+.74^2 = .55$.

This section has introduced the idea that the total variance in a set of scores can be broken down into two parts—a portion of the variance that is predicted by another variable and a remaining portion called unpredicted variance. The idea of partitioning variance into component parts is a very powerful one and sees many applications in statistics and related areas. We will encounter it in the next chapter on partial correlation and in the later chapters on *t* tests and analysis of variance (Chapters 10–15).

Reporting Regression Analyses

Reporting the results of regression analyses is fairly formulaic. We'll illustrate the approach to communicating the results of a regression analysis with a verbal summary of The Right Stuff analysis:

> Multiple regression analysis was used to evaluate the extent to which the four predictor variables of IQ, Interpersonal Intelligence, Intrapersonal Intelligence, and Problem Solving Intelligence were able to predict students' GPA. Two of the variables, *IQ* and Intrapersonal Intelligence, had significant B and ß values in the equation, collectively predicting over half of the variability in GPA.
>
> Although the variable of Problem Solving Intelligence showed a substantial bivariate association with the criterion ($r = +.39$), its inclusion in the regression equation along with *IQ* and Intrapersonal Intelligence produced little increase in the amount of variance accounted for by the equation. Of the two significant predictors, Intrapersonal Intelligence was approximately twice as strong in its association with *GPA* as traditional IQ.

The Usual Admonitions

The Linear Regression procedure in *JASP* is extremely feature rich. Take a few minutes to do a regression analysis on your *Big Five* data set from the Exercises at the end of Chapter 5 and explore all the statistics and options available with the procedure. You'll be amazed at what you can teach yourself. Better yet also do the regression problems in the exercises below.

As the sophistication of the statistical procedures has increased, so has the size of the data sets. Entering and checking data for accuracy is a lot of work, so you may want to find a partner to share in the work.

Chapter 6, Linear Regression
Chapter Exercises

Some questions are purely factual and have only one correct answer. Others require greater thought on your part and may have more than one possible answer.

1. Fill in the missing term(s) that makes the statement true.

   a. Linear regression allows one to assess the _____ of two or more predictor variables with the criterion.
   b. When there is only one predictor variable, it's called _____.
   c. Multiple linear regression is needed because important criterion variables are _____.
   d. In a linear regression equation, A is called the _____.
   e. In linear regression, B is called the _____.
   f. When there are two predictor variables, the linear equation generates a _____.
   g. A and B values for any data set are generated using the _____ criterion.
   h. That criterion minimizes _____.
   i. Using this criterion, the linear equation always predicts _____ when X is at the mean of X.
   j. Poor Joe College. Based on just his low *Intrapersonal* intelligence he is predicted to earn a _____ GPA.
   k. The technical term for a β value is _____.
   l. The advantage of β over B is that it uses _____ units.
   m. If a B value has an associated p value of .001, we say that it is _____.
   n. If the B value for a variable has a p value of .356, that variable should _____.
   o. _____ occurs when two predictor variables are strongly correlated and is _____.
   p. R is technically called the _____ and is the _____.
   q. R^2 is technically called the _____ and quantifies the _____.
   r. adj R^2 is preferred in research reports because it more accurately reflects _____.
   s. $S_{Y.X}$ is the symbol for the _____ and is an estimate of the _____ of the regression residuals.
   t. After making a lot of assumptions, $S_{Y.X}$ can be used to form _____ for Y'.
   u. For utterly unclear reasons, you were advised to remember that _____ are _____.
   v. In regression analyses, S_Y^2 is the sum of _____ and _____.
   w. Finally, r^2 is formally defined as _____ divided by _____.

2. Sarafina Snape teaches a section of the Psychology Department's required *Introduction to Statistical Methods*. As a demonstration of the usefulness of multiple linear regression, she collects one recent course evaluation form from each of 15 colleagues and herself for data analysis purposes. Some colleagues with less than stellar evaluations were reluctant to share, but Sarafina can be very persuasive. The forms contain the following summarized information for each class.

   a. Gender of the professor (*Gen*)

   b. Mean rating of the teacher by the class for the following traits using a 7-point scale (1= "Terrible", 4 = "Acceptable", 7 = "Outstanding"). For convenience in entering the data, the class means were rounded to the nearest integer value:

   Energy & Enthusiasm (*Energy*)
   Organization (*Organ*)
   Knowledge of the Field (*Know*)
   High Standards (*High S*)
   Flexibility (*Flex*)

   c. Mean rating (*Rating*) of the professor "as a teacher" on a 9-point scale (1= "Terrible", 5 = "Acceptable", 9 = "Outstanding"), again, rounded to the nearest integer value.

Here are the data for the 16 professors:

Prof	Gen	Energy	Organ	Know	High S	Flex	Rating
A	2	5	6	6	6	7	6
B	1	4	5	7	7	5	8
C	2	2	7	6	3	7	4
D	2	3	5	7	7	6	5
E	1	4	6	5	6	3	7
F	2	2	5	7	6	1	3
G	1	7	4	6	7	3	4
H	1	6	6	7	6	6	7
I	1	5	7	4	7	6	9
J	2	8	5	6	7	7	6
K	1	6	4	6	4	6	5
L	1	7	5	5	6	4	8
M	2	5	5	6	7	7	9
N	2	2	5	5	7	7	6
O	2	3	6	5	6	4	5
P	2	2	7	7	5	3	2

Do a complete descriptive and correlational / regression analysis to determine which trait ratings actually predict the final professor evaluation and write a thoughtful interpretive summary of your findings. In doing the regression analysis, make sure that *JASP* has all variables typed as numerical.

4. Mark Harried is the Office Manager for a new software start-up company whose intended product is an AI-assisted program to help high school students develop vocational plans and choose "best-fit" colleges and universities to further those plans. Because it involves the psychology of high-school students and their parents, the project is incredibly complex and, in many ways, tedious and frustrating. Low morale and attrition among new hires has been severe. To attempt to minimize the problem, the company CEO hired a high-powered psychological consulting firm to develop a battery of screening tests to identify people who would be good fits for the job. Alas, the preliminary findings were not encouraging, and the firm was scandalously expensive.

Mark believed that the bosses were over-thinking the problem and based on his experiences with the hiring process, believed that certain easily available markers could provide better predictors of worker success. Specifically, he thought that five factors were predictive of employee success: the person's age (*Age*), the quality of the person's handwriting in filling out the lengthy employment application (*Writing*), the appropriateness of the person's dress during the interview (*Dress*), the precision and clarity of the person's language during the interview (*Speech*), and the "friendliness" of the person during the interview (*Friendly*). As all interviews were videorecorded, this information was readily available to him. He scored the next 24 hired employees on those dimensions using a 7- point scale (1 = "lousy" to 4 "OK" to 7 "Excellent"). His criterion was composite supervisor ratings of the employee's level of performance at the end of a 3-month probationary period (*Ratings* 0–25 scale). Here are his data, which his partner offered to analyze and interpret:

Person	Age	Writing	Dress	Speech	Friendly	Ratings
1	20	1	4	4	2	15
2	26	5	2	3	6	23
3	21	6	2	5	3	22
4	22	3	4	4	3	18
5	20	4	4	2	3	14
6	27	4	4	4	6	22
7	20	3	5	4	4	16
8	32	3	2	5	3	25
9	24	7	2	5	2	20
10	19	4	3	7	4	17
11	24	5	4	3	3	20
12	22	4	3	2	5	23
13	23	5	4	4	7	22

14	20	2	7	4	1	13	
15	21	1	4	3	4	18	
16	18	3	4	2	5	16	
17	23	2	3	5	4	10	
18	40	4	3	6	5	24	
19	19	3	4	7	4	15	
20	24	4	2	6	5	17	
21	18	2	4	4	6	13	
22	20	4	4	6	4	22	
23	26	3	2	5	5	9	
24	20	5	6	2	2	19	

Do a complete descriptive and correlational / regression analysis of the data and write a thoughtful interpretive summary of your findings.

5. The dreaded "Freshman 15" is the tendency for first year college students to gain weight. It isn't really 15 lbs.—more like 8 lbs. on average. In a brief study, 35 incoming students were weighed and completed a Health Attitudes Questionnaire (*Health*) which was scored 0–50. They also rated the importance of personal physical attractiveness (*PA* 1–9 rating scale). During the year they kept daily calorie counts from meals and snacks, and the number of hours of moderate to hard exercise they engaged in. At the end of the year their weight gain/loss was computed. Here are the data:

	Health	PA	Cal / Meals	Cal / Snacks	Weekly Exercise	Gain/ Loss		Health	PA	Cal / Meals	Cal / Snacks	Weekly Exercise	Gain/ Loss
1	31	2	2000	810	2.0	15	19	42	5	1870	390	3.1	9
2	18	1	2150	1520	3.2	22	20	33	6	2680	100	9.3	18
3	41	8	1470	400	8.9	6	21	33	9	2550	260	6.7	2
4	34	3	2280	1060	1.0	18	22	19	3	1600	700	3.1	7
5	12	9	1750	0	2.0	4	23	46	8	1990	120	8.8	-5
6	23	6	2340	525	7.7	7	24	40	8	3210	95	12.4	-12
7	19	5	2500	670	3.0	19	25	30	6	2150	185	3.0	8
8	49	9	2880	410	9.2	-3	26	27	7	2240	850	8.5	13
9	24	3	1870	1270	6.4	9	27	8	4	1880	1000	3.0	26
10	43	7	2380	470	11.1	6	28	39	2	1980	800	8.5	15
11	27	5	2210	880	5.7	11	29	48	5	1150	710	13.5	5
12	40	3	2050	675	8.4	2	30	36	8	1970	0	2.2	7
13	38	2	1670	1375	2.8	27	31	30	8	1770	380	2.0	19
14	31	9	1925	415	3.9	4	32	44	7	1780	200	7.5	3
15	19	9	2280	50	9.7	-6	33	49	3	1975	760	11.3	-13
16	14	4	1460	1740	3.5	23	34	13	5	2450	1110	2.8	12
17	22	4	2000	790	1.4	14	35	26	4	2000	990	8.3	14
18	23	9	2010	590	2.0	8							

Do a complete descriptive and correlational / regression analysis of the data and write a thoughtful interpretive summary of your findings. This problem is interesting in terms of causal pathways!

6. Small college newspapers vary enormously in quality. Some are sloppy little affairs that appear irregularly and have minimal impact on the community. At other institutions, the student newspaper is a major force in campus politics, and an important source of news and entertainment. What factors influence the quality of a student newspaper? To attempt to answer this question, a national organization of college newspaper editors randomly selects 50 smaller college and university newspapers for study. Because the focus of the study is on smaller institutions, the population is defined as colleges and universities having a total enrollment of less than 5000 students and a weekly, rather than daily, newspaper. For purposes of the study, the printed version of the paper was used unless it was only available on-line. The research group evaluates two years' worth of issues of each newspaper and assigns weighted scores for general appearance, level of writing, diversity of topics, depth of coverage, and the like. The weighted component scores were

then summated to a total Quality Score (QS) that could range in value from 0 to 150. To determine whether certain institutional factors are related to the quality of the newspaper, the research group collected data on such factors as institution size, amount of tuition, overall quality of the school, size of the Journalism Department, etc. The data below are a subset of the larger body of information collected. The variables are (a) selectivity of the institution (*Sel*, 1-5, with 1 = open admissions policy and 5 = highly selective admissions policy), (b) the number of faculty members in the Journalism Department (*#JD*), (c) age of the school in years (*Age*), and (d) the Quality Score (*QS*) for each newspaper (the criterion variable).

Inst	Sel	#JD	Age	QS	Inst	Sel	#JD	Age	QS	Inst	Sel	#JD	Age	QS
01-	1	2	47	78	18-	2	3	71	103	35-	4	0	210	140
02-	3	1	26	126	19-	2	3	59	101	36-	2	2	81	84
03-	2	2	126	130	20-	1	4	51	110	37-	2	3	63	72
04-	3	3	55	100	21-	5	0	33	105	38-	3	3	51	90
05-	2	2	168	121	22-	2	5	43	114	39-	2	2	63	89
06-	1	6	32	120	23-	1	3	56	58	40-	2	5	45	114
07-	4	0	210	132	24-	3	2	39	76	41-	4	1	29	130
08-	2	2	52	85	25-	3	1	137	129	42-	2	3	80	106
09-	2	1	68	94	26-	2	0	50	81	43-	3	3	115	89
10-	2	2	39	68	27-	1	2	30	65	44-	2	5	19	83
11-	3	0	97	84	28-	3	1	39	106	45-	1	2	38	78
12-	2	3	46	95	29-	2	1	46	72	46-	5	1	106	109
13-	1	2	72	94	30-	1	3	65	90	47-	4	2	57	96
14-	2	3	51	130	31-	2	3	51	106	48-	1	3	41	84
15-	3	0	33	90	32-	2	5	83	100	49-	2	3	69	95
16-	2	3	102	110	33-	3	2	96	142	50-	2	5	71	92
17-	4	2	140	121	34-	2	0	29	65					

Do a complete descriptive and correlational / regression analysis of the data and write a thoughtful interpretive summary of your findings. This problem is interesting in terms of an odd discrepancy that exists between the correlation and the regression results.

Chapter 7
Partial Correlation

What's In This Chapter?

Partial correlation is a statistical technique for computing the Pearson correlation between a predictor and a criterion variable while controlling for (removing) the effects of one or more other correlated variables. Controlling for one variable produces a **first-order partial correlation**, controlling for two variables, a **second-order partial**, and so on. **Zero-order partials** are regular Pearson correlation coefficients.

Partial correlation uses a standard terminology. The expression $r_{12.3}$ means the **first-order partial correlation** between the criterion variable 1 and the predictor variable 2, with the effect of variable 3 controlled for or removed. The expression $r_{12.34}$ is the **second order partial** between variables 1 and 2, with variables 3 and 4 controlled. And so on.

Partial correlation is based on linear regression. Removing the effect of a third variable on the correlation between a predictor and criterion variable involves the following conceptual steps. First, a linear equation is derived to predict the values of the predictor variable from values of the third, to-be-controlled variable. The predicted value (Y') is subtracted from the actual value (Y) for all cases in the sample, leaving a **residual score** for each case. These residual scores for the predictor are now completely independent of (uncorrelated with) the values of the third variable. Next, a second linear equation is used to predict values of the criterion variable from values of the third variable and residual scores derived in a similar manner. Again, these residual scores for the criterion are also now completely uncorrelated with the third variable. Finally, the Pearson correlation is computed between the two sets of residual scores, yielding the partial correlation. Higher-order partials are computed in the same manner using multiple linear regression to remove the effects of two or more other variables from the predictor and criterion variables.

Part or **semi-partial correlation** is similar in logic to partial correlation except that linear regression is used to remove the effects of one or more variables from only one of the two variables. The residual scores for that variable are correlated with the unadjusted scores for the other variable. The expression $r_{1(2.3)}$ means the correlation between variable 1 and the **residualized values of variable 2**.

Introduction

We have seen that one of the major uses of multiple linear regression is to assess the **unique contribution** that each of several predictor variables makes in predicting a criterion variable. The relative sizes of the β values for the predictor variables included in the equation give us a direct measure of the strength of the relationship between each predictor variable and the criterion, independently of the effects of the other predictor variables. In that sense, multiple linear regression allows one to look at the correlational relationship between a predictor and criterion variable, while statistically holding other variables constant.

Another way to achieve the same objective is a technique called **partial correlation**. A partial correlation is the Pearson r between a predictor and criterion variable, with the effects of one or more other variables statistically removed. Because partial correlations are just modified Pearson correlations, they require the same assumptions about the data as do Pearson correlations, specifically:

1. **All the variables are numerical.**

2. **The relationship among all variables is linear in the population.**

To understand how partial correlation works, we'll begin with the simplest situation—two predictor variables and a criterion variable. As an example, assume that we are interested in the relationship between personal income and political conservatism, i.e., as people get richer, do they become more politically conservative? To study this, we assess the personal income and level of political conservatism of a sample of adult men. We also measure the age of each man, believing correctly that age is also related to political conservatism. **Figure 1** on the next page diagrams this situation.

Figure 1

Correlations Among the Three Variables of Income, Age, and Political Conservatism.

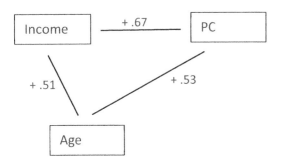

In the diagram, rectangles represent the variables. The criterion variable is political conservatism (*PC*) and the two predictor variables are labeled *Income* and *Age*. The labeled lines connecting the rectangles stand for the Pearson correlations between the variables. Note that *Income* and *PC* correlate +.67, indicating a strong linear relationship between the two variables. We might interpret this relationship causally to mean that as you become wealthier, you tend to also become more politically conservative in the interest of hanging on to as much of your income as possible.

The problem here is that *Age* also correlates fairly strongly with both *PC* (+.53) and with *Income* (+.51). As we learned in Chapter 5, when two predictor variables correlate strongly with one another, they both correlate with the criterion at least in part because they are measuring the same thing. Said somewhat differently, part of the correlation between *Income* and *PC* is due to the fact that *Income* also correlates positively with *Age*, which in turn correlates with *PC*. It's plausible to believe that as you get older, you tend to become (take your choice) either wiser or more traditional in your thinking. What we need to do is to somehow remove the effects of *Age* from the correlation between *Income* and *PC*. Technically we want to "partial out" the effects of *Age* from the relationship. This is called a **first-order partial correlation** because we are removing the effects of a single variable. A first-order partial correlation is symbolized as $r_{12.3}$. This is read as the Pearson correlation between variables 1 and 2, with the effects of variable 3 removed.

It is possible to remove the effects of more than one variable. When more than one variable is partialled out, we use the general term **higher-order partialling**. If two variables are partialled out, we call it a **second-order partial correlation** and symbolize it as $r_{12.34}$. In words, this is the Pearson correlation between variables 1 and 2 with the effects of variables 3 and 4 removed. For example, suppose that in addition to measuring income, age, and political conservation, the study had also measured the number of children in the family—on the assumption that the responsibilities of having children tend to make one more socially and politically conservative. Here, you might want to look at the correlation between income and conservatism when the effects of both age and number of children are partialled out (see below).

In principle, there is no limit to the number of variables that can be partialled out, although in practice it's rare to see more than **third-order partial correlations**. Generally, once you have partialled out the effects of two or three variables, removing the effects of additional variables has no further effect on the size of the correlation. The explanation for this is exactly the same one offered to account for the fact that multiple regression equations rarely contain more than three or four statistically significant predictor variables. Many predictor variables tend to intercorrelate substantially, and once two or three of them have been entered into the regression equation, the predictive association is "used up," so additional variables have no remaining unique association with the criterion. The same thing happens in partial correlation.

One last bit of terminology. A **zero-order partial correlation** is a Pearson correlation in which no variables are partialled out—in other words, a zero-order partial correlation is just a regular Pearson *r*.

Let's return to the original problem now and ask what should happen to the correlation between *Income* and *PC* once we partial out the effects of *Age*. Here is one way to think about the problem. Begin by assuming that *Income* correlates positively with *PC* for two reasons—first, because *Income* has a unique association with *PC* and second, because it correlates positively with another variable, *Age*, which in turn correlates positively with *PC* Thus, part of the observed correlation between *Income* and *PC* is due to this third-variable pathway. If we remove the pathway, the correlation between *Income* and *PC* should decrease.

As was the case with regression, although there are formulas for manually computing lower-order partial correlation coefficients, they are tedious and not very informative (Problem 3). Consequently, we will rely on *JASP* to do the computations for us, using the REGRESSION – Correlation procedure with the option of Partial. **Figure 2** on the next page shows the partial correlation table produced by this procedure.

Figure 2

JASP Partial Correlation Table Showing the Correlation Between Income and Political Conservatism While Partialling Out Age.

Pearson's Partial Correlations

Variable		Income	PC
1. Income	Pearson's r	—	
	p-value	—	
2. PC	Pearson's r	0.543	—
	p-value	0.016	—

Note. Conditioned on variables: AGE.

The entries in a partial correlation table are organized similarly to those in a correlation matrix. The first number is the value of the partial correlation itself (.543). The second entry is the *p* value for the correlation (.016), which, recall, is the probability of getting a correlation of this magnitude or greater by chance, when the correlation in the population is actually zero. The phrase under Note, "Conditioned on Variables:" is another way of saying, "Partialling out:".

According to the table, partialling out *Age* reduces the correlation between *Income* and *PC* from +.67 to +.54, i.e., $r_{12.3}$ = +.54. This is an appreciable decrease in the size of the correlation between *Income* and *PC*, and reflects the fact that *Age* had a fairly strong correlation with both the predictor and criterion variables, thus contributing substantially to the correlation between the two. Had *Age* been less strongly correlated with either the predictor or criterion, its removal would have produced much less of an effect.

For example, if the correlation between *Age* and *Income* had been +.31 instead of +.51, then $r_{12.3}$ would equal +.63—a decrease of only .04. On the other hand, the more strongly correlated *Age* is with either the predictor or criterion variables, the greater the effect of partialling it out. If *Age* and *Income* had correlated +.71, partialling out *Age* would have reduced $r_{12.3}$ to +.49—a decrease of .19.

Throughout this example we have treated *Age* as the variable to be controlled. But we could turn the question around and determine the correlation between *Age* and *PC* when the effects of *Income* are controlled for. Using the terminology of partial correlation, we could compute $r_{13.2}$, which turns out to be equal to +.30. Partial correlation works in any direction and the decision about which variable or variables to partial out has to be made on logical rather than statistical grounds.

Partial Correlation and Linear Regression

By now, you should be curious how partial correlation actually works, i.e., how variables are "partialled out" or controlled for. It turns out that the mechanism for doing this is linear regression. Let's return to our example of partialling out the effects of age on the correlation between income and political conservatism. We'll use this example to show how partial correlations are derived from linear regression. Understand that the steps below are given to show the conceptual relationship between partial correlation and linear regression. You would not actually carry out the steps, but instead would simply use the *JASP* REGRESSION – Correlation with the option Partial to do the work for you.

1. Our first step is to have *JASP* REGRESSION – Linear Regression derive the linear equation for predicting *Income* from *Age*. It takes this form:

$$Y'_{Income} = -6.771 + 1.722\ (Age)$$

2. Next, we use the equation to compute the predicted value of *Income* (Y') for each case in the sample. This gives us both the actual value of *Income* (Y) and its predicted value Y').

3. The third step is to subtract the predicted value Y' from the actual value Y for each case, i.e., ($Y_{Inc} - Y'_{Inc}$). Recall that this is termed the residual. The important thing about these residual values for *Income* is that they are completely uncorrelated with *Age*, i.e., **we have removed that portion of the variability in *Income* that is associated with variation in *Age*.**

4. Now, we repeat this entire process with the criterion variable *PC*. First, we would derive the linear equation for predicting *PC* from *Age*:

$$Y'_{PC} = 2.971 + .189\ (Age)$$

5. We then compute the predicted value of *PC* for each case and subtract it from the actual value of *PC*, yielding the residual values for *PC*, i.e., ($Y_{PC} - Y'_{PC}$). As was the case with *Income*, these residual value for *PC* are completely uncorrelated with *Age*. There is still variation in the values of the residual scores, but none of that variation is associated with variability in *Age*.

6. The final step is to compute the Pearson correlation between the residualized values for *Income* and *Age*. This correlation reflects any remaining association between *Income* and *PC* once the influence of

Table 1

Original Data and Computed Residual Scores for Hypothetical Study of the Relationship Between Income and Political Conservatism.

Age	Income	P.C.	Resid. Income	Resid. P.C.
47	28	5	-46.18	-6.85
35	56	14	2.49	4.42
32	65	8	16.66	-1.02
28	24	10	-17.45	1.74
44	95	16	25.99	4.72
75	125	18	2.60	.86
47	152	9	77.82	-2.85
41	32	11	-31.84	.28
25	21	5	-15.29	-2.69
24	45	8	10.44	.49
52	165	19	82.21	6.20
50	50	8	-29.35	-4.42
32	41	9	-7.34	-.02
53	36	11	-48.51	-1.99
48	78	12	2.10	-.04
27	86	15	46.27	6.93
54	49	16	-37.23	2.83
50	99	14	19.65	1.58
34	19	4	-32.79	-5.40
36	35	5	-20.23	-4.77

Age has been removed. It is, in fact, **the partial correlation between *Income* and *PC***

The six steps above are fairly abstract, so it's useful to see them actually carried out in practice. Look at **Table 1**. The first three columns show the original data that gave rise to the correlations between *Income*, *PC*, and *Age* shown in **Figure 1**. The fourth column labeled "Resid Income" contain computed values of $(Y_{Inc} - Y'_{Inc})$. The fifth column labeled "Resid PC" contains the computed values of $(Y_{PC} - Y'_{PC})$. Notice that some of the values in both columns are negative in sign. That's the nature of residuals—sometimes the predicted value of Y is larger than the actual value of Y, resulting in positive values of $(Y - Y')$. Sometimes Y' is smaller than Y, leading to negative values. **The fact that the residuals are signed has no effect on their ability to correlate with other variables**.

Figure 3 shows the correlation matrix produced by JASP REGRESSION – Correlation for these five variables. Three parts of the matrix are highlighted. The first part highlighted by the upper rectangle contains the correlations for the three original variables. Satisfy yourself that they agree with the correlations presented originally in **Figure 1**. Now look at the correlations highlighted by the lower rectangle. They are the correlations between *Age* and the resdiualized values of *Income* and *PC*. The correlations are exactly zero indicating that the two variables are now completely independent of *Age*. Finally, look at the lower right rectangle. It highlights the correlation between residualized *Income* and residualized *PC*. The value is +.543, which is exactly the value for $r_{12.3}$ given by the REGRESSION – Correlation – Partial procedure.

Higher-Order Partials

We can partial out more than one variable using the same logic. Let's assume that our hypothetical study had measured the number of children in the family (*Children*) in addition to *Age*, *Income*, and *PC*. **Figure 4** diagrams the pattern of correlations that exist for that situation. The first thing to notice from the diagram is that the number of correlations to be considered grew geometrically with the addition of another variable. There are now six

Figure 3

JASP Correlation Matrix Containing Pearson Correlations Among Age, Income, Political Conservatism, and the Residualized Scores for Income and PC.

Pearson's Correlations

Variable		Age	Income	PC	R Income	R PC
1. Age	Pearson's r	—				
2. Income	Pearson's r	0.507 *	—			
3. PC	Pearson's r	0.528 *	0.665 **	—		
4. R Income	Pearson's r	0	0.862 ***	0.461 *	—	
5. R PC	Pearson's r	0	0.468 *	0.849 ***	0.543 *	—

* p < .05, ** p < .01, *** p < .001

Figure 4

Correlations Among the Four Variables of Income, Age, Children, and Political Conservatism.

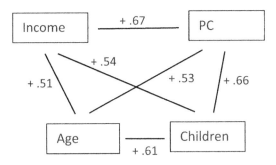

correlations, and their interpretation is fairly complex. As hypothesized, the number of children in the family correlates strongly with *PC*—being the same as the correlation between Income and *PC*. *Children* also correlates strongly with *Age* and *Income*.

What should happen to the correlation between *Income* and *PC* if we partial out both *Age* and *Children*? Because *Children* is strongly correlated with *PC*, it potentially provides another pathway by which *Income* can correlate with *PC*. The effectiveness of this pathway is quite strong, given that *Children* has a stronger correlation with PC than Age.

The situation is further complicated by the fact that *Age* and *Children* also correlate strongly, adding "higher order" pathways by which *Income* and *PC* could correlate. Generally, however, these higher-order pathways contribute very little to a correlation. Overall, we might expect that partialling out both *Age* and *Children* should substantially reduce the correlation between *Income* and *PC* even further.

Here are the conceptual steps involved in partialling out both *Age* and *Children*. We first derive the multiple regression equation to predict *Income* from *Age* and *Children*:

$$Y'_{Income} = +4.105 + .969\ (Age) + 7.748\ (Children)$$

Using the equation, we would get the predicted *Income* score for each case in the sample and subtract it from the actual *Income* score to get the residual score ($Y_{Inc} - Y'_{Inc}$).

Next, we would derive the multiple regression equation to predict *PC* for *Age* and *Children*. It is:

$$Y'_{P.C.} = +4.633 + .074\ (Age) + 1.184\ (Children)$$

We would use the equation to compute residual scores for *PC* ($Y_{PC} - Y'_{PC}$). Last, we would compute the Pearson correlation between the two sets of residuals. This would be $r_{12.34}$.

Again, we would not actually carry out the computations as described. Instead, we would simply use the JASP REGRESSION – Correlation – Partial procedure to give us the result directly. No surprise! JASP gives +.46 as the value for $r_{12.34}$. Recall that when we partialled out just *Age*, the correlation was +.54. The inclusion of *Children* substantially reduced the value of the correlation thanks to the very strong association between *Children* and *PC*.

Part Correlations

In the example that we have been using, it makes sense to partial out the influence of *Age* (and *Children*) on both the predictor (*Income*) and criterion (*PC*) variables. This is because we believe that *Age* is somehow causally affecting both, i.e., as you get older, you simultaneously get wealthier and more politically conservative. In certain situations, however, it makes sense to remove the influence of a variable only from the predictor or criterion variable. This is called **part correlation** or **semi-partial correlation**.

A good example of this is the "Right Stuff" data used in the previous two chapters. Recall that Problem Solving Intelligence (*PS*) was correlated with *GPA* and also with traditional *IQ*. Suppose that we wanted to use partial correlation to see whether *PS* had any unique association with *GPA*, apart from its association with traditional *IQ*. We already know that it doesn't, as shown by the fact that the *B* and beta values for *PS* were not significant when *IQ* was also in the equation. For the moment, let's pretend that we don't know that. In this case, what we want to do is to remove traditional *IQ* from the *PS* measure to see whether *PS* has a unique ability to predict academic performance. But we don't want to remove *IQ* from the criterion variable of *GPA*. The reason for this is that *GPA* realistically *should* reflect the influences of *IQ*. If we removed *IQ* from *GPA*, we would have a variable that doesn't exist in the real world, which would be counter to the purposes of the study. Thus, we need to use part correlation.

Conceptually, what we would do is to use linear regression to predict *PS* from *IQ* and get residual scores for *PS* by subtracting the predicted *PS* score from the actual *PS* score for each case. We would then correlate these residuals with the actual *GPA* scores. Assuming that

GPA is symbolized as "1", PS as "2", and IQ as "3", the resulting correlation would be expressed as $r_{1(2.3)}$. In words, this is the Pearson correlation between variable 1 and the residualized variable 2.

JASP includes a part correlation procedure as a statistic option under the REGRESSION – Linear Regression procedure. The correlation between PS and GPA is .194 with IQ fully partialled out. We have already argued on logical grounds that the partial would not be of theoretical interest to us. Instead, partialling IQ only out of PS yields a value of .174, which is the correlation between GPA and residualized PS. Thus, $r_{1(2.3)}$ = .17.

There are two things to notice about the values of the partial and part correlations. First, there is very little difference in the size of the two correlations, and second, the part correlation is smaller than the partial correlation.

The fact that the partial and part correlations are similar in value is a consequence of the fact that PS and GPA had very little in common to begin with (r = +.39), except that each had a modest correlation with IQ. Removing IQ from either variable dropped the correlation so low that it makes little difference whether it is removed from the other variable. Had PS and GPA been more strongly correlated to begin with, the difference between the partial and part correlation would be greater.

The second observation—the fact that the part correlation is smaller than the partial correlation—is not unique to these data. The part correlation will *always* be smaller than the partial correlation. The explanation for this is somewhat complex and depends upon the concept of **shared variance**. Shared variance means that the correlation between two variables is directly determined by the proportion of variation in each variable caused by common factors. For example, suppose that 30% (.30) of the variation in PS scores was due to IQ differences and 20% (.20) of the variation in GPA due to IQ differences. If the two variables had nothing else in common, the correlation between PS and GPA would be given by the following formula:

$$r_{PS \times GPA} = \sqrt{(.30)(.20)} = \sqrt{.06} = .25$$

The more that two variables share the same variance, the more strongly they will correlate. Conversely, the more that they contain non-shared variance, the lower will be the correlation. If IQ is partialled out of both variables, then any remaining correlation between them will be due to whatever else they share in common. But, if IQ is partialled out of only one variable, the other variable still contains variance due to IQ. This remaining IQ variance is no longer shared and acts as error to reduce the correlation.

By way of example, let's see how we would describe the results of our partial correlation analysis of the relationship between income and political conservatism, controlling for age:

> For the sample of adult men, political conservatism was strongly predicted by reported income (r = .67). Because age was also correlated with both variables, its influence was removed using partial correlation. When age was controlled, the correlation between income and political conservatism dropped to .54.

An Important Caution

One final comment. Partial correlation is a more limited procedure than multiple linear regression. It is used primarily when the purpose of the study is to focus on the relationship between a single predictor variable and a criterion, while eliminating the effects of other correlated variables. This is most appropriate when the variables to be controlled have no great theoretical significance. When the controlled variables are also potentially causally related to the criterion, you are, in a sense, throwing away information about the role that they play in influencing the criterion variable. In the present example, we should run multiple linear regression using *Income*, *Age*, and *Children* to see which variables best predict political conservatism.

Figure 5 shows the Coefficients table in which all three predictors have been entered as Covariates and the "Backward" option selected under Method. Recall the Backward option allows JASP to select variables for removal based on the *p* values. In the initial table with all three predictors entered (Model 1) the *p* value for *Age* is nowhere near statistically significant (.687), while the other two variables are "marginally" significant (.055 and .102, respectively).

In the next step (Model 2) JASP has eliminated *Age*, leaving *Income* and *Children* as significant predictors, with large and approximately equal β values (.439 and .420, respectively).

The Model Summary Table in the lower part of **Figure 5** shows that Model 2 does a very respectable predicting Political Conservatism, R = .753 and R^2 = .568. We might conclude from the regression analysis that increasing personal wealth and the responsibilities of having children are the real causes of increased political conservatism.

Figure 5

JASP Regression and Model Summary Tables with Age, Income, and Children Included as Initial Predictors of Political Conservatism Using the Backward Selection Option Under Method.

Coefficients

Model		Unstandardized	Standard Error	Standardized	t	p
1	(Intercept)	4.453	2.625		1.697	0.109
	Age	0.031	0.076	0.088	0.410	0.687
	Income	0.044	0.021	0.417	2.070	0.055
	Children	0.845	0.487	0.379	1.735	0.102
2	(Intercept)	5.364	1.365		3.929	0.001
	Income	0.046	0.020	0.439	2.321	0.033
	Children	0.936	0.422	0.420	2.219	0.040

Model Summary - PC

Model	R	R^2	Adjusted R^2	RMSE
1	0.756	0.572	0.492	3.231
2	0.753	0.568	0.517	3.151

"If you are depressed you are living in the past, if you are anxious you are living in the future. If you are at peace, you are living in the present."

Lao Tzu

Chapter 7, Partial Correlation
Chapter Exercises

Some questions are purely factual and have only one correct answer. Others require greater thought on your part and may have more than one possible answer.

1. Fill in the missing term(s) that makes the statement true.

   a. $r_{12.3}$ is the symbol for a _____.
   b. A zero-order partial is _____.
   c. $r_{12.34}$ is a _____.
   d. In the Income – Age – PC example, partialling out Age had the effect of _____.
   e. Conceptually, linear regression partials out a variable from two others by _____.
   f. Once Age is partialed out of Income and PC, it will correlate 0 with _____.
   g. A part or semi-partial correlation is one in which a third variable has been _____.
   h. If 30% of the variation in Variable X is due to Variable Z and 40% of the variation in Variable Y is due to Variable Z, and they share nothing else in common, X and Y will correlate _____.
   i. A part correlation is always _____ than a partial correlation because _____ (long answer required).
   j. Partial correlation is most useful in removing the influence of _____ variables.
   k. In one sense, partial correlation throws away information compared with _____.

2. Joe is doing a senior thesis project at Little Angels Elementary School, a private church-run school. He is interested in factors that predict conflict and aggression in elementary age children. He gathers the following data on the 35 children at the school: child's gender (*Gender*, coded "1" for boys and "2" for girls), age in years (*AGE*), social maturity score (*SM*, score on a standardized teacher rating scale, ranging from 5–25, with higher scores indicating greater social maturity), and the criterion variable, number of discipline calls made to the parents during the current school year (*CALLS*).

   Here are his data:

Gender	AGE	SM	CALLS	Gender	AGE	SM	CALLS
2	6	9	6	2	8	18	4
2	6	12	3	2	8	21	1
1	6	11	11	2	8	23	0
1	6	20	8	2	9	14	6
2	6	15	0	2	9	19	4
1	6	8	9	1	9	12	13
1	6	10	3	1	9	18	11
1	6	12	5	2	9	20	7
2	7	14	1	2	9	23	3
1	7	12	9	1	9	18	9
1	7	19	4	1	10	15	14
1	7	8	12	1	10	22	4
1	7	10	8	2	10	25	0
2	7	11	7	1	10	13	13
1	8	16	9	1	10	19	11
1	8	18	10	2	11	21	8
1	8	15	5	1	11	16	21
2	11	22	10				

How well do social maturity ratings predict discipline calls once the age of the child is partialled out? What about partialling out *Gender*? *Gender* is coded 1 = boys and 2 = girls. Nominally, gender is a categorical variable, but as we know in correlational and regression research it can be treated as a numerical variable. Recall that to do that in *JASP* you need to click on the type symbol in the variable name *Gender* in the Data Display window and change it to a numerical variable. Once you've done a partial correlation analysis, do a complete regression analysis and you'll be surprised by the findings.

3. Assume that you are doing a study of the relationship between psychological empathy and favorable attitudes toward abortion among a large sample of working women, ages 19 to 65. You predict that the more empathic a woman is, the more she will favor giving another woman the option of having an abortion. To your surprise, the correlation between a standardized measure of empathy and favorable attitudes toward abortion is only +.25. When you include age in your analysis, you get the following pattern of correlations.

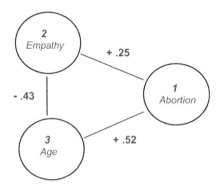

a. Explain what would happen to your correlation of + .25 if you were to partial out Age from the relationship, i.e., explain what $r_{12.3}$ would be. Why would this happen?

b. The following formula is used to manually compute 1st order partial correlations:

$$r_{12.3} = \frac{r_{12} - r_{13}r_{23}}{\sqrt{(1 - r_{13}^2)(1 - r_{23}^2)}}$$

Manually compute $r_{12.3}$ to see how well your prediction was confirmed. Why did this happen?

Chapter 8
Basic Probability

What's In This Chapter?

Probabilities are numbers between 0 and 1.00 that quantify the likelihood that an event will occur. **Odds** are non-standard ways of expressing probabilities. There are three types of probability. **Theoretical** or **equal-likelihood probability** is appropriate in situations in which a **simple random experiment** gives rise to equally likely outcomes. An **event** is made up of one or more **outcomes**, and the probability of an event is defined as the number of outcomes that produce the event divided by the total possible number of outcomes. **Empirical probability** defines the probability of an event as the relative frequency with which the event has occurred over a series of trials or observations. **Subjective probability** characterizes most real-world decision making. The subjective probability of an event is the subjective interpretation of one or more sources of probability information.

Computation of theoretical probabilities is facilitated using **outcome trees**, which are techniques for identifying all of the possible outcomes of a simple random experiment. The use of such techniques is limited, however, by the fact that the number of outcomes grows geometrically large as the number of elements comprising the random experiment increases.

To facilitate probability computations in these more complex situations, mathematicians have developed a number of probability theorems. The two simplest of these are the **addition** and **multiplication theorems**. The addition theorem is used to compute the probability that either one or the other (or both) of two events will occur. The multiplication theorem is used to compute the probability that two events will both occur together. In situations in which there are only two possible outcomes – each having known probabilities – the **binominal theorem** is used to compute how often an event will occur across any number of trials.

Probability distributions specify the probability of obtaining specific values or ranges of values for an event. **Discrete probability distributions** are arrays of discrete event values, each with its associated probability. **Continuous probability distributions** specify the probability of ranges of values for continuous events in terms of the relative area under the mathematical function that describes the event distribution. Probability distributions are at the heart of inferential statistics because they tell you what random sampling (chance) will produce.

Introduction

Most of us have a "rough and ready" understanding of **probability** as being the likelihood or chance that something will happen. This definition has been adequate for understanding the topics we have studied so far, but before we go on, we need to develop a more complete understanding of basic concepts of probability. That is the purpose of this chapter. We will begin by defining some important terms in probability and distinguishing among three different types of probability. We will then explore some of the basic concepts associated with one of those types of probability called theoretical probability and learn how to solve simple problems in theoretical probability. Finally, we will consider the concept of a "probability distribution," which will lead us into the next chapter concerned with the basics of sampling distributions and statistical inference.

Probability Terminology

As indicated above, we all understand probability to be the likelihood that something will occur. The "something" that we are talking about is formally called an **event**. An "event" can be anything for which we can compute the probability—from rolling a seven with a pair of dice to getting an "A" in statistics.

Probabilities are expressed as numbers that range in value from 0 (no chance) to 1.00 (must happen). We express the probability of an event by first identifying the event with some letter like "E" or "B" and then stating that "$p(E)$" or "$p(B)$" is equal to some value. For example, let the event be "Winning the Lottery" with one ticket, which we will abbreviate "W." From a knowledge of the number of possible values that the Lotto number can take, we can state the approximate probability of winning the lottery as

$$p(W) = .000000044$$

The probability of not winning the lottery is abbreviated $p(\neg W)$, where the symbol "$\neg$" means "Not." Because a single ticket will either win or lose (assuming only one winning ticket), and nothing else, the sum of $p(W)$ and $p(\neg W)$ must add up to 1.00, i.e.

$$p(W) + p(\neg W) = 1.00$$

From this, we know that the probability of not winning the lottery with a single ticket is

$$p(\neg W) = 1.00 - p(W)$$
$$= 1.00 - .000000044$$
$$= .999999956$$

Given that the probability of winning the lottery is so low, you may wonder why millions of people buy lottery tickets each week. The answer has to do with what's called the "utility" of the ticket. Assume that the tickets cost $3.00 each and this week's prize is $50,000,000, the expected value of a lottery ticket is about $.80. Despite a theoretical loss of $2.20 for every ticket you buy, the "happiness" that it buys you in terms of fantasizing about what you'd do with the money is actually greater than its cost. In that sense, people are being very "rational" in buying a lottery ticket!

The term "odds" is sometimes used to express probabilities. This can be confusing, depending upon how the term "odds" is used. For example, if you hear that the odds of something happening is "1 in 4," figuring the probability is easy—it's just 1 divided by 4, which is .25. On the other hand, if someone says that the "odds are 5 to 2" against some event happening, the probability is not so obvious. The phrase "5 to 2" means that you have 5 "chances" of losing and 2 "chances" of winning. To compute the probability of the event not happening, i.e., $p(\neg E)$, you divide the number of losing chances by the sum of the losing and winning chances:

$$p(\neg E) = \frac{5}{(5+2)} = \frac{5}{7} \approx .714$$

Conversely, you can reverse the odds and talk about the likelihood of the event happening. Here the odds would be 2 to 5 for the event happening, and the probability would be

$$p(E) = \frac{2}{(5+2)} = \frac{2}{7} \approx .286$$

Types of Probability

There are three types of probabilities depending upon the way that one generates probability numbers. They are referred to as "theoretical", "empirical", and "subjective."

Theoretical Probability

The first type is called **theoretical** or **equal likelihood** probability. In theoretical probability, the probability of an event E, $p(E)$, is given by:

$$p(E) = \frac{Number\ of\ Outcomes\ Producing\ E}{Total\ \#\ of\ Outcomes}$$

This definition is based on the assumptions that some random process will produce two or more **outcomes**, and that each outcome is equally likely to occur. An "outcome" is simply **the most fundamental thing that a random process can produce**. Consider the random process of flipping a coin. We will pretend that the coin can't land on edge (although it can), so there are only two outcomes to our "experiment." Assume that the coin is a fair one, i.e., each of the two outcomes is equally likely to occur. Because the assumptions are met, we can use the definition of theoretical probability to arrive at the probability of any event—say the probability of getting "heads." We will abbreviate the event heads with "H." Using this definition of probability, we can state that:

$$p(H) = \frac{1\ Outcome\ Produces\ H}{2\ Outcomes} = \frac{1}{2} = .50$$

In this example, there was a one-to-one correspondence between an outcome and an event, i.e., "head outcome" = "head event." This is not usually the case as we will see below. We will have more to say about theoretical probability in the next section, so let's pass on to the next type of probability.

Empirical Probability

The second type of probability is called the **empirical** or **relative frequency** definition of probability. According to this definition, the probability of an event E, $p(E)$, is the **relative frequency with which the event has occurred across past trials**. The phrase "relative frequency" means just what we have meant it to be in the past—the proportion of times that something occurs. The term "trials" can literally mean individual trials like discrete flips of a coin, or it can refer to samples of observations, like counting the number of A's given in a class of 35 students. Clearly, it is "empirical" in the sense that probabilities are based on evidence accumulated over time, rather than upon some theory of occurrence of equally likely outcomes.

Here is an example of empirical probability. Being of an observant and forward-looking cast of mind, you notice that Professor Snape seems to favor alternative "C" on his multiple-choice tests in Introductory Sociology. Sure enough, examination of the answer sheets from your last three tests reveals that of the 225 questions asked, 91 had alternative C as the correct answer. The proportion works out to be 91/225 ≈.404. On the next exam, you resolve that on those questions for which you haven't a clue as to the answer, you will always select alternative C. You reason that you will have about a .40 probability of being correct. Good reasoning! If Professor Snape remains oblivious to his bias, past probability may well reflect future probabilities.

Subjective Probability

The third definition of probability is called **subjective probability**. Subjective probability is more difficult to define than the other two, but it is the type of probability that makes our non-statistical world go 'round. Here is one definition of subjective probability. The probability of some event E, p(E), is the **assignment of a probability number based upon subjective interpretation of one or more sources of probability information**. The term "subjective" means from the person's unique perspective or viewpoint. The sources of probability information could include both formal empirical probabilities and casual frequency observations.

Despite the unsatisfactory vagueness of its definition, most important decisions in the world are made based on subjective probabilities. For example, a President's decision to send American troops to some troubled location in the world is in part determined by his (hopefully, soon, her) subjective probability that the military mission will be "successful." The evidence on which the decision is made includes such things as formal risk estimates made by the President's military advisors (these are themselves subjective probabilities), the President's feelings about how other countries will react, the tolerance of the American people for this adventure, and so on. The President may never sum it all up by saying, "My subjective probability is .85 that the mission would succeed," but in principle he or she could.

Frequently, empirical and subjective probabilities are systematically combined by Bayesian statisticians to predict a broad range of important events. Sadly, one not-so-important area where this has become widespread is sports betting, mostly because enormous sums of money are involved. Professional football is a particularly good example because there exist long game, team, and player histories on which to develop stable empirical probabilities, which then can be combined with expert current information to generate probabilities about game and score outcomes.

Basic Concepts of Theoretical Probability

The world assumes that if you have taken an elementary course in statistics, you will have some formal knowledge of theoretical probability. Indeed, some statistics courses spend up to 6 or 8 weeks on this topic. We won't do so, simply because most statistical applications in the social and physical sciences are not dependent upon an extensive knowledge of this type of probability. What we'll do instead is to give a "bare bones" introduction to the basic concepts of theoretical probability and the procedures for solving simple probability problems.

The starting point for theoretical probability is the **simple random experiment**. This is not really an experiment at all, as you now think of the term "experiment." Instead, a simple random experiment is **any process in which the outcome is entirely due to chance or random forces**. Some examples of simple random experiments include our old standby of flipping a coin, plus other things like rolling dice, drawing cards from well-shuffled decks, and so on. The result of a simple random experiment is two or more outcomes that are all equally likely to occur. The complete set of all outcomes of a simple random experiment comprises what is called the **outcome space**. **Figure 1** shows in diagrammatic form the outcome space for the simple random experiment of rolling a single die (one-half of a pair of dice).

Figure 1

Outcome Space for a Simple Random Experiment of Rolling a Single Die One Time.

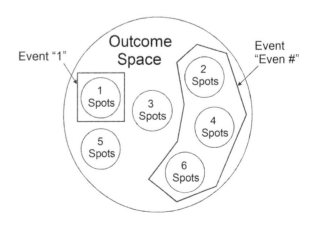

Within the large circle representing the outcome space are six smaller circles representing the outcomes one spot, two spots, three spots, four spots, five spots, and six spots. Single outcomes or collections of outcomes form **events**. Events can be simple or complicated. Each outcome is associated in a one-to-one fashion with what is called an **elementary event**. The square placed around the one spot outcome stands for the elementary event "1." Events in general are more interesting than elementary events. Notice the irregular polygon drawn around for the two spots, four spots, and six spots outcomes. That represents the event "even numbered spots," which we can abbreviate "E."

The probability of this event is given by the defining formula for theoretical probability:

$$p(E) = \frac{3 \ outcomes \ produce \ E}{6 \ possible \ outcomes} = \frac{3}{6} = .50$$

This seems pretty simple. Let's take another example. We have a leather bag containing five balls of the same size and weight. Two of the balls are black, two are red, and one is white. We are going to draw a ball out of the bag at random, record its color, and then return it to the bag. Then, we are going to draw a second ball and record its color. This is our simple random experiment. Let the event be "neither ball is red." Technically, we are asking for the probability of the event ($\neg R_1$ and $\neg R_2$), where the subscripts refer to the first and second balls. What is the probability of this event? According to the definition of theoretical probability, to figure the probability of this event, we need to know the total number of outcomes possible in the experiment and the number of outcomes that produce the event of interest. Unfortunately for this simple random experiment, it's not immediately obvious how many outcomes are possible, nor how many outcomes produce the event of interest. Happily, however, there is a simple device for identifying the complete set of outcomes in situations like this one. It is called an **outcome tree**.

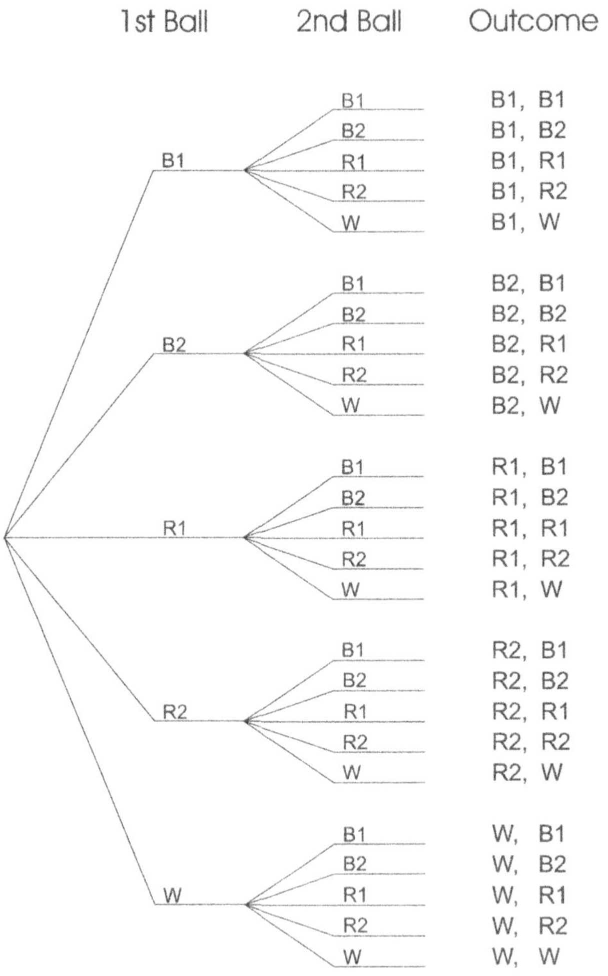

Figure 2 shows the outcome tree for this simple random experiment. The "tree" has two points where "branches" occur. The first branching point is at the first draw, and the five branches refer to the five different balls that could be drawn. Notice that there must be a separate branch for each of the two black balls and each of the two red balls. Although of the same color, they are different balls and must be kept track of separately. The branches are each labeled to correspond to the five colored balls. At the end of each of the branches, five more branches "sprout," corresponding to the second draw. These branches represent the five different balls that could be drawn, after drawing any one of the five balls on the first

Figure 2

Outcome Tree for the Random Experiment of Selecting Two Balls from Bag, Sampling with Replacement.

1st Ball	2nd Ball	Outcome
B1	B1	B1, B1
	B2	B1, B2
	R1	B1, R1
	R2	B1, R2
	W	B1, W
B2	B1	B2, B1
	B2	B2, B2
	R1	B2, R1
	R2	B2, R2
	W	B2, W
R1	B1	R1, B1
	B2	R1, B2
	R1	R1, R1
	R2	R1, R2
	W	R1, W
R2	B1	R2, B1
	B2	R2, B2
	R1	R2, R1
	R2	R2, R2
	W	R2, W
W	B1	W, B1
	B2	W, B2
	R1	W, R1
	R2	W, R2
	W	W, W

draw. Each of these branches is labeled for a particular ball. The terminus of each of the second-stage branches is labeled with the two balls making up the completed branch. There are 25 completed branches. These identify all the different combinations of balls that could be drawn, i.e., all the outcomes of the experiment.

What one needs to do now is to count how many of these outcomes lead to the event "neither ball is red." These include the first complete branch (B_1, B_1), the second (B_1, B_2), the fifth (B_1, W), the sixth (B_2, B_1), and so on. In all, there are nine outcomes that don't include a red ball. According to our definition, the probability of not getting a red ball is

$$p(\neg R_1 \ and \ \neg R_2) = \frac{9}{25} = .36$$

The advantage of the outcome tree should be obvious to you now. "Simple" random experiments can rapidly grow quite elaborate in terms of all the outcomes that they can produce. The tree is a device to make sure that you identify and keep track of these outcomes.

There is a simple formula for computing the total number of outcomes in these types of experiments. Let E stand for the number of "elements" or "stages" in the experiment and O stand for the number of outcomes associated with *each* element or stage. The total number of outcomes is O^E. If you were to draw *three* balls in the manner described in the five-ball experiment, the number of outcomes would be $5^3 = 125$. If you were to flip five pennies in the air and record the number of heads produced, how many outcomes are possible? Here there are five elements corresponding to the five pennies. Each penny produces two outcomes, so $2^5 = 32$. For the purposes of our simple random experiment, flipping five pennies all at once is equivalent to flipping one penny five times.

In the ball experiment, we replaced the first ball after drawing and recording its value. This is called **sampling with replacement**. The expression means just what it says. After an element is sampled, it is returned to the experimental pool of elements. We will generalize this concept to all types of sampling, including the general statistical procedure of drawing samples from populations.

One can also conduct simple random experiments in which sampling is done **without replacement**. Let's repeat the five-ball experiment, except that we will simply draw two balls out at random, one at a time. Under this condition of sampling without replacement, what is the probability of not getting a red ball? Our outcome tree will still work. We just need to be more attentive in drawing it. **Figure 3** shows the tree generated when sampling is without replacement.

The first stage still has five branches, because all balls are available for selection. But at stage two, only four branches can sprout from each of the first branches. This is because only four balls are left in the bag after the first one is drawn. There are 20 outcomes; five have been eliminated because the same balls now cannot be drawn twice. Of the 20 outcomes, six produce the event "neither ball is red," $(B_1, B_2), (B_1, W), (B_2, B_1), (B_2, W), (W, B_1)$, and (W, B_2). The probability of this event is given by:

$$p(\neg R_1 \text{ and } \neg R_2) = \frac{6}{20} = .30$$

There is a simple computational rule to determine the number of outcomes when you are sampling without replacement. For O outcomes for the first element E, the number of total outcomes will be $O!$ (read "O factorial") carried out for E elements. The factorial of a number is a series of products of the following type: Number x (Number -1) x (Number -2) x ... x (1). Thus, $5! = 5 \times 4 \times 3 \times 2 \times 1 = 120$. In the ball drawing experiment, it would be $5!$ carried out twice, yielding 20 outcomes.

Figure 3

Outcome Tree for the Random Experiment of Selecting Two Balls from Bag, Sampling without Replacement.

1st Ball	2nd Ball	Outcome
B1	B2	B1, B2
	R1	B1, R1
	R2	B1, R2
	W	B1, W
B2	B1	B2, B1
	R1	B2, R1
	R2	B2, R2
	W	B2, W
R1	B1	R1, B1
	B2	R1, B2
	R2	R1, R2
	W	R1, W
R2	B1	R2, B1
	B2	R2, B2
	R1	R2, R1
	W	R2, W
W	B1	W, B1
	B2	W, B2
	R1	W, R1
	R2	W, R2

Simple Probability Theorems

The procedures described above will work, in principle, to solve any problem in theoretical probability. The difficulty in practice is that the number of outcomes in simple random experiments grows exponentially large with the number of elements comprising the experiment. Consider the following scenario. The Biology Club of a small college has 30 members, 15 of whom are male and 15 of whom are female. The president of the club wants to select four members in addition to herself to organize the Spring Fund Raiser. She selects four names at random, without replacement. What is the probability that all four members selected will be women, thereby creating charges of favoritism and inadequate political correctness? To use our outcome tree here would be tedious to say the least. It turns out that there are 570,024 possible outcomes to this not-so-simple random experiment, i.e., O = 29 and E = 4, so O! carried out to four products is 29 x 28 x 27 x 26 = 570,024. The sheet of paper required to draw the tree would have to be about four feet wide by 780 feet high. Clearly, there must be a better way to solve this type of problem. There is. Mathematicians have come up with shortcut procedures for determining the probability of certain kinds of events. These are called probability theorems, and we will study three of the simpler of these theorems. By the way, the probability of getting four female members is about .05.

Addition Theorem

The first theorem is known as the **addition theorem** of probability. It is used to compute what are called **disjunctive events**. A disjunctive event is the occurrence of either one event or another event, or both events. Disjunctive events are usually symbolized "A or B," where A and B refer either to elementary events or more complex events. The theorem states that

$$p(A \text{ or } B) = p(A) + p(B) - p(A \text{ and } B)$$

In words, this expression says that the probability of *either A or B or both* occurring is equal to the separate probability of A plus the separate probability of B, minus the probability of A and B occurring together. The apparently contradictory necessity of subtracting out the probability of A and B occurring together will be made clear below.

We can best understand the operation of this theorem by using a specific example. We will invoke a favorite statistician's example—the card drawing experiment. Assume that we have a well-shuffled deck of 52 playing cards. We draw one card at random from the deck. Let F stand for the event "A Face Card" and let S stand for the event "A Spade." What is the probability that the card will be either a face card or a spade (or both), i.e., what is p(F or S)? According to the addition theorem,

$$p(F \text{ or } S) = p(F) + p(S) - p(F \text{ and } S)$$

The probability of a face card, p(F), is 12/52 (three face cards of each of four suits). The probability of a spade, p(S), is 13/52 (there being 13 cards of each suit). But what is the probability of a card being *both* a face card *and* a spade at the same time, p(F and S)? A moment of thought should tell you that it is 3/52 (the king, queen, and jack of spades). The expression thus becomes:

$$p(F \text{ or } S) = \frac{12}{52} + \frac{13}{52} - \frac{3}{52} = \frac{22}{52} \approx .42$$

Let's clarify why the expression contains the "- p(A and B)" component. **Figure 4** diagrams the card drawing problem. The probability of drawing a face card is illustrated by the box that surrounds the jack, queen, and king of clubs, diamonds, hearts, and spades (12 cards). The probability of drawing a spade is also indicated by a box the surrounds the deuce through the ace of spades (13 cards). The problem occurs because three of the cards that are surrounded by both boxes (jack, queen, and king of spades) were counted twice—once when the face cards were tabulated and again when spades were counted. To correctly reflect the probability of (F or S), they should only be counted once, so it is necessary to "subtract out" the duplicate count.

Figure 4

Diagram of Card Sampling Experiment Showing Double Counting of the Jack, Queen, and King of Spades.

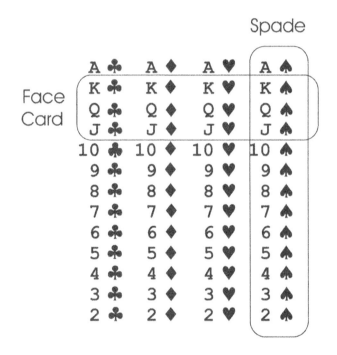

One more example. What is the probability that from one well-shuffled deck you will draw either an odd numbered card or a face card? Here, let O = "odd numbered card" and F = "face card," so

$$p(O \text{ or } F) = p(O) + p(F) - p(O \text{ and } F)$$

The $p(O)$ = 16/52 (the 3, 5, 7 and 9 of each suit, not counting aces as 1's) and $p(F)$ = 12/52 (three face cards of each suit). What is $p(O$ and $F)$? The answer is 0, because no card can simultaneously be an odd number *and* a face card. Thus,

$$p(O \text{ or } F) = \frac{16}{52} + \frac{12}{52} - \frac{0}{52} = \frac{28}{52} \approx .54$$

When two events cannot both occur at the same time, we say that they are **mutually exclusive** events. When events are mutually exclusive, the addition theorem simplifies to

$$p(A \text{ or } B) = p(A) + p(B)$$

because $p(A$ and $B)$ = 0.

Multiplication Theorem

The second probability theorem is concerned with the computation of the probability of the joint occurrence of two events. When two events occur together, we say that they are **conjunctive**. Conjunction is indicated by the word "and." The theorem is called the **multiplication theorem** of probability. The multiplication theorem is stated as follows:

$$p(A \text{ and } B) = p(A) \times p(B \mid A)$$

In words, the expression says that the probability of events A and B occurring together is equal to the separate probability of A, times the probability of B, *given that A has already occurred*. The latter part of the expression, $p(B \mid A)$, is called a **conditional probability**. It is the probability that a particular event will occur after some other event has already occurred. This idea of conditional probability will become clearer after we solve a couple of example problems.

Let's do some more card playing probability. What is the probability that if you draw two cards at random without replacement from a well-shuffled deck, that both cards will be clubs? Let C_1 stand for a club on the first card and C_2 a club on the second card. According to the multiplication theorem

$$p(C_1 \text{ and } C_2) = p(C_1) \times p(C_2 \mid C_1)$$

The probability of getting a club on the first card, $p(C_1)$, is 13/52 (13 cards in a standard deck are clubs). The probability of getting a club on the second card, given that you have already drawn a club, $p(C_2 \mid C_1)$, is 12/51. This is true, of course, because there are only 12 clubs left in the deck and only 51 cards remaining from which to draw. So,

$$p(C_1 \text{ and } C_2 \mid C_1) = \frac{13}{15} \times \frac{12}{51} = \frac{156}{2652} \approx .069$$

Here is another example. While playing around, you shuffle the deck and deal yourself three cards—8 of clubs, 8 of spades, and the jack of diamonds. What is the probability that the next two cards that you deal yourself will be two black aces, thereby completing the "dead man's hand." Three cards have been dealt, so the probability of dealing a black ace on the fourth card $p(A_{B,4})$, is 2/49. If you get a black ace on the fourth card, the probability of getting a black ace on the fifth card, $p(A_{B,5} \mid A_{B,4})$ will be 1/48. Thus,

$$p(A_{B,4} \text{ and } A_{B,5}) = \frac{2}{49} \times \frac{1}{48} = \frac{2}{2352} \approx .0009$$

Fortunately, you probably will not draw the dead man's hand. (The interested student will no doubt Google "dead man's hand.")

In the two examples involving drawing clubs, you "sampled" cards without replacement. What would happen to the probabilities if you used a replacement strategy for sampling? Sampling with replacement, what is the probability of drawing two clubs? The probability of the first club, (C_1), is 13/52 and, because you replaced the card and reshuffled, the probability of a club on the second card, (C_2), is also 13/52. Thus,

$$p(C_1 \text{ and } C_2) = \frac{13}{52} \times \frac{13}{52} = \frac{169}{2704} \approx .063$$

Recall above that when you sampled without replacement, the probability of two clubs was about .069. The value of .063 that was obtained when you sampled with replacement really isn't much different. Why do you suppose this is so? The answer lies in the fact that sample size (two cards) was small compared with the entire set of cards (52). Under these conditions, a card here or there doesn't make too much difference. In our other example, the probability of drawing the "dead man's hand" if the fourth and fifth cards are sampled with replacement is 2/49 x 2/49 = 4/2401 = .002, compared with .0009 when sampling without replacement. Still pretty safe odds.

The difference between sampling with and without replacement can make a bigger difference, however, when the total number of elements comprising the random experiment becomes smaller relative to the sample size. Recall in the ball drawing experiment, the probability of not drawing a red ball in two draws was .36 when sampling with replacement and .30 when sampling without replacement.

Sampling with replacement causes two events *A* and *B* to be **independent** of one another. We say that two events are independent if the occurrence of one event in no way changes the probability of the other event. In probability notation, two events, *A* and *B*, are independent if

$$p(B|A) = p(B)$$

That is, the conditional probability of *B* given *A* is the same as the simple probability of *B*. If two events *A* and *B* are independent, then the multiplication rule can be rewritten as

$$p(A \text{ and } B) = p(A) \times p(B)$$

Many (perhaps most) events in the real world are independent—not because we are "sampling with replacement," but simply because they are caused by processes that have nothing to do with one another. Whether it will rain today is most likely independent of whether you select lemon or blueberry yogurt for breakfast. Such independence simplifies probability calculations.

By the way, we could use the multiplication theorem to solve the ball drawing problem (2 Black, 2 Red, 1 White), both with and without replacement. Let $\neg R_1$ stand for "not red on the first draw" and $\neg R_2$ for "not red on the second draw." Sampling with replacement, the probability of neither ball being red is

$$p(\neg R_1 \text{ and } \neg R_2) = p(\neg R_1) \times p(\neg R_2)$$
$$= 3/5 \times 3/5$$
$$= 9/25$$
$$= .36$$

Sampling without replacement,

$$p(\neg R_1 \text{ and } \neg R_2) = p(\neg R_1) \times p(\neg R_2 | \neg R_1)$$
$$= 3/5 \times 2/4$$
$$= 6/20$$
$$= .30$$

It may be that you don't spend much of your time drawing balls from bags or cards from well-shuffled decks. Are there any more realistic applications of the addition and multiplication theorems of probability? Happily, the answer is yes. Consider new cars. A particular dealer's records reveal that 20% of the cars that she receives from the factory have a defect that requires significant alteration or repair in the dealer's service department. We can take this to mean that the probability of any car having a defect is .20, i.e., $p(D) = .20$. These, by the way, are empirically rather than theoretically derived probabilities, but they will work just fine in our probability theorems. We will assume that cars are independent of one another with respect to requiring dealer service.

Salesperson John sells two cars on Monday. What is the probability that neither car will have a defect requiring service? Let $\neg D_1$ stand for the first sale being free of defects and $\neg D_2$ for the second sale being free of defects. The probability of any car being free of defects is given by the expression $p(\neg D) = 1.00 - p(D)$), which is $1.00 - .20 = .80$. This, of course, is true because the two events *D* and $\neg D$ are all that can happen and the sum of their two probabilities must be 1.00. So, according to the simplified form of the theorem for independent events,

$$p(\neg D_1 \text{ and } \neg D_2) = p(\neg D_1) \times p(\neg D_2)$$
$$= .80 \times .80$$
$$= .64$$

What is the probability that both of John's sales will be defective? The probability of each car being defective is .20, and the events are independent; therefore,

$$p(D_1 \text{ and } D_2) = p(D_1) \times p(D_2)$$
$$= .20 \times .20$$
$$= .04$$

What is the probability that only one of the two cars would have a defect. This can happen two ways—the first car is defective and the second is not, or the first car is not defective, and the second car is. The probability of the first situation is:

$$p(D_1 \text{ and } \neg D_2) = p(D_1) \times p(\neg D_2)$$
$$= .80 \times .20$$
$$= .16$$

The probability of the second situation is:

$$p(\neg D_1 \text{ and } D_2) = p(\neg D_1) \times p(D_2)$$
$$= .20 \times .80$$
$$= .16$$

Because the two situations are mutually exclusive, the probability of one or the other happening is the sum of these two separate probabilities, i.e., $.16 + .16 = .32$. Thus, the probability of having exactly one defective car is .32.

Probability Distributions

This last section introduces the concept of **probability distributions**, which are mathematical entities that specify the probability of obtaining different values or ranges of values of an event. Probability distributions may be derived empirically, but the probability distributions that we are concerned with in this text are known through mathematical theory. The concept of probability distributions is central to statistical inference. In this section, we will consider probability distributions in general, using examples drawn from simple random experiments. In Chapter 9, we will focus on a special type of probability distribution called a "sampling distribution."

Discrete Probability Distributions

Consider the simple random experiment of tossing four fair coins in the air and letting them land on a flat surface. Each element of the experiment can produce two outcomes and there are four elements, so we know that the experiment will generate $O^E = 2^4 = 16$ possible outcomes. Our event of interest is the number of heads showing for the four coins.

Figure 5 is the outcome tree with which we can tally the 16 outcomes. At the terminus of each branch in the tree is the outcome for that branch and the *value* of the event, "Number of Heads," produced by the outcome. The top branch, for example, is the outcome *HHHH*, which yields "4" as the value for the event. The next branch is the outcome *HHHT*, which yields "3" for the event. The third branch is *HHTH*, which also yields the value "3" for the event, and so on. If you scan down the values listed, it is obvious that some values of the event "Number of Heads" are more likely to occur than others. We can compute the probability that the event "Number of Heads" will take on each of the different values by using the defining formula for theoretical probability. The value "4" has a probability of 1/16 = .063, because only 1 of the 16 outcomes yields this value. The value "3" has a probability of 4/16 = .25, because four outcomes produce that value.

Table 1 lists the five different values possible for the event "Number of Heads" and the probability of each value. This listing looks very much like the relative frequency distribution that we studied in Chapter 2. Indeed, it is a distribution, and the probabilities are in one sense relative frequencies. There is, however, an important difference. The relative frequencies that appear in a relative frequency distribution are empirically derived. The "relative frequencies" in this distribution are theoretically derived probabilities. This type of distribution is called a **discrete probability distribution**. Technically, a discrete probability distribution is defined as **an array of values,**

Figure 5

Outcome Tree for Random Experiment of Flipping Four Fair Coins.

Coin 1	Coin 2	Coin 3	Coin 4	Outcome	Event
H	H	H	H	H H H H	4
			T	H H H T	3
		T	H	H H T H	3
			T	H H T T	2
	T	H	H	H T H H	3
			T	H T H T	2
		T	H	H T T H	2
			T	H T T T	1
T	H	H	H	T H H H	3
			T	T H H T	2
		T	H	T H T H	2
			T	T H T T	1
	T	H	H	T T H H	2
			T	T T H T	1
		T	H	T T T H	1
			T	T T T T	0

each with its associated probability. The qualifier "discrete" means that we are dealing with events that can take on only a finite number of values. A bit later in this section we'll consider what are called "continuous" probability distributions.

Table 1

Probability Distribution for the Event "Number of Heads" for a 4-Coin Random Experiment.

Value of Event "Number of Heads"	Probability of Value Occurring
0 Heads	1/16 = .0625
1 Head	4/16 = .2500
2 Heads	6/16 = .3750
3 Heads	4/16 = .2500
4 Heads	1/16 = .0626

Binomial Theorem

By now it may have struck you that drawing outcome trees is quite laborious. Happily, there is a much more efficient way to derive probabilities for different outcomes in simple experiments that have binary outcomes. It is called the **binomial theorem**. The theorem has a long history dating back to the Middle Ages, but Isaac Newton is credited with developing the modern form of the equation. When a simple random experiment produces two outcomes (usually called successes and non-successes) having known probabilities, the formula becomes:

$$p(X) = \frac{N!}{(N-X)!(X!)} p^X q^{(N-X)}$$

Where: N = number of trials,
X = number of successes (Heads),
p = probability of success on a trial, and
q = probability of non-success on a trial.

In our four-coin flipping experiment, we can compute the probability of 0, 1, 2, 3, or 4 coins coming up heads (successes) using the theorem. Computing the probability of 0 Heads becomes:

$$p(0) = \frac{4!}{(4-0)!(0!)} .5^0 \, x \, .5^{(4-0)}$$

$$p(0) = \frac{24}{(24)(1)} 1 \, x \, .0625$$

$$p(0) = 1 \, x \, 1 \, x \, .0625 = .0625$$

Note that any number raised to the 0 power is defined as equaling 1. The result agrees with the probability in **Table 1** derived from the outcome tree. How about computing the probability of 2 Heads?

$$p(2) = \frac{4!}{(4-2)!(2!)} .5^2 \, x \, .5^{(4-2)}$$

$$p(2) = \frac{24}{(2)(2)} .25 \, x \, .25$$

$$p(2) = 6 \, x \, .25 \, x \, .25 = .375$$

Using this theorem we can compute the remaining probabilities for 1 Head, 3 Heads, and 4 Heads and finish the probability distribution table.

If the outcome of a random process yields only two states, the binomial theorem can be used to calculate the frequency (probability) of different events based on those outcomes. The theorem is widely used in probability calculations in the real world and more generally in solving a variety of higher mathematical problems.

Let's do one more problem in which $p \neq q$. Assume you are rolling a four-faced die, i.e., an equilateral pyramid (tetrahedron) with the numbers 1 – 4 on the four faces. Each of the faces is equally likely to occur and the face down is the outcome of the roll. The probability of each of the four numbers is .25. We'll define "success" as getting a 1. Its probability is $p = .25$, so the probability on a non-success is $q = .75$. You roll the die four times and count the number of 1's that occur. What is the probability of getting 0, 1, 2, 3, and four 1's?

The probability of getting zero 1's across the four rolls is given as:

$$p(0) = \frac{4!}{(4-0)!(0!)} .25^0 \, x \, .75^{(4-0)}$$

$$p(0) = \frac{24}{24(1)} x \, (1)(.3614)$$

$$= 1 \, x \, 1 \, x \, .361 = .3614$$

The probability of one 1 is:

$$p(1) = \frac{4!}{(3)!(1!)} .25^1 . 75^3$$

$$= \frac{24}{6} x \, .25 \, x \, .4419$$

$$= 4.00 \, x \, x \, .25 \, x \, .4219 = .4219$$

In a similar manner, the computed probabilities for 2, 3, and 4 ones are .2109, .0469, and .0039. Notice that when $p \neq q \neq .50$, the probability distribution is not symmetrical. **Table 2** shows the probability distribution for the four rolls of the pyramidal die out of 256 possible outcomes ($4^4 = 256$).

Table 2
Probability Distribution for the Event "Number of Ones" for the Four Rolls of a Pyramidal Die Random Experiment.

Value of Event "Number of Ones"	Probability of Value Occurring
0 Ones	81/256 = .3164
1 Ones	108/256 = .4219
2 Ones	54/256 = ..2109
3 Ones	12/256 = .0469
4 Ones	1/256 = .0039

What does one do with probability distributions? Actually, you already know. Recall that in Chapter 3 we considered the problem of "The (Perhaps) Crooked Coin."

In that problem our task was to decide whether a coin was fair or biased towards "heads." Our strategy was to flip the coin 16 times and compare the number of heads obtained in this sample to the theoretically derived distribution of numbers of heads that would occur in an unbiased coin. **Probability distributions tell us what "chance" or random processes will produce.**

Continuous Probability Distributions

The second type of probability distribution is called a **continuous probability distribution**. The "continuous" means that we are dealing with a variable that can assume infinitely graded values. For our purposes, a continuous probability distribution is a **distribution defined by a continuous mathematical function in which the probability of a range of values is given by the relative area under the function**. This is a mouthful, but, again, you are already familiar with the concept. In Chapter 3 where we were considering the problem of estimating the probability of different ranges of IQ scores, we noted that the histogram display of the discrete probability distribution of the IQ scores could be well-approximated by a normal distribution having $\mu = 100$ and $\sigma = 15$. The normal distribution was used as a continuous probability distribution to give approximate results for the discrete histogram distribution of IQ scores.

Let's return to "The (Perhaps) Crooked Coin" example also discussed in Chapter 3. **Figure 7** shows the relationship between the histogram display of the discrete probability distribution for the 16–flip random experiment and the continuous normal curve that approximates the discrete distribution. The normal distribution curve that approximates the histogram has $\mu = 8.00$ and $\sigma = 2.00$. We are already familiar with the methods of solving normal curve problems, so we'll just do one continuous probability distribution problem.

What is the probability of obtaining "7," "8," or "9" heads from the simple random experiment of flipping a fair coin 16 times? Recall that because the continuous function is used to describe a discrete probability distribution, we need to express discrete values as ranges of values, i.e., "7" is given by the *range* between the real limits of 6.5 and 7.5, 8 by the range between the real limits of 7.5 and 8.5, and 9 by the range between 8.5 and 9.5. Thus, our question must be expressed as the probability of obtaining a value between the lower real limit of 6.5 and the upper real limit of 9.5.

In a normal distribution having $\mu = 8.00$ and $\sigma = 2.00$, the value of 6.5 lies $-1.5 / 2.00 = -.75$ standard deviations below the mean, while 9.5 lies $+1.5 / 2.00 = +.75$ standard deviations above the mean. From the unit normal distribution in **Table A** in the Appendix we note that $\mu \pm .75$

Figure 7

Approximation of a Discrete Probability Distribution by a Continuous Normal Probability Distribution.

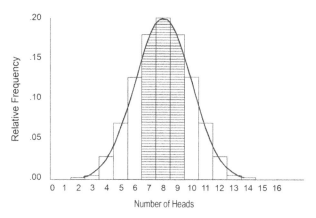

(σ) includes .5568 of the area under the function. The exact probability obtained from summing the probabilities for 7, 8, and 9 heads for the discrete probability distribution is +.1746 + 1964 + 1746 = .5456 (refer to **Table 5** in Chapter 3). Clearly, the continuous normal probability distribution is a very good approximation of the discrete probability distribution.

Final Comment

This chapter has given you the basics of probability terminology and introduced three important theorems used in discrete probability computations. This has been the "barest of the bones" introduction because the remainder of this text is concerned with a variety of continuous probability distributions that are used in many inferential statistics. If you found the topic interesting, you should take an "Introduction to Probability" or a "Probability and Statistics" course from the math department.

"It is not true that people stop pursuing dreams because they grow old, they grow old because they stop pursuing dreams."

Gabriel García Márquez

Chapter 8, Basic Probability
Chapter Exercises

Some questions are purely factual and have only one correct answer. Others require greater thought on your part and may have more than one possible answer.

1. Fill in the missing term(s) that makes the statement true.

   a. Probabilities are numbers between _____ and _____.
   b. If the odds for an event are 3 to 5, the event has a probability of _____.
   c. If $p(A) = .40$, then _____ = .60.
   d. The "happiness" that you get imaging what you would do if you won the lottery is called _____.
   e. In theoretical probability, $p(A)$ = _____ divided by _____.
   f. Flipping three coins is an example of a _____.
   g. All the possible outcomes that can occur in flipping the three coins is called the _____.
   h. Getting all Heads from the coin flip is called an _____.
   i. When an outcome equals an event, we call the event an _____.
   j. _____ are useful devices for keeping track of outcomes.
   k. $p(A \text{ or } B)$ is computed using the _____ theorem.
   l. $p(B|A)$ is called a _____ and is a component of the _____ theorem.
   m. If $p(B|A) = p(B)$, the events A and B are _____.
   n. Discrete probability distributions are _____, each with its associated _____.
   o. The more complex computational theorem introduced in the chapter is called _____.
   p. It is used to compute the probability of _____.
   q. In a continuous probability distribution, probabilities are computed by finding _____.

2. For each of the following, indicate what type of probability is involved (theoretical, empirical, or subjective). If you believe that two different types of probability may be possible, justify your answer.

   a. Drawing two cards to an inside straight in five-card poker
   b. Getting a pop quiz in a particular class on any given day
   c. Buying a defective blender from Amazon
   d. Wining the Florida lottery with one ticket
   e. The weather report predicting rain for this afternoon being correct

3. Card drawing questions. Standard deck of 52 cards.

   a. Two Aces without replacement
   b. An Ace or a Face Card with replacement
   c. Two cards of different suits with replacement
   d. Two even numbered cards without replacement

4. You flip a fair coin 5 times. Use the Binominal Theorem to compute the probability of getting three Heads.

5. You roll a die three times. A "Success" is defined as 1 or 2 spots. A "non-success" is 3, 4, 5, or 6 spots. Use the Binominal Theorem to compute the probability of getting exactly two successes.

Chapter 9
Sampling Distributions & Statistical Inference

What's In This Chapter?

A **sampling distribution** is a probability distribution of values of a statistic computed from random samples drawn from the population of interest. Sampling distributions can be **discrete** when the population size is small and enumerable, but most sampling distributions used in statistics are **continuous functions** derived mathematically from knowledge of the parameters of populations. A very important theorem in statistics is the **Central Limit Theorem (CLT)**, which describes the characteristics of the continuous sampling distribution of the mean.

Sampling distributions are used in inferential statistics to test hypotheses about population parameters and to establish confidence interval estimates of population parameters. In **hypothesis testing**, sample data are used to decide whether a population parameter is equal to a particular value or range of values. The value to be tested is expressed in the **Null Hypothesis (H_0)**, which is pitted against its logical complement called the **Alternative Hypothesis (H_A)**. If the obtained value of the sample statistic is one that is *likely* to occur if H_0 is true, there is no basis for **rejecting** H_0. If the obtained value of the sample statistic is one that is *unlikely* to occur if H_0 is true, then H_0 is rejected, and H_A accepted. The criterion of unlikeliness for rejecting H_0 is a probability value that is called the α **level**, and it is usually set at .05 or .01.

When testing hypotheses about the population mean (μ), H_0 and H_A can be formulated in either a non-directional or directional manner. **Non-directional hypotheses** establish **areas of rejection** symmetrically in both tails of the sampling distribution of interest. The test is therefore called a **2-tail test** of H_0. In a **directional hypothesis**, the entire area of rejection is placed in one tail of the sampling distribution, leading to a **1-tail test** of H_0.

In **parameter estimation**, knowledge of the sampling distribution allows you to construct a **confidence interval** around the value of the statistic that has a known **prior probability** of including the value of the corresponding population parameter. Two confidence intervals commonly used are the **90%** and the **95% confidence intervals.**

When sample data are used to test a null hypothesis, two types of **inferential errors** are possible. If H_0 is true, but the sample value leads you to reject the hypothesis, you have made a **Type I Error**. If H_0 is false, but the sample value causes you to fail to reject it, you have made a **Type II Error**. The **power** of a statistical test is the probability of correctly rejecting H_0, i.e., rejecting H_0 when H_0 is false.

Two important considerations involved in parameter estimation are estimator **bias** and **precision**, both of which contribute to the **accuracy of estimation**. A statistic is an unbiased estimator of its population parameter if the **expected value** (average value) of the statistic equals the parameter estimated. The precision with which a statistic estimates its population parameter refers to the consistency of the estimate. Precise statistics have narrow sampling distributions, which means that most of the sample values will be close to the parameter value.

Introduction

This chapter introduces the key concept of a **sampling distribution** and how sampling distributions are used in the inferential processes of **hypothesis testing** and **parameter estimation**. We will also consider the closely related concepts of **inferential error** and **statistical power**. If you fully understand this chapter, then you will have mastered the essentials of inferential statistics. The focus of this chapter is on the sampling distribution of the mean, but most of the concepts introduced here apply to the other sampling distributions that we will consider in later chapters.

Sampling Distributions

A sampling distribution is a **probability distribution in which the event of interest is a statistic, and the random experiment is sampling with replacement from a**

population of values. Sampling distributions can be either discrete or continuous. We will begin with an example of a discrete sampling distribution because such distributions are easy to discuss and to understand. We will then quickly move on to continuous sampling distributions, because virtually all the inferential statistical procedures in this textbook are based on continuous sampling distributions.

Discrete Sampling Distributions

Imagine a population consisting of just three scores: 3, 4, and 5. This is a very small population, but it will suffice for our purposes. In technical terms, we have a small, finite population defined as {3, 4, 5}. The μ of our little population is 4.00 and σ is approximately .816, i.e.,

$$\mu = \frac{\sum X}{N} = \frac{12}{3} = 4.00$$

$$\sigma = \sqrt{\frac{\sum(X-\mu)^2}{N}} = \sqrt{\frac{2}{3}} \approx .816$$

We are going to construct the discrete sampling distribution of the mean for sample size $n = 3$ for this miniature population. Note that for every different sample size, there will be a different sampling distribution of the mean. We will sample with replacement, so no matter how big the sample size, we would never "use up" the population.

We'll begin by drawing the outcome tree to determine the sampling distribution for all the possible samples of size $n = 3$ that could be drawn from this population. This is shown in **Figure 1**.

The outcome tree for sampling three numbers with replacement has 27 branches ($3^3 = 27$ outcomes). At the end of each branch is the sample outcome and the value of the event "Mean of the Sample." The first branch yields the outcome 3,3,3 with a value of 3.00 for the event, the second branch yields 3,3,4 with 3.33 for the value of the event, and so on. Scanning down the values for the mean, note that the value 3.00 occurs only once and, thus, has a probability of 1/27 = .037. The value 3.33 occurs 3 times, with a probability of 3/27 =.111. **Table 1** displays the probability distribution for the different values of the mean for this random sampling experiment.

The discrete probability distribution in **Table 1** is technically called a **random sampling distribution**, or just a **sampling distribution**. In this example, it is the **sampling distribution of the mean**. But had we computed the range for each of 27 samples, we could have constructed the probability distribution for the event "Range of the Sample." That probability distribution would be the sampling distribution of R. In fact, we could construct a sampling distribution for any statistic that could be computed on samples of size $n = 3$, e.g., Mo, Md, S, , etc.

Figure 1

Outcome Tree Showing the 27 Possible Samples of Size n = 3 Drawn with Replacement from the Population {3, 4, 5}.

Draw 1	Draw 2	Draw 3	Outcome	Mean
3	3	3	3 3 3	3.00
		4	3 3 4	3.33
		5	3 3 5	3.67
	4	3	3 4 3	3.33
		4	3 4 4	3.67
		5	3 4 5	4.00
	5	3	3 5 3	3.67
		4	3 5 4	4.00
		5	3 5 5	4.33
4	3	3	4 3 3	3.33
		4	4 3 4	3.67
		5	4 3 5	4.00
	4	3	4 4 3	3.67
		4	4 4 4	4.00
		5	4 4 5	4.33
	5	3	4 5 3	4.00
		4	4 5 4	4.33
		5	4 5 5	4.67
5	3	3	5 3 3	3.67
		4	5 3 4	4.00
		5	5 3 5	4.33
	4	3	5 4 3	4.00
		4	5 4 4	4.33
		5	5 4 5	4.67
	5	3	5 5 3	4.33
		4	5 5 4	4.67
		5	5 5 5	5.00

Table 1

Probability Distribution of Values of the Sample Mean for Sample Size n = 3 from the Population {3, 4, 5}.

Value of the Mean	Probability of Value
3.00	1/27 = .037
3.33	3/27 = .111
3.67	6/27 = .222
4.00	7/27 = .259
4.67	6/27 = .222
4.33	3/27 = .111
5.00	1/27 = .037

$$\mu_{\bar{X}} = 4.000$$
$$\sigma_{\bar{X}} = .471$$

With this idea in mind, we offer now a "generic" definition for a discrete sampling distribution. The **discrete sampling distribution** of a statistic is **a discrete probability distribution of values of a statistic arising from repeated random samples of a given size drawn with replacement from a defined population.**

From this "generic" definition, it is possible to generate the definition of the sampling distribution for any statistic. Here is the definition of the discrete sampling definition of the mean for the random sampling experiment described above. The sampling distribution of the mean is **the discrete probability distribution of values of the mean for repeated random samples of size n = 3, drawn with replacement from the population of scores {3,4,5}.**

Continuous Sampling Distributions

What are sampling distributions good for? Simple. They do what all probability distributions do—**they tell you the probability of obtaining different values of a statistic arising from random sampling.** For our modest experiment, the sampling distribution lets us know the probability of obtaining a sample mean of any value or range of values. We know, for example, that the probability of obtaining a sample mean greater than 4.00 through random sampling, is approximately .37, i.e., the sum of the probabilities for the means of 4.33, 4.67, and 5.00, i.e., .222 + .111 + .037 = .370.

The problem with small discrete sampling distributions like this is that they are not good for much except teaching statistics. Real world populations are so large and usually so indefinite that it is either impractical or impossible to construct a discrete sampling distribution for any statistic. To compute a discrete sampling distribution, you need to know all of the values comprising the population. In practical situations, however, we never know the precise numerical composition of the population. At most, we know or hypothesize what the population mean and standard deviation are, and perhaps something about the distribution shape of the population.

The solution to this problem lies in using a **continuous sampling distribution** to approximate the underlying (and unknowable) discrete sampling distribution. **The mathematical functions that define continuous sampling distributions can be derived from a knowledge of population parameters and population shape, rather than from a knowledge of the exact composition of the values comprising the population.**

Here is the generic definition of a continuous sampling distribution. The continuous sampling distribution of a statistic is a **continuous probability distribution for values of a statistic obtained from an infinitely large number of random samples of size n drawn with replacement from a defined population.**

The balance of this chapter will concentrate on the **continuous sampling distribution of the mean**. The nature of this distribution was worked out many years ago by statisticians and is embodied in a very important mathematical theorem called the **Central Limit Theorem (CLT)**. The CLT as it applies to the sample mean is stated as follows:

1. If a population of scores is normally distributed, then the continuous sampling distribution of the mean will itself be normally distributed. If the population of scores is not normally distributed, the sampling distribution will still tend to be normally distributed, and this tendency will increase with the size of the sample.

2. The mean of the sampling distribution of the mean ($\mu_{\bar{X}}$) is equal to the mean of the population, i.e.,

$$\mu_{\bar{X}} = \mu$$

3. The standard deviation of the sampling distribution of the mean—also called the **standard error of the mean**—is abbreviated $\sigma_{\bar{X}}$ and is equal to the standard deviation of the population divided by the square root of the sample size, i.e.,

$$\sigma_{\bar{X}} = \frac{\sigma}{\sqrt{n}}$$

According to this theorem, **if you know or hypothesize what μ and σ of the population are, you also automatically know what the sampling distribution of the mean will be like for any sample size.** The tendency for the sampling distribution to be normal in shape is quite marked, even if the population distribution is not particularly normal in shape. If the population distribution is at least unimodal, the continuous sampling distribution of the mean will be well approximated by a normal curve even with sample sizes as small as n = 10–15. Regardless of the precise shape of the sampling distribution, however, $\mu_{\bar{X}}$ will always equal μ and $\sigma_{\bar{X}}$ will always be given by the expression $\sigma/\sqrt{n}$.

The description of the CLT above focuses on the sampling distribution of the mean, but the theorem is actually much broader and applies to a variety of linear combinations of numbers drawn from normally distributed populations. We'll see in Chapter 11 that it applies to the sampling distribution of **differences between pairs of sample means**.

> A dynamic demonstration of the operation of the Central Limit Theorem is included in the end-of-chapter problems under *JASP* Learn Stats.

Let's apply the CLT to our miniature population of three scores {3,4,5}. Recall that the population of these scores had μ = 4.00 and σ = .816. According to the central limit theorem, $\mu_{\bar{X}}$ will equal the mean of the population. If we compute the mean of the 27 sample means shown in **Table 1**, we find it to be exactly 4.00. The CLT says that $\sigma_{\bar{X}}$ will equal $\sigma/\sqrt{n}$ = .816/√3 = .816/1.732 = .471. The computed standard deviation of the 27 sample means shown in **Table 1** is .471.

Sampling distributions are at the heart of statistical inference. Knowledge of the sampling distribution of a statistic allows you to test hypotheses about the value of a population parameter from sample data and use sample data to estimate values of population parameters. We will develop the logic of both procedures using the continuous sampling distribution of the mean.

Sampling Distributions & Hypothesis Testing

On this page is a sample problem "The Dean's Dilemma" that we will use to explain both hypothesis testing and parameter estimation. **Read through it carefully before continuing**.

In this problem, the Dean of course hopes that students at her college are made to work at least as hard as students at prestigious institutions. What she wants to do is to use her sample data to determine whether it is reasonable to believe that the population mean for the college is the same as the national mean of 10 hours. Technically, she wants to test the hypothesis that μ_{Col} = 10.0 hr.

A "hypothesis" is **a testable assertion about some aspects of reality**. In statistics, the hypothesis to be tested is given a special name. It is called the **Null hypothesis**. The null hypothesis is abbreviated H_0 and it is **always a statement about one or more characteristics of a population or populations**. Usually, it is a statement about the value of one or more population parameters.

The word "null" in logic means "empty," which might lead you to believe that the null hypothesis is empty or meaningless. Nothing could be farther from the truth. The null hypothesis is the starting point for statistical inference. It is a provisional assertion about a population (or populations) whose truth or falsity is to be tested with sample data. Our Dean's null hypothesis would be formally stated as

H_0: μ_{Col} = 10.0 hr.

The Dean's Dilemma

One of the questions on the student course evaluation form at a college asks how many hours per week on average the student spent studying for the course, including time in class. The Dean of the college is extremely sensitive to student responses to this question because she is concerned that the courses be maximally challenging, consistent with the college's very stiff tuition and fees. Presumably, the more challenging the course, the more time students must spend studying.

Suppose one day, the Dean reads in the *Chronicle of Higher Education* that in a survey of prestigious private colleges and universities, students reported spending a mean of 10 hours per week studying for each of their courses (including in-class time). The Dean wonders whether students at her institution spend a comparable amount of time in class preparation. She secures a random sample of 100 course evaluation forms drawn from the thousands completed at the end of the semester and computes the sample mean and standard deviation for the "hours studying" question. She obtains the following:

$$\bar{X} = 9.3 \; hr$$
$$S = 3.0 \; hr$$

In classical hypothesis-testing statistics, two things can happen. First, the sample data may be found to be consistent with the hypothesis, in which case we will conclude **that H_0 cannot be rejected**. Second, the sample data may make it unlikely that H_0 is true, and we will decide **to reject the hypothesis**. Oddly enough, although we may decide that the hypothesis is false, for reasons to be discussed below, we will *never* be able to conclude that H_0 is true!

Rejection of H_0 causes us to accept what statisticians call the **Alternative Hypothesis**, which is abbreviated "H_A" (or frequently "H_1"). The H_A is also a statement about characteristics of one or more populations. The H_A is *always* the logical complement of H_0—it is a statement about what *must* be true if H_0 is false. In our example, H_0 states that μ_{Col}= 10.0 hr., so H_A must state that $\mu_{Col} \neq 10.0$ hr. ("≠" means "does not equal"). The H_0 and H_A are always presented together, i.e.,

$$H_0: \mu_{Col} = 10.0 \; hr$$
$$H_A: \mu_{Col} \neq 10.0 \; hr$$

There are other ways to formulate H_0 and H_A, which we will consider later in the chapter. For the moment, we'll just say that the hypotheses as currently formulated are called **non-directional hypotheses**, because they don't specify that the μ_{Col} is above or below some value—only that it is either equal to or not equal to some value. "Directional" H_0 and H_A are quite possible.

The sample mean obtained by the Dean was 9.3 hr. How do we decide whether the sample value is consistent with H_0 that μ_{Col} = 10.0 hr.? We now come to the heart of the process of inferential hypothesis testing. We reason in the following way:

> If H_0 were true and the μ_{Col} really did equal 10.0 hr., is a sample mean of 9.3 hr. for a sample of 100 cases likely to occur through random sampling (by chance)? If it is a value that readily could occur by chance, then we have no basis for rejecting H_0. If, on the other hand, it is a value that is *unlikely* to occur through random sampling if the population mean were 10.0 hr., then we will conclude that H_0 is not correct and reject it.

It is *extremely* important that you understand the logical process outlined in the above paragraph. If it is not clear to you, read it again and think about it until it is clear.

The crux of the problem is to know what values of the sample mean are likely to occur through random sampling. Here is where the sampling distribution of the mean comes into the process. Remember, that's what a sampling distribution is good for—it tells you the probability of getting different values or ranges of values of a statistic through random sampling. What we do is this. We *tentatively* assume that H_0 is true and construct the sampling distribution that would result for the statistic in question. We then compare the value of our sample statistic to the sampling distribution. If our value is one of the "likely" ones, we conclude that there is no basis for rejecting H_0. On the other hand, if our sample value is one that would rarely occur by chance, we decide that H_0 is false.

What would the sampling distribution of the mean look like if the Dean's H_0 were true, i.e., μ_{Col} = 10.0 hr.? Here is where the CLT comes into play. Recall that the theorem tells us that the continuous sampling distribution of the mean will be a normal distribution if the population is normally distributed or it will tend toward normality for large samples, even if the population is not normally distributed. A sample of 100 is enormous as far as the normalizing tendency is concerned, so we are safe in assuming that the sampling distribution of the mean will be extremely close to a normal distribution.

The CLT says that the mean of the sampling distribution will equal the μ of the population. So, tentatively assuming H_0 to be true,

$$\mu_{\bar{X}} = \mu_{COL} = 10.0 \; hr$$

Finally, the CLT says that the standard deviation of the sampling distribution ($\sigma_{\bar{X}}$) will be given as $\sigma/\sqrt{n}$. Here we have a slight problem. We haven't said anything about the population σ. To use the CLT, we need to know the value of the population σ. As we will see in the next chapter, the solution is to use $\hat{s}$ as an estimate of σ. But that also introduces an additional complication that we want to ignore for the moment. Right now, we are going to pretend that we know that the standard deviation for the college is the sample value of 3.0 hr., i.e., σ_{Col} = 3.0 hr. Based on this assumption, the standard deviation of the sampling distribution will be:

$$\sigma_{\bar{X}} = \frac{\sigma_{Col}}{\sqrt{n}} = \frac{3.0}{\sqrt{100}} = .30 \; hr$$

We now can actually construct the sampling distribution of the mean based on the tentative assumption that H_0 is true. **Figure 2** shows what the sampling distribution of the mean would look like. It will be a normal distribution having $\mu_{\bar{X}}$ = 10.0 hr. and $\sigma_{\bar{X}}$ = .30 hr. The X axis has been scaled to show the values that correspond to a range from minus two standard deviations to plus two standard deviations.

Figure 2

Continuous Normal Sampling Distribution of the Mean Expected if the College Mean Equals 10.0 Hr.

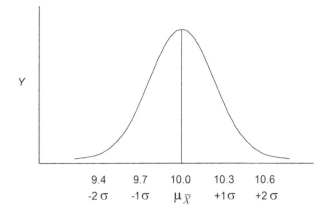

Now we're almost ready to decide about H_0. We need to compare our sample mean of 9.3 hr. to our sampling distribution. We will reject H_0 and accept H_A if our sample mean is not very likely to have occurred if H_0 were true. Here is a catch. What shall we consider "likely" and "unlikely" to mean? How unlikely does our sample value have to be before we reject H_0? In statistical parlance, what we need is a *criterion of unlikeliness*. Naturally, there is such a criterion. It is called the **alpha level** and is symbolized by the lower-case Greek "α." **The α level**

defines an area of the continuous sampling distribution that includes all the values of the statistic that will cause us to reject H_0.

The α level is expressed in the form of a probability or proportion. Two values are commonly used —.05 and .01. The .05 α level means that H_0 will be rejected if the value of the sample statistic is among the most deviant 5% of sample values that would occur if H_0 were true. The .01 level means that H_0 will be rejected if the statistic falls among the most extreme 1% of sample values. In terms of the sampling distribution, the most deviant (unlikely) values are out in the "tails" of the distribution.

We will adopt an α level of .05 to make our decision about H_0. **Figure 3** shows the sampling distribution of the mean for our problem with an area equal to 2.5% of the total area under the curve marked in each tail of the sampling distribution. The two areas combined constitute the most deviant and, therefore, unlikely 5% of sample values that would occur through random sampling if the population μ really were 10.0 hr. These areas are called the **areas of rejection** for the sampling distribution.

Figure 3

Sampling Distribution of the Mean when μ_{Col} =10.0 Hr. Showing Areas of Rejection for α = .05.

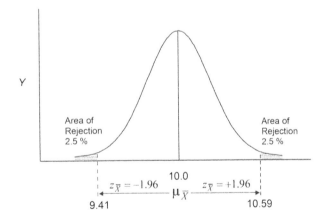

Why is the 5% area equally divided between the two tails of the distribution? The answer hinges on how H_0 and H_A are formulated. In our present example, H_0 simply stated that μ_{Col} = 10.0 hr. Recall that we called this a "**non-directional**" hypothesis. In a non-directional H_0, an *unlikely* sample value for the statistic—either *below* or *above* the hypothesized value of the parameter—leads to rejection of H_0. Because the areas of rejection are established on both tails of the sampling distribution, the test of a **non-directional** H_0 is usually called a **2-tail test** of the hypothesis. As we will see below, a directional statement of H_0 and H_A would cause us to put the entire area of rejection of one tail of the sampling distribution, leading to (you guessed it) a **1-tail test** of H_0.

The values defining the two areas of rejection are 9.41 hr. for the lowest 2.5% and 10.59 hr. for the highest 2.5%. The value 9.41 hr. was obtained by finding the z score corresponding to the lowest 2.5% of the area under the left-hand side of the normal distribution curve and converting the z score back to a raw score. If you consult the unit normal distribution table in the Appendix (**Table A**), you find that you need a z score of -1.96 to get to the lowest 2.5% of the distribution. Recall from Chapter 3, to recover the score value corresponding to a z score of −1.96, you multiple the z score times the standard deviation of the distribution and subtract that value from the distribution mean, i.e.,

$$Lower\ Value = \mu_{\bar{X}} - (1.96)(\sigma_{\bar{X}})$$
$$= 10.0\ hr - (1.96)(.30\ hr)$$
$$= 10.0\ hr - .59\ hr$$
$$= 9.41\ hr$$

The same process was carried out to find the extreme upper 2.5% of the curve, this time adding 1.96 $\sigma_{\bar{X}}$ to the mean, yielding 10.59 hr., i.e.,

$$Upper\ Value = \mu_{\bar{X}} + (1.96)(\sigma_{\bar{X}})$$
$$= 10.0\ hr + (1.96)(.30\ hr)$$
$$= 10.0\ hr + .59\ hr$$
$$= 10.59\ hr$$

Comparison of our sample mean value of 9.3 hr. with the areas of rejection for the sampling distribution reveals that 9.3 hr. falls in the area of rejection in the lower tail of the distribution. This means that our sample mean is among the most unlikely 5% of sample values that would occur if H_0 were true. This outcome is so unlikely that we will conclude the H_0 is false and the population mean must be some other value.

The process described above illustrates the logic of hypothesis testing. In practice, however, researchers don't go to the trouble of actually computing the score values that define the areas of rejection. Instead, they compute the z score value for the sample mean and compare it to the value of z required for an α level of either .05 or .01, whichever they are using. Recall that in Chapter 3, a z score was defined as

$$z = \frac{X - \mu}{\sigma}$$

When you are dealing with the sampling distribution of the mean, the sample mean is treated as a score, so the formula for converting it to a z score is

$$z_{\bar{X}} = \frac{\bar{X} - \mu_{\bar{X}}}{\sigma_{\bar{X}}} = \frac{\bar{X} - \mu_{H_0}}{\sigma_{\bar{X}}}$$

where: μ_{H_0} = population μ assumed under H_0.

Substituting our sample mean of 9.3 hr. into the equation gives:

$$z_{\bar{X}} = \frac{9.30\ hr - 10.0\ hr}{.30\ hr} = \frac{-.70\ hr}{.30\ hr} = -2.33$$

This obtained z score of -2.33 exceeds the value of -1.96 that defines the lower area of rejection on the sampling distribution, so we know that the sample mean falls in the area of rejection. **Figure 4** diagrams the result of our test.

Figure 4

Sampling Distribution of the Mean when μ_{Col} =10.0 Hr. Showing the Location of the Sample Mean of 9.3 hr.

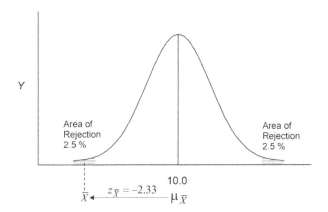

The z score used here to test H_0 is given a special status when it is employed in this manner. It is called a **test statistic. Test statistics have no descriptive function; instead, they are used solely to test null hypotheses.** The special status is indicated by the subscript "$\bar{X}$" meaning that $z_{\bar{X}}$ tests hypotheses using the sample mean. We will encounter numerous other test statistics in subsequent chapters.

Our sample value meets our α criterion of unlikeliness, and we will reject H_0 that μ_{Col} = 10.0 hr. Seemingly, students at her college do not work quite as hard as those attending more prestigious institutions.

In statistical terms, we will conclude that the sample mean differs **significantly** from the hypothesized value of 10.0 hr. But here is an important point that you should not miss. The phrases "statistical difference" and "significant result" always mean that H_0 in question has been rejected. The use of the term "significant" in this regard does **not** mean, however, that the result is practically or theoretically important. As we will see below, if you make a statistical test sufficiently powerful, you will be able to detect even very small and practically unimportant differences. Some statisticians urge the use of the term "statistically unlikely" to avoid this possible confusion of meaning. Unfortunately, the term "significant difference" has the weight of many years behind it and will be hard to change. Let's all agree that a "significant difference" is something that is unlikely to have occurred by chance—whether it is also significant (important) in a practical or theoretical sense depends upon other considerations.

In this example we formulated H_0 and H_A. as "non-directional" hypotheses, i.e., μ_{Col} either was or was not equal to some value. The advantage of non-directional hypotheses is that deviant sample values of the statistic either larger or smaller than the parameter value being tested can lead to rejection of H_0. In the present example, a non-directional set of hypotheses would make sense if the Dean cherished the secret hope that her students might be found to work *harder* than the average of 10 hr. for students at prestigious colleges and universities. Suppose, for example, that the sample mean had been 10.6 hr.? This value is 2.00 standard deviations *above* the H_0 value of 10.0 hr. and falls in the area of rejection in the upper tail of the sampling distribution (see **Figure 3**). Such a result would lead the Dean to conclude that students at her college studied *significantly more* on average than do students at more prestigious institutions. This would make her very happy.

Suppose on the other hand, that the Dean wanted *only* to find out if her students were significantly *below* the average for prestigious colleges and universities. Here, a directional formulation of H_0 and H_A. would make sense. The hypotheses would look like this:

$$H_0: \mu_{Col} \geq 10.0\ hr$$
$$H_A: \mu_{Col} < 10.0\ hr$$

The symbol "≥" should be read "greater than or equal to." When formulated in this manner, H_0 can only be rejected if the sample mean is numerically *smaller* than 10.0 hr. A value larger than 10.0 hr.—no matter how much larger—cannot lead to rejection of H_0, because H_0 encompasses *all* values greater than 10.0 hr. Such an outcome is necessarily ambiguous.

Perhaps you are probably wondering what the fuss is all about. Is there any practical consequence to using a directional formulation of H_0 and H_A? The answer is yes. Here's why. **Figure 5** shows the sampling distribution of the

Figure 5

Sampling Distribution of the Mean when μ_{Col} =10.0 Hr. Showing All 5% Area of Rejection Set in Lower Tail of the Sampling Distribution.

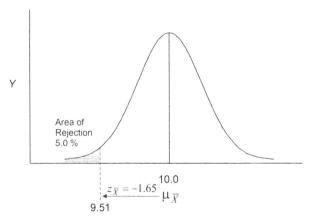

mean redrawn to reflect the directional hypotheses stated above. The α level remains at .05, but now the entire area of rejection has been put in the lower tail of the sampling distribution. This area corresponds to the extreme 5% of sample means that would cause rejection of H_0. Notice the magnitude of the z score that defines the beginning of the area of rejection. It is -1.65 rather than the -1.96 that was required with a 2-tail test of the hypothesis. The effect of this difference is that a *less extreme* value of the sample mean is required to reject H_0. Specifically, the minimum value of the mean required to reject H_0 would be 9.51 hr. using the directional hypothesis, compared with the more extreme 9.41 hr. with the non-directional hypothesis, i.e.,

Non- Directional Hypothesis:

$$\mu_{\bar{X}} - (1.96)(\sigma_{\bar{X}}) = 9.41 \; hr$$

Directional Hypothesis:

$$\mu_{\bar{X}} - (1.65)(\sigma_{\bar{X}}) = 9.51 \; hr$$

Technically, the use of a directional hypothesis (1-tail test) makes the hypothesis test more **powerful**. Statistical "power" will be explained more fully below. Right now, we'll simply say that the power of a statistical test is its ability to correctly reject H_0.

Sampling Distributions & Parameter Estimation

Knowledge of the sampling distribution of the mean has enabled us to use our sample data to reject H_0 that μ_{Col} = 10.0 hr. We will now use the sampling distribution to help us obtain an estimate of what μ_{Col} actually is. Recall from Chapter 1 that parameter estimates come in two types—point estimates and confidence interval estimates. In a point estimate, the value of the sample statistic is taken as the estimated value of the population parameter. In our example, we would estimate the mean for the college to be 9.3 hr. A confidence interval estimate, on the other hand, provides a range of values about the point estimate that has a known prior probability of including the parameter. Construction of this range depends upon knowledge of the sampling distribution of the statistic in question.

Let's assume that we want to construct the **90% confidence interval** estimate for the μ of the college. By this we mean that we want to define **an interval around the sample mean constructed in such a way that you are 90% confident that the interval contains the population mean.**

Figure 6 shows how the interval is constructed. At the middle of the X axis is the value of the sample mean, 9.3 hr. Placed symmetrically below and above this point are two sampling distributions of the mean. The lower one is centered at 8.81 hr., which is 1.65 standard deviations below the sample mean. The upper one is centered at 9.80 hr., which is 1.65 standard deviations above the sample mean. Notice that exactly 5% of the area of the lower sampling distribution lies *above* the sample mean of 9.3 hr. and exactly 5% of the area of the upper sampling distribution lies *below* the value of the sample mean.

The two values, 8.81 hr. and 9.80 hr., represent the "worst case" scenario for the 90% confidence interval. If the population μ were any farther away from the sample mean in either direction, then our sample value would have been among one of the most unlikely 10% of sample outcomes to occur through random sampling. We are betting that this sample was not one of those unlikely occasions. We could be wrong, of course. The population μ might be lower than 8.81 hr. or greater than 9.80 hr., in which case we just happened to have drawn a very unlikely sample. We have a 90% chance, however, that this did not happen and the population μ is between 8.81 and 9.80 hr.

The formula for constructing the 90% confidence Interval estimate of μ is:

$$90\% \; CI = \bar{X} \pm (1.65)(\sigma_{\bar{X}})$$

We can extend this logic to constructing other confidence intervals. Suppose that we wished to construct the 95% confidence interval estimate of μ. Here we want an interval that in 95 out of 100 random samples would

Figure 6

Diagram of the Logical Process for Constructing the 90% Confidence Interval Estimate of $\mu_{Col.}$

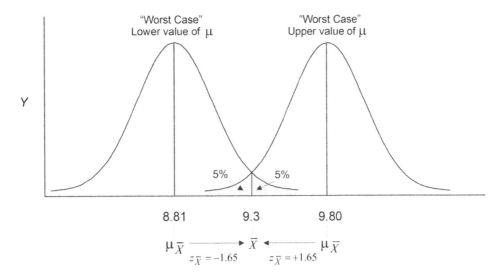

encompass the population μ. The formula for constructing the 95% confidence interval is:

$$95\% \ CI = \bar{X} \pm (1.96)(\sigma_{\bar{X}})$$

Figure 7 diagrams what this situation would look like. Now the "worst case" values of μ are 8.71 hr. and 9.89 hr., and each associated sampling distribution has exactly 2.5% of its area *beyond* the sample mean of 9.3 hr. If the population μ were less than 8.71 hr. or greater than 9.89 hr., then our sample mean would have been among the most unlikely 5% of sample values that could occur by chance. We're betting that this sample is not one of those 5% outcomes and that the μ actually is less extreme than 8.71 and 8.89 hr.

This leads us to the general formula for any desired confidence limit for estimating μ:

$$CI = \bar{X} \pm (z_{1-CI})(\sigma_{\bar{X}})$$

where: z_{1-CL} = value of z for a 2-tail significance test at the (1-CI) α level.

Figure 7

Diagram of the Logical Process for Constructing the 95% Confidence Interval Estimate of $\mu_{Col.}$

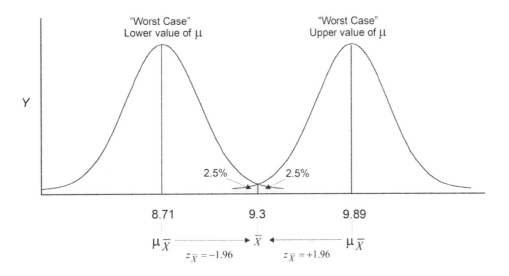

For example, assume that you wanted to construct the 50% CI. The α level for this test is (1 − 50%) = 50% or .50. In a 2-tail test with α = .50, you need a z score that defines the most extreme 25% of each tail of the distribution. **Table A** gives the value as ± .67. Thus, z_{1-CI} = .67.

Let's make a couple of points about terminology. First, it is incorrect to say that μ "falls" inside or outside the confidence interval. The population mean isn't moving around from sample to sample; the value of the sample mean does the moving. Usually, you say that the confidence interval either does or does not "bracket" or "include" the value of μ. Second, once a sample has been selected and the interval computed, we don't say that there is some probability that the interval brackets or includes μ—it either does or doesn't at this point. The probability exists *in advance of drawing the sample*. That's why we call it a "confidence" interval rather than a "probability" interval.

In the present example, we pursued the process of parameter estimation only after rejecting the hypothesis that μ = 10.0 hr. Frequently, however, research is directed solely toward estimating some parameter. Suppose, for example, that the Dean had never read the article describing the average number of hours of studying per class of students at prestigious colleges and universities. Instead, she simply may have been curious what the situation was like at her institution and selected the sample purely to estimate the institution μ.

Political polling research is always of this type. When pollsters ask a sample of registered voters what they think about some issue, they are not testing some hypothesis, but are trying to obtain an accurate estimate of the *proportion* of registered voters who hold some opinion. Here the parameter being estimated is a population proportion (*p*), rather than μ, but the principles are exactly the same. You will learn how to establish a confidence interval estimate for a population proportion in Chapter 10 on Single-Sample Inference.

A dynamic demonstration of Confidence Interval Estimation is included in the end-of-chapter problems under *JASP* Learn Stats.

Inferential Error

When you test a H_0 about a population parameter, you run the risk of coming to an erroneous conclusion—of making a mistake. In our example above, we tested H_0 that μ_{Col} = 10.0 hr. Based upon the value of the sample mean, we decided to reject H_0. We did so because examination of the sampling distribution of the mean showed that a sample value of 9.3 hr. was a rather unlikely event if the population μ were equal to 10.0 hr. But here is the key point. The sample value was unlikely, *but possible*. Just by the whim of the fates, we may have gotten one of those unlikely samples. If so, we falsely rejected H_0. Statisticians call this a **Type I Error**. A Type I Error occurs **when H_0 is true, but your sample value leads you to conclude that it is false**. The probability of making a Type I Error is set by the α level that we use. If H_0 is true, and we are employing an α level of .05, then by the very nature of the process, 5% of the samples will lead to incorrect rejection of H_0. In our example, we rejected H_0 that μ_{Col} = 10.0 hr. We could be making a Type I Error. Formally, the likelihood of making a Type I Error is designated by α and is expressed as a conditional probability:

$$\alpha = p(\text{Rejecting } H_0 \mid H_0 \text{ is True})$$

If this were the whole story, the problem would not be a difficult one to solve. We could reduce the probability of making a Type I Error to any desired level by selecting a smaller and smaller value for α. If, instead of using a .05 α level, we selected a .001 α level, we would falsely reject H_0 only .1% of the time.

Unfortunately, there is more to the story. There is a *second* type of error that you can make in testing H_0. Suppose that H_0 is false, but you *failed to reject it*. This is called a **Type II Error**. Formally, a Type II Error occurs **when H_0 is false, but your sample result does not permit you to reject H_0**. The probability of making a Type II Error is designated β (Greek lower case "**beta**") and is given by:

$$\beta = p(\text{Not Rejecting } H_0 \mid H_0 \text{ is False})$$

Type II Errors are the real scourge of hypothesis testing because β, the probability of making such an error, is *unknown*! We can see why this is so by modifying our example slightly. Suppose that the Dean's sample mean had been 9.6 hr. instead of 9.3 hr.

Now the sample mean of 9.6 hr. is closer to the hypothesized value of μ_{Col} and *does not* fall in the area of rejection for a 2-tail test with α set at .05. The obtained value of z of −1.33 does not approach the critical value of −1.96 required for a 2-tail test at the .05 α level:

$$z_{\bar{X}} = \frac{\bar{X} - \mu_{\bar{X}}}{\sigma_{\bar{X}}} = \frac{9.6\ hr - 10.0\ hr}{.30\ hr} = \frac{-.40\ hr}{.30\ hr} = -1.33$$

We failed to reject H_0, so it is possible that we have made a Type II Error. What is the likelihood that we have done so? The way to look at this is to imagine drawing a very large number of random samples and carrying out the

Figure 9

Likelihood of Making a Type II Error when Testing H_0 that μ_{COL} = 10.0 hr., when Actual Value of μ_{COL} = 9.6 hr.

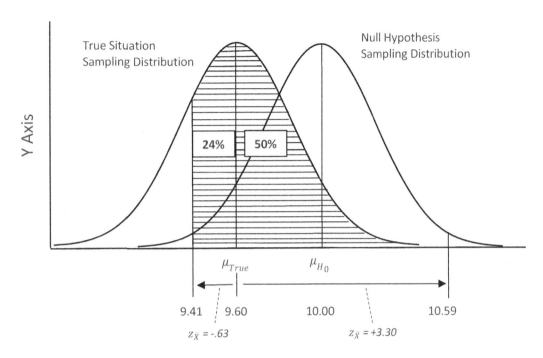

test of the hypothesis for each sample. The question would be: how many of those samples would lead to a Type II Error? If, for example, 30% of the samples result in a Type II Error, we would say that the probability of a Type II Error is .30.

Let's imagine that the College mean actually is 9.6 hr., i.e., μ_{Col} = 9.6 hr. What would be the probability of making a Type II Error if this were the true value of μ? **Figure 9** diagrams this situation. There are two sampling distributions shown. The one labeled "Null Hypothesis" is the sampling distribution that is used to test the Null Hypothesis that μ = 10.0 hr. The second distribution labeled "True Situation" is the sampling distribution that would actually occur given that the population mean is really 9.6 hr. Basically, what we're going to do is to compute what fraction of samples that would actually occur would fall in the areas of acceptance under the null hypothesis sampling distribution, thereby causing a Type II Error.

Recall that for a 2-tail test of the null hypothesis that μ = 10.0 hr., the critical vales required to reject H_0 were 9.41 hr. and 10.59 hr. These are shown for the "Null Hypothesis" sampling distribution. Now we switch our attention to the "True situation" sampling distribution. This distribution shows the actual probability of getting different values of the mean **given that μ_{Col} really is equal to 9.6 hr.** With respect to the true situation, what proportion of the samples will yield means between 9.41 hr. and 10.59 hr.? To get the area under the true situation sampling distribution that lies between 9.41 hr. and 10.59 hr., we find the area between μ_{True} and 9.41 hr. and between μ_{True} and 10.59 hr., and then add the two areas together.

a. Proportion of means falling between μ_{True} and 9.41:

$$z_{\bar{X}} = \frac{\bar{X} - \mu_T}{\sigma_{\bar{X}}} = \frac{9.41\ hr - 9.60\ hr}{.30\ hr} = \frac{-.19\ hr}{.30\ hr} = -.63$$

Area ≈ .24

b. Proportion of means falling between μ_{True} and 10.59:

$$z_{\bar{X}} = \frac{\bar{X} - \mu_T}{\sigma_{\bar{X}}} = \frac{10.59\ hr - 9.60\ hr}{.30\ hr} = \frac{.99\ hr}{.30\ hr} = 3.30$$

Area ≈ .50

The total area between 9.41 and 10.59 hr. under the "True situation" sampling distribution is thus .24 + .50 = .74 = 74%. This means that about 74% of the sample means that will actually occur will fall between 9.41 and 10.59 hr., causing us not to reject H_0. That is, the probability of a Type II Error in this situation is about .74. Rather discouraging!

Let's carry our reasoning a little bit further. Look at **Figure 9** and in your mind, slide the "True Situation" sampling distribution to the right, so that it gets closer to the "Null Hypothesis" sampling distribution. That is, make the true value of μ_{Col} closer and closer to the H_0 value of 10.0 hr. As you do so, consider what proportion of the actual sample means will fall between 9.41 and 10.59 hr. If you are good at visualizing, it should be clear that as the true μ_{Col} approaches the H_0 value of 10.0 hr., this proportion will approach .95—the area for the H_0 sampling distribution that leads to acceptance of H_0.

Statisticians say technically that as the difference between the true μ and the hypothesized mean approaches 0, the probability of a Type II Error with $\alpha = .05$ approaches .95. Notice that if the true mean exactly equals the hypothesized mean, H_0 becomes true and you can no longer make a Type II Error. More generally, the limit for the probability of making a Type II Error in this situation is $1 - \alpha$. In our example α was set at .05, so $1 - .05 = .95$. Had the Dean used a .01 α level, the probability of a Type II Error would approach .99 as the true μ approached the H_0 μ. If this is not clear to you, draw a diagram like that in **Figure 9**, but setting the α at .01.

Suppose that you mentally slide the "True Situation" sampling distribution to the left, away from the H_0 value of μ. As the μ_{True} gets further and further away from the H_0 value of μ, less and less of the "True Situation" distribution falls between the values of 9.41 and 10.59 hr, so fewer and fewer Type II Errors are made. In the limit, the probability of making a Type II Error approaches 0.

> A dynamic demonstration of the operation of Type I and II Error is included in the end-of-chapter problems under *JASP* Learn Stats.

Statistical Power

Perhaps because "correctness" is more appealing than "error," statisticians focus on the ability of a statistical procedure to correctly reject H_0 rather than upon its failure to do so. This is called the "power" of a statistical procedure. Technically, **power is the probability of rejecting H_0, given that H_0 is false**. It is symbolized as $1 - \beta$. This makes sense, because β is the probability of making a Type II Error, and power is the logical compliment of a Type II Error. Formally, power is defined as

Power = p(Rejecting H_0 | H_0 is False) = $1 - \beta$

In the discussion that follows, we are going to consider five different factors that determine the power of a statistical procedure. Keep in mind that because power and Type II Error are inversely related to one another, anything that increases the power of a procedure correspondingly reduces the probability of a Type II Error.

Factors Influencing Power

In testing the H_0 about the value of a population μ, the power of the statistical procedure is determined by the following five factors:

1. The difference between the value of μ tested in H_0 and the actual value of μ, i.e., how "wrong" H_0 is.

2. The amount of score variation present in the population.

3. The size of the sample employed in the procedure.

4. The α level employed in the procedure.

5. Whether a 1- or 2-tail test is employed.

We'll see how each of these factors operates using the approach introduced in the last section on Type II Error.

How Wrong H_0 Actually Is

There's a saying in statistics that it's easier to find big differences than small differences. As the actual population mean gets further and further away from the value of μ under H_0, the two sampling distributions will overlap less and less, and a greater percentage of the sample means will fall in the area of rejection for the H_0 sampling distribution. Power will go up dramatically.

Size of the Population Standard Deviation

The second factor that influences power is the amount of score variation present in the population, i.e., the size of σ. One might think that this factor is outside the control of the investigator, but often in social science research the "random variation" that influences the size of σ arises from sloppy or inefficient data collection. Where it is possible, reducing such random variation reduces the size of σ, thereby making $\sigma_{\bar{X}}$ smaller. This reduces the "width" of the H_0 and true situation sampling distributions, leading to less overlap and more of the true situation samples falling in the area of rejection.

Sample Size

The third factor that influences power is the size of the sample. Sample size exerts its effect in the same way that population variability does—by influencing the width of the H_0 and true situation sampling distributions. As sample size increases, $\sigma_{\bar{X}}$ decreases, making the two distributions "skinnier" with less overlap.

The role that sample size plays in influencing power is particularly important because, unlike the first two factors, sample size is always under the control of the investigator. It is always possible—in principle at least—to increase the power of a test to any desired level by increasing sample size. Unfortunately, the relationship between sample size and power is not a simple linear function. For all statistical procedures, power increases as a negatively accelerated function of sample size. That means that if your sample size is small to begin with, adding 10–15 cases can make a big difference in power. But as sample size grows, adding the same number of cases has less and less of an effect on power. As a result, it can be costly to increase sample size enough to achieve very high levels of power.

Also recall that we made a distinction between "statistically significant" and "practically significant" differences. The use of very large samples makes it possible to detect very small and practically unimportant differences. By making the sample large enough, we could have a high probability of rejecting the Dean's H_0 even if the true mean at the college were 9.8 hr.—which is only a difference of .2 hr. from the H_0 value. Really, is a .2 hr. difference from the national average of prestigious colleges and universities really worth getting upset about?

The α level Selected

The fourth factor that influences power of a test is the α level selected by the researcher. Like sample size, choice of an α level is also under the researcher's control. In the description of the sample problem, we have left α at .05. What happens to power if we adopt a more rigorous α level, say .01? Recall that you might adopt the .01 α level in preference to the .05 level if you were seeking to reduce Type I Errors. But also recall that we said there was a cost to such a strategy. Highly stringent α levels reduce Type I Errors *if* H_0 is true but *increase* Type II Errors if H_0 happens to be false, which means diminished power.

Making α more rigorous means that the areas of rejection for the H_0 sampling distribution extend further out in the tails, thereby increasing the number of actual sample means that will fall into the area of acceptance (non-rejection) for the H_0 sampling distribution.

The choice of .05 as the α level represents a balance between permitting too many Type I Errors if H_0 is true and reducing power too much if H_0 happens to be false.

1- vs. 2-Tail Test

The final factor that can influence power in this situation is the choice of a 1-tail vs. a 2-tail test. We say "in this situation" because in some statistical procedures that we will consider later, the area of rejection is always set in one tail of the relevant sampling distribution, i.e., the test is always 1-tail. Going from a 2-tail test of H_0 to a 1-tail test can have diametrically opposite effects on the power of the statistical procedure—power can either plummet to near 0 or increase substantially.

If the area of rejection for a 1-tail test is placed on the tail of the H_0 sampling distribution opposite the true situation sampling distribution, power will be effectively zero. On the other hand, if the area of rejection is placed on the tail of the H_0 sampling distribution in the direction of the true situation sampling distribution, a larger percentage of sample means will fall in the area and lead to correct rejection of the null hypothesis.

An apparent solution to this dilemma that frequently occurs to students is to wait until the data are collected and then state H_0 and H_A in a manner consistent with the values of the sample mean. This way you always have the power advantage of using a 1-tail test, without running the risk of an "incorrect" formulation of H_0 and H_A. Unfortunately, this is not a legitimate procedure, because it sneakily doubles the α level without letting anyone know about it. If your stated α level is .05, use of this strategy will result in a true α level of .10. The reason for this is that you have effectively placed a .05 area of rejection at both tails of the sampling distribution. If across a large series of tests, H_0 were always true and sample means occurred randomly on either side of μ, your strategy would cause you to falsely reject H_0 10% of the time, instead of the stated 5%.

A variant of this clearly illegitimate strategy is a more subtle, but still shady, technique that is actually employed by some researchers. What they do is to "predict" the direction of the statistical outcome from theory or intuition and formulate H_0 and H_A accordingly. If they have predicted correctly, well and good. If, however, the result comes out in the opposite direction, they mumble something about the "unpredictability of nature" and switch to a 2-tail test. Why is this illegitimate? Basically, for the same reason that the "look and state" approach is illegitimate. The researcher always has 7.5% of the sampling distribution available to reject H_0, so the effective α level is .075 rather than the stated .05.

Estimation Bias & Precision

Much of inferential statistics is concerned with estimating population parameters. **Accurate estimates** are obtained when the statistics used to estimate the

Table 2

Sampling Distributions of $\hat{S}^2$, S^2, and $\hat{S}$ for Sample Size n = 3 From the Population {3, 4, 5}.

Population Parameters	Value of $\hat{S}^2$	Prob.	Value of S^2	Prob.	Value of $\hat{S}$	Prob.
$\mu = 4.00$	.000	3/27 = .111	.000	3/27 = .111	.000	3/27 = .111
$\sigma = .816$	.333	12/27 = .444	.222	12/27 = .444	.577	12/27 = .444
$\sigma^2 = .666$	1.000	6/27 = .222	.667	6/27 = .222	1.000	6/27 = .222
	1.333	6/27 = .222	.889	6/27 = .222	1.155	6/27 = .222
	$\bar{X}_{\hat{S}^2} = .666$		$\bar{X}_{S^2} = .444$		$\bar{X}_S = .735$	

parameters are both **unbiased** and **precise**. We will discuss each of these concepts in turn.

Bias in Estimation

Technically, a statistic is an unbiased estimator of its corresponding parameter **if the expected value of the statistic equals the parameter estimated**. The expected value of a statistic is the mean value of the statistic across an infinite number of samples. We have already said that the sample mean is an unbiased estimator of its parameter, μ, in our small, finite population. This is the case because the mean of the sampling distribution of the mean ($\mu_{\bar{X}}$) *does* equal the population mean (μ). This is embodied in the second component of the CLT.

On the other hand, the sample standard deviation (S) is *not* an unbiased estimator of the population standard deviation (σ), which is a consequence of the fact that the sample variance (S^2, the sample standard deviation squared) is not an unbiased estimator of the population variance (σ^2). We can illustrate this point with a "miniature population" example. Recall at the beginning of this chapter, we constructed the discrete sampling distribution of the mean for n = 3 from a population consisting of the numbers 3, 4, and 5. There were 27 possible samples that could be drawn with replacement. We computed the mean for each sample and constructed the sampling distribution of the mean. To show that S^2 is a biased estimator of σ^2, we will construct the sampling distribution of S^2 based on the 27 samples. We will then compare the mean of that sampling distribution with the value of the population σ^2, which equals .666 (**i.e., σ^2 = .816² = .666**). By way of contrast, we will also use the same samples to generate the sampling distribution for a related statistic $\hat{S}^2$, which is called the unbiased estimate of the population variance. Recall that $\hat{S}$ was defined in Chapter 2 as:

$$\hat{S} = \sqrt{\frac{\sum(X - \bar{X})^2}{n - 1}}$$

so,

$$\hat{S}^2 = \frac{\sum(X - \bar{X})^2}{n - 1}$$

The sampling distributions for $\hat{S}^2$ and S^2 and are given in **Table 2**. Notice that, unlike the sampling distribution of mean, the sampling distribution for $\hat{S}^2$ and S^2 and are both positively skewed. The mean of the sampling distribution for $\hat{S}^2$ = .666, which equals the σ^2 of the population. The mean of the sampling distribution of S^2 = .444, which underestimates the σ^2 of .666, showing that S^2 is a biased estimator of σ^2.

Because $\hat{S}^2$ is an unbiased estimator of σ^2, you might reasonably assume that $\hat{S}$ is an unbiased estimator of σ. Curiously, because of the non-linear nature of square root transformations, this is not the case. The population σ = .816, but the mean of the sampling distribution of $\hat{S}$ (shown in the third column) equals .735. We say that $\hat{S}$ is a "reasonably unbiased" estimator of σ. Although slightly biased in estimating σ, it does a much better job than S. The mean of the sampling distribution of S (not shown) equals .600 (way smaller than the population σ of .816).

Precision in Estimation

Recall that in Chapter 2 we discussed the "sampling stability" of different statistics. Another name for sampling stability is "precision." The **precision** with which a statistic estimates its corresponding population parameter is defined in terms of the sampling distribution of the statistic. An estimator is said to be precise if the **standard deviation (or variance) of the sampling distribution of the estimator is relatively small**. Narrow sampling distributions mean that there is relatively little sample-to-sample variation in the value of the statistic, so any particular sample is likely to yield a relatively accurate estimate of the parameter. **The standard deviation for the sampling distribution of $\bar{X}$ is equal to .471**, indicating that the sample mean is a precise estimator of its corresponding population parameter.

But not all statistics are equally precise estimators. By way of comparison, let's construct the sampling distribution of the median for size $n = 3$ for our miniature population. This distribution is shown in **Table 3**. The mean of this sampling distribution of medians (μ_{Md}) is 4.00, which exactly equals the median of the population of three scores. This indicates that the sample median is an unbiased estimator of the population median. **The standard deviation of this sampling distribution (σ_{Md}) is equal to .720.** Recall that the standard deviation of the sampling distribution of the mean equals .471. The larger value of σ_{Md} shows that the median "bounces around" more in value from sample to sample, thus making it a less precise estimator than the mean.

Table 3

Sampling Distribution of Values of the Sample Median for Sample Size n = 3 from the Population {3, 4, 5}.

Value of the Median	Probability of Value
3.00	7/27 = .259
4.00	13/27 = .481
5.00	7/27 = .259

$\bar{X}_{Md} = 4.000$
$S_{Md} = .720$

Final Comment

Virtually all students find this to be a very challenging chapter and for many, the chapter marks the beginning of a drop in examination grades. Do not let this happen to you. To repeat a statement made in the introduction, the material in this chapter is critical for understanding everything that follows. You need to adopt a very active approach to studying the information. Work in a study group and make up and solve sample power problems.

"Don't just say you have read books. Show that through them you have learned to think better, to be a more discriminating and reflective person. Books are the training weights of the mind. They are very helpful, but it would be a bad mistake to suppose that one has made progress simply by having internalized their contents."

                                              Epictetus

Chapter 9, Sampling Distributions & Statistical Inference
Chapter Exercises

Some questions are purely factual and have only one correct answer. Others require greater thought on your part and may have more than one possible answer.

1. Fill in the missing term(s) that makes the statement true.

   a. The key to statistical inference is the concept of _____.
   b. {5, 7, 9} symbolizes a _____.
   c. The sampling distribution of $\bar{X}$ for the above for size n = 4 has _____ samples.
   d. For {5,7,9}, μ = _____ and σ = _____.
   e. The characteristics of continuous sampling distributions are derived _____ from population _____.
   f. CLT stands for _____.
   g. According to the CLT, $\sigma_{\bar{X}}$ = _____.
   h. The symbol H_0 is called the _____ and is the _____ for hypothesis testing.
   i. The logical complement of H_0 is _____.
   j. In the "Dean's Dilemma" example, the Dean's H_0 is _____.
   k. Hypothesis tests use a criterion of "unlikeliness" called the _____.
   l. That criterion is usually set at _____ or _____.
   m. The most unlikely events are those that fall in the _____ of a sampling distribution.
   n. Sample outcomes that meet the criterion of unlikeliness lead to _____ of H_0.
   o. $z_{\bar{x}}$ is called a _____ and is used to make inferential decisions about _____.
   p. H_0 that $\mu \geq 10.0$ hr. is called a _____.
   q. $\bar{X} \pm 1.65\, \sigma_{\bar{x}}$ defines the _____ estimate of _____.
   r. $p(\text{Rejecting } H_0 \mid H_0 \text{ is True})$ is called a _____ and p is given by _____.
   s. If α = .05, $z_{\bar{x}}$ must exceed _____ to reject a non-directional H_0.
   t. $p(\neg\text{Rejecting } H_0 \mid H_0 \text{ is False})$ defines a _____, is symbolized by _____ and p is usually _____.
   u. $1 - \beta$ defines the _____ of a statistical test.
   v. There are _____ factors that influence β in a hypothesis test of the value of μ.
   w. The three factors under the control of the investigator are _____, _____, and _____.
   x. An unbiased estimator is one whose _____ equals the parameter in question.
   y. Sadly, S^2 is a _____ of _____.
   z. Happily, _____ is an _____ of _____.
   aa. The precision of a statistic refers to the "narrowness" of the _____ of the statistic.
   bb. The Md is an _____ estimator of _____, but not as _____ as the $\bar{X}$.

2. A right-wing "reliable" news source reports that the average college student who has an account with X (formerly Twitter) "tweets" an average of 25 times a day. Jason, hurting for a stat project for his Introductory Statistics class, decides to see whether students at his college really waste that much of their time. He sends a blanket text to all currently enrolled students asking for volunteers who have X accounts. Horrifyingly, 358 students reply that they would love to participate in a study of tweeting. Jason randomly selects 120 students from the volunteers and sends them a text asking their gender and an estimate of how many tweets they send during the average day, including weekends. Remarkably, 100 reply with the information. Here are the findings, broken down by gender. Because we haven't done Chapter 10 yet, assume that the sample values of $\hat{s}$ equal the population values for σ.

Females	Males
$\bar{X}_{Tw}$ = 21.5 Tw	$\bar{X}_{Tw}$ = 27.3 Tw
$\hat{s}_{Tw}$ = 7.2 Tw	$\hat{s}_{Tw}$ = 9.6 Tw
n = 64	n = 36

   a. Write the non-directional Null and Alternative hypotheses to be tested for each gender.
   b. With α set at .05, test the Null hypothesis for each gender that the college mean is 25 Tw.
   c. Find the 90% Confidence Interval estimate of μ for both genders.

d. Suppose that the college mean was actually 22.0 Tw for both genders. What would be the power of tests of the Null Hypothesis that µ = 25 for the female and male college students for the given sample sizes and σ's? Draw the sampling distributions carefully.

JASP "Learn Stats" Simulations

JASP has a large number of additional statistical routines in addition to the core seven that are shown across the top of the JASP Home Screen. Most of the routines are more advanced than what we are covering in this course, but one of them, "Learn Stats", is designed specifically for showing students important concepts in an interactive manner. We're going to do three of them to illustrate important concepts introduced in this chapter: the Central Limit Theorem, confidence intervals, and Type I and Type II Error. At the JASP Home Screen, click on the ✚ sign at the far right of the display. That will open a display of all of the additional statistical routines available in JASP. The 11th option is "Learn Stats." Click on the inclusion box. That will put it up on the list of available routines next to Factor. Click anywhere to close the list.

Central Limit Theorem

Many statisticians are very taken with the Central Limit Theorem because it magically solves so many problems in inferential statistics (see the rest of the text!). JASP has a nifty module that shows how powerful the theorem actually is. Click on the Learn Stats procedure at the top of the home screen and select "Central Limit Theorem" from the drop-down menu. This screen will appear:

Because we are already familiar with IQ scores, enter 100 for the Mean and 15 for the Std. dev values. The Results window will then show a beautiful normal distribution having $\mu = 100$ and $\sigma = 15$. Now click on Sample options. There you'll see that the default is 30 observations per sample (n = 30) with 100 samples, along with the options "Show samples" and "Display rug marks." The "Set seed" value is any arbitrary seed value for the random number generator used to select samples. You can set it to anything—like 23! The default is to display the first seven of the 100 random samples in histogram form. Here's what the third one looked like for me when I has ran the simulation:

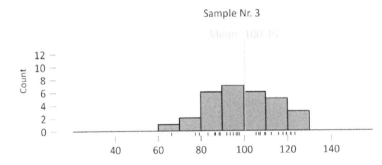

The mean of that sample of 30 cases was 100.46. Recall that the population μ is 100, so our sample mean very accurately estimated the population value. The little ticks at the bottom of the histogram are the "rug marks," the actual sample values.

If you scroll down the Results window, you'll see a histogram display of the sampling distribution of the mean for the 100 samples drawn by default as shown in the Sampling distribution options section of the Statistics Panel. A normal distribution curve is superimposed, showing that the sampling distribution of the mean tends toward normality as sample size increases.

Now, play around with the settings—like increasing sample size and number of samples to get a feel for the operation of the Central Limit Theorem!" Each time you change a parameter and hit enter the whole display resets with a new set of samples.

Confidence Intervals

Click on Learn Stats and select "Confidence intervals." This is the initial display in the Statistics Panel:

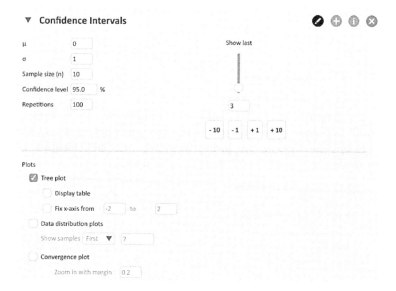

This simulation shows the accuracy of sample estimates of the population mean for samples of different sizes and different confidence intervals. For consistency, change μ to 100 and σ to 15. Slide the "Show last" lever up so that "Tree plot" shows all 100 of the random samples. The Results Table at the top of the Results window shows what percentage of the 100 samples using the default 95% confidence interval actually included the population mean of 100. The Tree plot shows the results for each of the 100 samples. Your results should be very close to five "misses" per 100 samples! Again, play around with the sample size (intervals get wider or narrower!) and value for the confidence interval. Check out some of the other options as well.

P Values

Click on Learn Stats and select "p Values." The initial display in the Statistics Panel is shown on the next page, where I have already selected "Plot theoretical distribution," "Highlight critical region," and under Output, "Plot test statistics" and "Frequency table". This module really should be called Type I and II Error because it shows dynamically how Type I and II Errors occur across repeated random sampling.

Look at the Results window. The first thing shown is the distribution of $z_{\bar{x}}$ values that would occur if the Null Hypothesis is true and the population of scores is normally distributed. (It doesn't say so, but $\mu = 0$ and $\sigma = 1.00$ for this population) The areas of rejection for $\alpha = .05$ and a 2-tail test are shown. Below that is a sampling distribution of values of $z_{\bar{x}}$ for 100 samples of any size that would occur is H_0 is true. And below that is a Results table showing the number and percentage of samples that incorrectly lead to rejection of the H_0, i.e., a Type I Error. Run the simulation several times for different numbers of "studies" (samples), clicking on the "Reset" button between simulations.

Now, click on the "Simulation under the alternative hypothesis" Option. That's also shown on the next page, where I have already changed the mean to 1.33 and selected several other display options. This option is basically exploring the

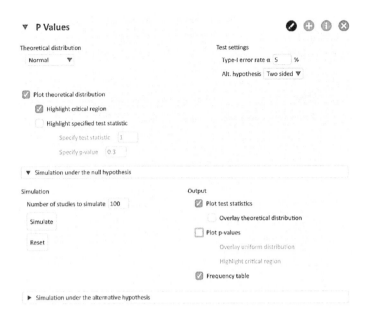

probability of a Type II Error for different "True Situation" values for the population mean. Recall that the distribution under the H_0 had $\mu = 0$ and $\sigma = 1.00$. By changing the mean to 1.33, I have said that the actual population mean is 1.33 standard deviations away from the Null Hypothesis mean. This is like the example shown in **Figure 9** on page 119 in the chapter where the True Situation mean was 9.60, which was 1.33 standard deviations (.30) away from the Null Hypothesis mean of 10.0.

If you look at the Results window, you'll see a graph like the one that I got (below) that is constructed similarly to that shown in **Figure 9**. The "Null dist. on the left is the sampling distribution of values of $z_{\bar{X}}$ for 1000 samples. The distribution to the right is the sampling distribution of $z_{\bar{X}}$ under the True Situation. The Results Table for this particular simulation (not shown) had 73% Type II Errors, which is close to the value of 74% that we calculated on page 119. Again, play around with this simulation, changing values for the true situation mean and see what happens.

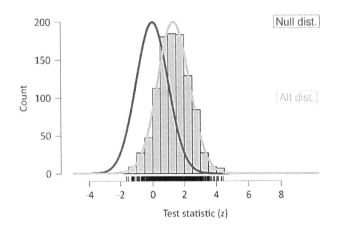

Chapter 10
Single-Sample Inference

What's In This Chapter?

This chapter explains how you can use data collected on a single sample to estimate and test hypotheses about population means and proportions. The logic underlying the statistical procedures is the same for tests of both parameters.

Inference about individual population means uses the **t distribution** to evaluate the probability of obtaining different sample means for any hypothesized value of μ. Such tests are called "**t tests**." The *t* distribution must be used in place of the normal distribution because the standard error of the mean ($\sigma_{\bar{X}}$) must be estimated from sample data. The *t* distribution is a family of distributions whose exact shape depends upon the **degrees of freedom** (**df**) for the estimate of $\sigma_{\bar{X}}$. For small *df*, the *t* distribution is markedly **leptokurtic** in shape compared with the normal distribution. As *df* increase, the *t* distribution more and more closely approximates a normal distribution.

A proportion of cases is the decimal fraction of a sample or population that shows one of two values of a dichotomous variable. A **sample proportion** (**P**) can be used to estimate and test hypotheses about the proportion (**p**) that exists in a population. The procedures for doing so use the normal distribution to approximate the discrete sampling distribution that is actually followed by a sample proportion. The two procedures are the **binominal test of a proportion** and the **large sample test of a proportion**.

Introduction

This chapter will teach you how to estimate and test hypotheses about population means and proportions. This chapter is titled "Single-Sample Inference" because the procedures focus on making inferences about individual parameters. The next chapter called "Two-Sample Inference" deals with drawing conclusions about the differences between *pairs* of population means and proportions. As you study these different procedures, it is important to keep in mind that they all employ the same logical model developed in Chapter 9. The details will differ a bit, but the following steps are always involved in testing hypotheses about any parameter or pair of parameters.

1. State H_0 and H_A for the parameter(s) of interest and decide on an α level to be used in testing the H_0.

2. Select random or at least representative sample(s) from the population(s) of interest.

3. Use the sampling distribution of the statistic of interest to determine the probability of obtaining the sample value by chance if H_0 were true.

4. If the probability of obtaining the sample value is less than the established α level, reject H_0; otherwise, fail to reject H_0.

Similarly, each of the procedures used in this and the next chapter for establishing confidence interval estimates of population parameters employs the following logical steps:

1. Select random or representative sample(s) from the population(s) of interest and choose a percentage value for the confidence interval to be established (e.g., 90% CI, 95% CI, etc.).

2. From the sample(s), compute the statistic that yields an unbiased estimate of the parameter of interest.

3. Use the sampling distribution of the statistic to establish the confidence interval for your estimate.

Inference About Population Means

As we saw in the last chapter, the Central Limit Theorem (CLT) allows you to know the characteristics of the sampling distribution of the mean from knowledge of the mean and standard deviation of the population from which the sample was drawn. The CLT states that the continuous sampling distribution of the mean (a) will be normal or close to normal in shape, (b) will have a mean ($\mu_{\bar{X}}$) equal to the mean of the population (μ), and (c) will have a standard deviation ($\sigma_{\bar{X}}$) given by the expression:

$$\sigma_{\bar{X}} = \frac{\sigma}{\sqrt{n}}$$

It is this last part of the CLT that causes us problems. Almost never will we know what σ is for the population of interest. If we did, we also would probably know what μ was and would not be spending our time estimating or testing hypotheses about it. Our solution to this problem is to estimate the value of σ from the sample data—specifically, from the statistic $\hat{s}$, which is the relatively unbiased estimate of the population standard deviation. Recall that $\hat{s}$ is given by:

$$\hat{s} = \sqrt{\frac{\sum(X - \bar{X})^2}{n - 1}}$$

When we do that, the formula for the estimated standard deviation of the sampling distribution of the mean is:

$$s_{\bar{X}} = \frac{\hat{s}}{\sqrt{n}}$$

The symbol "$s_{\bar{X}}$" is used in place of "$\sigma_{\bar{X}}$" to indicate that we are dealing with an estimate of the standard deviation of the sampling distribution. We call "$s_{\bar{X}}$" the "**estimated standard deviation of the sampling distribution of the mean**" or for brevity, the "**estimated standard error of the mean.**"

If the sample is large ($n > 25$), $\hat{s}$ is likely to be a very accurate estimate of σ, and substituting $s_{\bar{X}}$ in place of $\sigma_{\bar{X}}$ doesn't present any problem. But consider what happens when the sample size is not so large, say when $n = 10$. Now the sample value of $\hat{s}$, while relatively unbiased in estimating σ, can easily yield an estimate of σ that is substantially in error, which means that the estimated value of $\sigma_{\bar{X}}$ also will be substantially in error. If we then try to use $s_{\bar{X}}$ to compute the z score for as particular sample $\bar{X}$, we can arrive at erroneous conclusions. We can understand why this is so by comparing the formula for $\sigma_{\bar{X}}$ with the formula that employs $s_{\bar{X}}$ in place of $\sigma_{\bar{X}}$.

Formula when $\sigma_{\bar{X}}$ is known:

$$z_{\bar{X}} = \frac{\bar{X} - \mu_{\bar{X}}}{\sigma_{\bar{X}}}$$

Formula when $\sigma_{\bar{X}}$ is estimated:

$$? = \frac{\bar{X} - \mu_{\bar{X}}}{s_{\bar{X}}}$$

Consider the upper formula for $z_{\bar{X}}$. According to the CLT, if the population is normally distributed or the sample size is large, **across an infinitely large number of samples**, the value of the mean will be a normally distributed variable. From each of these means we subtract $\mu_{\bar{X}}$, which is a constant for all samples, and then divide the difference by $\sigma_{\bar{X}}$, which is also a constant across all samples. We can rewrite the formula in more abstract terms to reflect the nature of the components. Let N.D.V. = a normally distributed variable, so:

$$z_{\bar{X}} = \frac{\text{N.D.V.} - \text{Constant}}{\text{Constant}}$$

Seen this way, it should be clear that the test statistic $z_{\bar{X}}$ must also be normally distributed. And, of course, it is. **The sampling distribution of the test statistic $z_{\bar{X}}$ is the unit normal curve**.

The lower formula using $s_{\bar{X}}$ presents a different picture. Across an infinitely large number of samples, the means will be normally distributed. From each mean, a constant ($\mu_{\bar{X}}$) is subtracted, but now the difference is divided by a fluctuating estimate of $\sigma_{\bar{X}}$—in other words, by a variable. The abstract formula reflecting this situation looks like this:

$$? = \frac{\text{N.D.V.} - \text{Constant}}{\text{Variable}}$$

The ratio will *not* be normally distributed, because each different sample yields a different estimate of $\sigma_{\bar{X}}$. Instead of following a normal distribution, across a very large series of samples, the values will distribute themselves in a different continuous sampling distribution called the **t distribution**. The details of this distribution were worked out mathematically in the early part of the 20th century by the statistician William Gossett. Gossett was forced by conditions of his employment as Head of Brewing at Guinness Brewery in Dublin to publish his solution under the pseudonym "Student," so the distribution is called "**Student's t distribution.**"

Unlike the unit normal curve, which is the sampling distribution of $z_{\bar{X}}$, the t distribution is a family of curves whose precise shape depends upon the size of the sample used to estimate $\sigma_{\bar{X}}$. **Figure 1** shows two different t distributions along with the normal curve for comparison.

Recall that the unit normal distribution of $z_{\bar{X}}$ has μ = 0 and σ = 1.00. The various t distributions also will have μ = 0, but their standard deviations are greater than 1.00. This is because t distributions are a bit "pointier" in the middle and have more of their area under the curve out in the tails of the distribution. These departures from a normal distribution are referred to as **leptokurtosis**. The degree of leptokurtosis is inversely related to the size of the sample. The bigger the sample, the closer will the t distribution approximate the normal distribution. The t distribution for $n = 5$, for example (see **Figure 1**), is markedly leptokurtic,

Figure 1

Comparison of a Normal Distribution with Two t Distributions Having the Same μ and Based on Samples of n = 10 and n = 5.

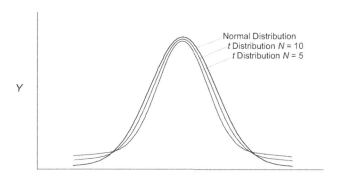

while that for *n* = 10 deviates more modestly from the unit normal curve.

There are two issues here. Why is the *t* distribution leptokurtotic, and why does it approach the normal distribution as sample size increases? The answer to the first question lies in the fact that across a large series of samples, $s_{\bar{X}}$ sometimes is an overestimate of $\sigma_{\bar{X}}$ and sometimes is an underestimate. For any given discrepancy of $(\bar{X} - \mu_{\bar{X}})$, when $s_{\bar{X}}$ is smaller than $\sigma_{\bar{X}}$, the resulting quotient of $(\bar{X} - \mu_{\bar{X}}/s_{\bar{X}})$ will be larger than it would have been had $(\bar{X} - \mu_{\bar{X}})$ been divided by $\sigma_{\bar{X}}$. Conversely, on those occasions when $s_{\bar{X}}$ overestimates $\sigma_{\bar{X}}$, the resulting quotient will be smaller than it would have been had $\sigma_{\bar{X}}$ been used. The result of this over- and underestimation is to pile up more values in the middle *and* in the tails of the distribution.

The answer to the second question lies in the fact that as sample size increases, the sample estimate of σ becomes more and more stable, i.e., larger sample sizes yield more precise and consistent estimates of a parameter. At the limit, when *n* grows infinitely large, $s_{\bar{X}}$ is always equal to $\sigma_{\bar{X}}$ and the *t* distribution is normal in shape. For practical purposes, with a sample size of 25, the *t* and normal distributions are virtually the same.

We said above that the precise shape of the *t* distribution depended upon sample size. That's not quite correct. Actually, it depends upon what is called the **degrees of freedom**, abbreviated **df**. The *df* is defined as **the number of elements in a sample that independently contribute to estimating a population parameter**. We can understand the concept by the following example. Consider a sample of size of *n* = 5. To estimate μ with this sample, we would compute the sample mean by summing all five values and dividing by five. The resulting quotient is our estimate of μ. In computing the mean, each case contributes its own independent "bit of evidence" concerning the value of μ. We would therefore say that the estimate of μ is based on all five cases and *df* = 5. In contrast, suppose that we want to estimate σ using $\hat{s}$. To compute $\hat{s}$ using its definitional formula, we would begin by computing the sample mean. Then we would subtract the mean from each score, square the difference, and sum up the squared differences yielding:

$$\sum (X - \bar{X})^2$$

There are five squared differences, but only four of the five give you independent "bits of evidence" concerning the value of σ. This is because the discrepancies are taken about the sample mean, and the sum of the discrepancies about the mean must equal 0, i.e.,

$$\sum (X - \bar{X}) = 0$$

Given that the mean must be some value for the sample, four of the $(X - \bar{X})$ discrepancies can assume any value. But once four discrepancies have been compiled, the fifth is fixed in value and is not independent of the first four. This fifth discrepancy contributes no new (independent) evidence about score variability beyond that given by the first four discrepancies. The estimate of σ is thus based on *n* – 1 or 4 *df*.

In estimating σ, *df* = *n* – 1, but this is not always the case. The *df* can be *n* – 2, *n* – 3, etc., depending upon what parameter is being estimated. Generally, the *df* for estimating a population parameter is the sample *n* minus the number of "linear restrictions" on the data. In the estimate of σ, one linear restriction existed because the quantity $\sum(X - \bar{X})$ had to equal 0. In the *t* distribution, σ (and therefore $\sigma_{\bar{X}}$) is estimated from sample data, so the particular shape of the distribution is a function of the *df* for the sample estimate, rather than the sample size *per se*.

With this background, we are now going to put the *t* distribution to work to test hypotheses about and estimate individual population means. The procedure described below is almost always called a **one-sample t test**. We will use the example problem, "Booked Up" to show how the *t* test works. **Read through the example problem on the next page before continuing**.

In the problem, "Booked Up", our interest lies in using the sample data to decide whether the Director is correct in her assertion that mean usage rates for the microcomputer labs are 70% on weekdays and 50% on weekends. Specifically, we are going to test H_0 that μ_{Usage}

Booked Up?

The Director of Computing Services at our college claims that the main microcomputer laboratory operates at an average of 70% of capacity during weekdays and 50% of capacity during weekends. This figure seems a bit high to me, because whenever I pass by the lab, I notice only one or two students on the machines. Of course, maybe I always go by the lab at the wrong time—whenever I'm not around, it's full.

How might I check up on the Director's figures in a more systematic manner? Here is one possibility. For two weeks, I send my work scholar, Andrea, over to the lab at randomly selected times during the day to count the percentage of machines that are being used. Andrea thinks that the assignment is a drag and only manages to make it over to the lab 14 times during weekday hours and eight times during the weekend. Here are her data:

Weekdays				Weekends	
35	45	30	70	0	30
85	95	50	80	20	25
60	65	35		80	40
30	50	85		35	10

on weekdays is 70% and on weekends is 50%. We will treat weekdays as one population and weekends as another population.

Assumptions of the Procedure

To use the *t* distribution to analyze these data, we need to make the following assumptions about each of the two populations of interest:

1. **The data are numerical in character.** This assumption is required because the mathematical operations required to find the mean, standard deviation, etc. assume numerical variables. The data meet this assumption.

2. **The sample is randomly selected from the population of interest.** If sampling is not random, then the details of the sampling distribution of the mean are unknown. Our time samples are random with respect to the 2-week period of the study, but their representativeness for the semester is unknown.

3. **The populations of usage rates are normally distributed.** This is required because the expression ($\bar{X} - \mu_{\bar{X}}/s_{\bar{X}}$) will follow the *t* distribution only if the sampling distribution of the mean is normal in shape *and* if the sample estimate of σ is statistically independent of the sample estimate of μ. **The latter will only be true if the sample is drawn from a normally distributed population.** The normalizing tendency of large sample sizes will help to insure an approximately normally distributed sampling distribution of the mean but will not help make the two estimates independent. There isn't much we can do about this assumption. We don't know what the populations are like. If they are unimodal and not too heavily skewed, the results of the procedure will be approximately correct.

We will use two *JASP* procedures to fully analyze these data. The DESCRIPTIVES procedure will be used to compute the descriptive statistics needed to summarize the data for each sample. The other *JASP* procedure for testing hypotheses about and estimating the population mean is called T-TESTS, with the One Sample T-Test sub procedure.

The first step in the analysis would be to summarize the data for each sample using the DESCRIPTIVES procedure. **Figure 2** shows the *JASP* Descriptive Statistics Table for the two samples. The estimated standard error of

Figure 2

JASP Descriptive Statistics Table for Usage Data for Weekdays and Weekends.

Descriptive Statistics for "Booked Up" Study

	Weekdays	Weekends
Valid	14	8
Mean	58.214	30.000
Std. Error of Mean	5.944	8.504
90% CI Mean Upper	68.741	46.112
90% CI Mean Lower	47.688	13.888
Std. Deviation	22.241	24.054
Minimum	30.000	0.000
Maximum	95.000	80.000

the mean ($s_{\bar{X}}$) and 90% confidence intervals have been included as one of the options under Statistics.

Our emphasis in this chapter is on developing the logic and procedures of hypothesis testing and parameter estimation, so we won't dwell on the descriptive analysis of the data. Suffice it to say that the data for both samples are highly variable and modestly positively skewed. The sample sizes are too small to meaningfully estimate shapes of the population distributions, so histograms are not given.

We will work through the analysis of the Weekday data by "hand" to give you a feel for the logic of using the t distribution to test hypotheses about and estimate μ from sample data. We'll use JASP to analyze the Weekend data.

Our first step is to formulate H_0 and H_A. We will use a directional H_0, because our only interest is whether the actual usage rate is *below* the asserted value of 70%. Here is how the hypotheses would be stated:

$$H_0: \mu_{Usage} \geq 70\%$$
$$H_A: \mu_{Usage} < 70\%$$

We will adopt an α level of .05 for our test.

Computational Procedures for Testing H_0

To carry out the analysis "by hand," we will use the following statistics from the JASP table: $n = 14$, $\bar{X} = 58.2\%$, and $\hat{s} = 22.2\%$. Our question is this. Does a sample mean of 58.2% fall among the least likely 5% of sample outcomes if μ_{Usage} really equals 70%? Because the population σ is unknown, we estimate the standard error of the mean as:

$$s_{\bar{X}} = \frac{\hat{s}}{\sqrt{n}} = \frac{22.2\%}{\sqrt{14}} = \frac{22.2\%}{3.74} = 5.94\%$$

Notice that JASP has already computed the standard error of the mean for you in the Descriptive Statistics table.

Because we are using the t distribution to test a hypothesis about a population μ, we will label the obtained value of t as $t_{\bar{X}}$ to indicate that it is serving as a test statistic for the sample mean. The $t_{\bar{X}}$ is computed as:

$$t_{\bar{X}} = \frac{\bar{X} - \mu_{\bar{X}}}{s_{\bar{X}}} = \frac{58.2\% - 70\%}{5.94\%} = \frac{-11.8\%}{5.94\%} = -1.99$$

The next step is to determine whether the obtained value of $t_{\bar{X}}$ falls in the area of rejection for the sampling distribution of t. Because the t distribution varies in shape as a function of the df on which $s_{\bar{X}}$ is based, it is necessary to consult the t distribution appropriate for our test. The df for the estimate of $\sigma_{\bar{X}}$ is $n - 1$. Our sample consists of 14 cases, so $df = 14 - 1 = 13$. If you were carrying out the procedures by hand, you would consult a table of what are called "critical values" of t that define areas of rejections for both 1- and 2-tail tests at different α levels for t distributions having different df.

Table B in the Appendix contains these critical values of t. To use the table, you run your finger down the first column labeled "df" until you find the correct df for your test. Then, you run your finger along the row until you find the column whose header is the α level that you are using for your test. The value tabled at the intersection of df and α level is the critical value of t.

For our 1-tail test with $\alpha = .05$ and $df = 13$, the critical value of t is -1.77. Notice that this value is larger than the normal curve value of -1.65 that would be required had we been using $z_{\bar{X}}$. This is because the t distribution has more area in the tails and, therefore, one must go further out in the tails to reach the most extreme 5% of the distribution. The obtained $t_{\bar{X}}$ of -1.99 exceeds this critical value, so we can reject H_0 that $\mu_{Usage} \geq 70\%$. **Figure 3** shows this situation graphically.

It is interesting to note that had we used a non-directional H_0, the 2-tail critical value of t would be ± 2.16 (see **Table B**), and our obtained value of -1.99 would not have led to rejection of H_0.

Figure 3

Area of Rejection for the t Distribution when df = 13, α = .05 and H_0 is Directional

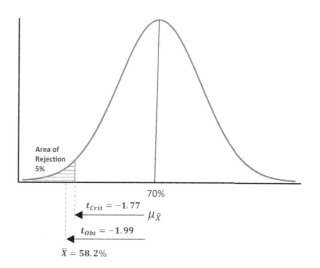

Computational Procedures for Making Parameter Estimates

Because we have rejected H_0 that $\mu_{Usage} \geq 70\%$, it is appropriate to use the sample data to establish an estimate of the actual value of μ_{Usage}. Our sample value of 58.2% is, of course, our point estimate of μ_{Usage}. Knowing the point estimate is likely to be somewhat in error, we are more interested in establishing a confidence interval estimate of the parameter. Let's assume that we want the 90% CI. Recall from Chapter 9 that the confidence interval estimate of μ was constructed using the following formula:

$$CI = \bar{X} \pm (z_{1-CI})(\sigma_{\bar{X}})$$

where: z_{1-CI} = value of z for a 2-tail significance test at the (1-CI) α level.

But, if $\sigma_{\bar{X}}$ must be estimated from the sample data, the unit normal curve is no longer appropriate for describing the probability of the sample mean falling any distance away from the hypothetical value of μ. To establish the "worst case" locations for μ for any confidence interval, we need to use the appropriate t distribution in place of the unit normal curve. Our formula thus becomes:

$$CI = \bar{X} \pm (t_{1-CI})(s_{\bar{X}})$$

where: t_{1-CI} = value of t for a 2-tail significance test at the (1-CI) α level.

The value of t_{1-CI} is obtained from Table B for 13 df and a 2-tail .10 α level. The value is 1.77. Using this formula, the 90% CI would be given by:

$$CI = \bar{X} \pm (t_{1-CI})(s_{\bar{X}})$$

$$= 58.2\% \pm (1.77)(5.94\%)$$

$$= 58.2\% \pm 10.52\%$$

$$= 47.7\% - 68.7\%$$

We can say, then, that we are 90% confident that the μ_{Usage} is included in the range between 47.7% and 68.7%, which are the same values given in **Figure 2**. Apparently, usage of the micro lab on weekdays is somewhat lower than the Director supposes. Notice the 90% CI does not include the value of μ_{Usage} (70%) stated under H_0. This is consistent with the fact that a 1-tail test of H_0 carried out at the .05 α level led to rejection of the hypothesis that $\mu_{Usage} \geq 70\%$.

JASP Analysis of the Data

Now that you have a feel for how the t distribution is used to test hypotheses and make parameter estimates, we'll analyze the Weekend data using *JASP*. We will test the following directional H_0 and H_A with the weekend data:

$$H_0: \mu_{Usage} \geq 50\%$$
$$H_A: \mu_{Usage} < 50\%$$

Figure 4 shows the results of the T-TEST – One-Sample T-Test. The Sample Statistics table (**Figure 2**) shows that the mean usage figure is 30%, with a standard deviation of 24.1%. The One-Sample Test Table gives a value of t of –2.35, the negative sign reflecting the fact that the observed sample mean of 30% lies *below* the H_0 value of 50%. The *p* value associated with this value of *t* is .025. Recall from Chapter 6 that the *p* value for a test statistic is **the probability of obtaining that value of the statistic plus all more extreme values by chance if H_0 is true.**

Figure 4

JASP One-Sample T-Test of Room Usage During Weekends with Directional Null Hypothesis.

One Sample T-Test

	t	df	p
Weekends	-2.352	7	0.025

Note. For the Student t-test, the alternative hypothesis specifies that the mean is less than 50.

JASP allows a good deal of customization of the One Sample T-Test procedure. **Figure 5** is a screen capture of the available statistics and options available.

Figure 5

Screen Capture of the Test Options and Statistics Available with the One Sample T-Test Procedure in JASP.

To carry out the test, the variable Weekends was moved into the active Variables box, the Test value was set at "50," and the Alternative Hypothesis set at "< Test value." The 90% CI for the estimated value of µ is given in **Figure 2**.

Again, the Director of Computer Services seems to have overestimated room usage. Our sample data make it unlikely that μ_{Usage} on weekends equals or exceeds 50% and we estimate that the actual usage is between about 13.9% and 46.1%.

Inferences About Population Proportions

Binominal Test of a Proportion

We are now going to extend the logical process of hypothesis testing and estimation to a new statistic—the **sample proportion (P)**. The sample P is an appropriate statistic whenever a variable can be expressed as a "dichotomy," i.e., the variable can take on only two values. Some categorical variables like *Hand Preference* (left vs. right) and *Club Membership* (belongs vs. doesn't belong) are naturally dichotomous, while other variables can be made dichotomous by collapsing several categories into two categories. An example of the latter is the variable *Religion*, where we could define one value of the variable to be "Protestant" and the other as "non-Protestant." Sometimes, we even elect to create a dichotomous classification from numerical data. We could, for example, define a dichotomous variable *Mastered Statistics*, in which each student was classified as "Yes" if his/her final course average ≥ 70% and "No" if the average < 70%.

For a dichotomous variable, P is defined as **the proportion of the sample falling into one of the two categories**, i.e.,

$$P = \frac{\text{Number in category}}{\text{Total number of cases}}$$

The category to which P is assigned is arbitrary; usually P stands for the category of greater interest to the researcher. The proportion of cases in the other category is symbolized **Q**. Because P + Q must equal 1.00, Q is usually defined in the following manner:

$$Q = 1 - P$$

Corresponding to the sample statistic P is the population parameter **p**. The parameter p refers to **the proportion of the population that falls into the category of interest for a dichotomous variable**. The proportion of the population in the other category is abbreviated **q** and $q = 1 - p$.

The value of the sample proportion P is used to test hypotheses about and/or make estimates of the value of the population proportion p. The first procedure that we will use for this purpose is a variant of our old friend called the Binominal Test of a Proportion and is found in *JASP* under FREQUENCIES — Binominal Test. The data for such a test are variables having just two values. **We will develop the steps of this procedure using the example titled "Some Depressing News." Read the example on the next page before continuing**.

Your basic question is whether the *population* of clients going through this program have a higher proportion of graduates who remain drug free at 6-week follow-up is greater than the national average of .50. To be conservative, you are going to test the H_0 that this population p is equal to the national average of .50. The sample value of P will be used to test the validity of this hypothesis. Because a significant difference, either above or below the population of .50 would be of interest, we'll use a 2-tail test. Naturally, the folks who developed the program hope that the H_0 will be rejected.

The Null and Alternative Hypotheses would be formulated as follows:

$$H_0: p_{Usage} = .50$$
$$H_A: p_{Usage} \neq .50$$

We'll set α at .05.

The data for the Depressing News study are simply 60 YES's or NO's entered for the 60 cases in the study. The procedure lets you enter the Test Value under the null hypothesis (.50), the nature of the Alternative Hypothesis (≠), and the desired confidence interval for the population estimate (90). The binominal test computes the exact probabilities associated with various outcomes when the Null Hypothesis is true.

The test assumes that:

1. The observations are independent of one another, i.e., the response of each person is unrelated to how other people respond.

2. The sampling distribution of P is approximately normal in shape.

Some Depressing News

People addicted to drugs lead horrible lives, cause untold misery to those around them, and cost society billions of dollars each year. As a result, enormous efforts have been made to develop therapeutic programs to "cure" drug addiction. Unfortunately, most drug rehabilitation programs are not very effective, probably because they are unable to alter most of the situational variables in a person's life that encourage drug addiction.

Suppose that you were doing a study of the effectiveness of a new drug rehabilitation program developed by physicians and social workers at a county hospital. You select as your sample 60 randomly selected "graduates" of the program and maintain contact with them for a period of six weeks. At the end of the 6-week follow-up, the number of clients remaining drug free is 42 (70%). Your review of the literature indicates that the average success rate of drug rehabilitation programs nationally is 50% of clients drug free at six weeks. Is this program doing better than the national average?

Here are the data for the 60 cases. YES means the person was drug free at the time of the 6-week follow-up. NO means that the person was not drug free at the 6-week follow-up.

YES	YES	NO	YES	NO	YES	YES	YES
NO	YES	YES	NO	YES	YES	YES	YES
YES	NO	YES	YES	YES	NO	NO	NO
NO	YES	YES	NO	YES	YES	YES	YES
NO	NO	YES	YES	NO	YES	NO	YES
YES	YES	NO	NO	YES	NO	YES	YES
YES	YES	YES	YES	YES	YES	NO	YES
YES	YES	YES	YES				

Figure 6

JASP Binomial Test of the Null Hypothesis that the Population Proportion Equals .50.

Binomial Test

Variable	Level	Counts	Total	Proportion	p	90% CI for Proportion Lower	Upper
Drugfree	NO	18	60	0.300	0.003	0.204	0.412
	YES	42	60	0.700	0.003	0.588	0.796

Note. Proportions tested against value: 0.5.

Figure 6 shows the table produced by the Binominal Test procedure. Our concern is the YES category in the second row of the table. The sample proportion of .70 is extremely unlikely to have occurred by chance (random sampling) if the population proportion were .50 ($p < .003$). The 90% confidence limits for the estimate of the actual population proportion are .588 to .796.

The Large Sample Test of a Proportion

An alternative to the binomial test is what's called a **large sample test of a proportion**. The term "large sample" is used because the procedure employs a continuous sampling distribution, the normal curve, to approximate the discrete sampling distribution followed by a dichotomous variable. The approximation by the normal distribution is adequate only for large samples.

The major advantage of the large sample test is that you do not need the actual case data to conduct the test—just the sample proportion (P) and the size of the sample (n).

The large sample test of a proportion assumes that the sample is sufficiently large that both ($n \times p$) and ($n \times q$) are greater than 5.0, where p = the population proportion tested under the Null Hypothesis and $q = 1 - p$. This is clearly the case:

$$n \times p = 60 \times .50 = 30.0$$
$$n \times q = 60 \times .50 = 30.0$$

For our present example of the "Some Depressing News" study, to test H_0 that $p = .50$, we need to know what the sampling distribution of P looks like when $p = .50$ and $n = 60$. As noted above, the exact sampling distribution of P is a discrete distribution, but with larger samples the distribution is well approximated by a normal distribution.

The Null and Alternative hypotheses would be stated as they were with the binominal test:

$$H_0: p_{Usage} = .50$$
$$H_A: p_{Usage} \neq .50$$

If H_0 is true, the normal curve approximation to the discrete sampling distribution of P will have μ and σ given by the following formulas:

$$\mu_P = p$$

$$\sigma_P = \sqrt{\frac{p \times q}{n}}$$

where: μ_P = the mean of the normal curve approximation to the sampling distribution of P, and

σ_P = standard deviation of the normal curve approximation to the sampling distribution of P.

The test statistic for the large sample procedure is z_P which follows the unit normal distribution. z_P is defined as:

$$z_P = \frac{P - \mu_P}{\sigma_P}$$

We are now ready to test the H_0 that in our 6-week follow-up population, the proportion of graduates who remain drug free is equal to .50. If H_0 were true, the sampling distribution of P would be a normal distribution having $\mu_P = .50$ and σ_P given by:

$$\sigma_p = \sqrt{\frac{p \times q}{n}} = \sqrt{\frac{.50 \times .50}{60}} = \sqrt{\frac{.25}{60}} = \sqrt{.0042} = .065$$

For the sample value of $P = .70$, the test statistic z_P is computed to be:

$$z_P = \frac{P - \mu_P}{\sigma_P} = \frac{.70 - .50}{.065} = \frac{.20}{.065} = 3.08$$

We recall that for a 2-tail test with α = .05, the areas of rejection for the unit normal curve begin at z = ± 1.96. Our value of the test statistic z_P is much larger than this critical value, indicating that our sample outcome is among the least likely 5% of sample Ps that would occur by chance if the H_0 were true. We will therefore reject H_0 and accept H_A that $p \neq .50$. Clearly, at 6-week follow-up, this rehabilitation program results in a higher proportion of graduates who have remained drug free compared with the national average.

Making Parameter Estimates

We are now able to use the large sample procedure to estimate what p is for our 6-week follow-up population. The sample P is an unbiased estimate of population p, so our point estimation of the population proportion is .70. To establish the confidence interval, we need to know the sampling distribution of P, and here we run into difficulty. The sampling distribution of P used to test the H_0 was generated based on the provisional assumption that H_0 was true and $p = .50$. We have now rejected that hypothesis and concluded that p is some value greater than .50.

Because σ_P is determined by the values of p and q, the exact value of σ_P is unknown. Our solution is to estimate σ_P from the sample data according to the following formula:

$$s_P = \sqrt{\frac{P \times Q}{n}}$$

where: s_P = the estimated standard deviation of the sampling distribution of P (the "estimated standard error of P").

The confidence interval estimate of p is then given by the formula:

$$CI = P \pm (z_{1-CI})(s_P)$$

where: z_{1-CI} = value of z for a 2-tail significance test at the (1-CI) α level.

Because p and q are being estimated from the sample values of P and Q, the accuracy of the estimation procedure is influenced by sample size. If n is too small, the estimates of p and q could be substantially in error, resulting in a confidence interval that is either too narrow or too wide. The rule of thumb for deciding whether sample n is large enough is to compute n x P and n x Q and require that both products be larger than 5.0

Let's use these formulas to establish the 90% confidence interval for estimating p for the 6-week follow-up population. First we note that both n x P and n x Q are larger than 5.0 (42.0 and 18.0, respectively), so our sample is large enough to justify the approach. The estimated standard deviation of the sampling distribution of P is computed as:

$$s_P = \sqrt{\frac{P \times Q}{n}} = \sqrt{\frac{.70 \times .30}{60}} = \sqrt{\frac{.21}{60}} = \sqrt{.0035} = .059$$

The 90% confidence interval is then computed as:

$$90\% \; CI = P \pm (z_{1-CI})(s_P)$$

$$= .70 \pm 1.65 \, (.059)$$

$$= .70 \pm .097$$

$$= .60 - .80$$

Thus, we are 90% confident that p for our population lies between .60 and .80. Note that these values agree closely with the *JASP* values given in **Figure 6**.

Analysis of the Data Using *Interactive Statistics*

It's clear that hand computing the necessary statistics to test hypotheses about population proportions and finding confidence interval estimates is a fair amount of work, which is best left to computers. In its current version JASP doesn't provide a procedure for working with proportions from already summarized data, but we're going to cheat a bit and use an alternative analysis package.

LibreTexts is a wonderful open-source foundation that produces a wide range of on-line texts and other educational resources. One such resource is **Interactive Statistics**, which is produced by:

Dr. Larry Green
Professor of Mathematics
Lake Tahoe Community College
South Lake Tahoe, CA 96150

Interactive Statistics can be found at Libretexts.org → Platforms → Libraries → Statistics → Learning Objects → Interactive Statistics. This is a collection of 52 modules that do all manner of interesting and useful things with statistics—demonstrations of important theorems and processes, as well as specific data analyses.

Figure 7 is a screen shot of Module 29: *Hypothesis Test for a Population Proportion Calculator* with the summarized data from the *Depressing News* study filled in and the results calculated. Notice that the actual values for the sample size and the number of "successes" have been entered rather than the proportion. The results are enormously more accurate than those calculated by hand.

A Final Word

This chapter has described just three of many procedures for testing hypotheses about population parameters. Statisticians have developed techniques for testing hypotheses about virtually any population parameter—correlation coefficients, standard deviations, medians, etc.—but all procedures follow the logic described above. If you encounter a situation where one of these procedures is required, consult a more advanced statistics text in the library or Google the topic. Hopefully, you will discover that you actually understand what you are reading!

Figure 7

Interactive Statistics Data Entry Screen and Analysis Results for the Depressing News Study.

hypothesis test for a population Proportion calculator

Fill in the sample size, n, the number of successes, x, the hypothesized population proportion p_0, and indicate if the test is left tailed, <, right tailed, >, or two tailed, ≠. Then hit "Calculate" and the test statistic and p-Value will be calculated for you.

n: 60
x: 42
p_0: .50

○ <
○ >
● ≠

[Calculate]

z: 3.098386676965933
p: 0.0019457736937391612

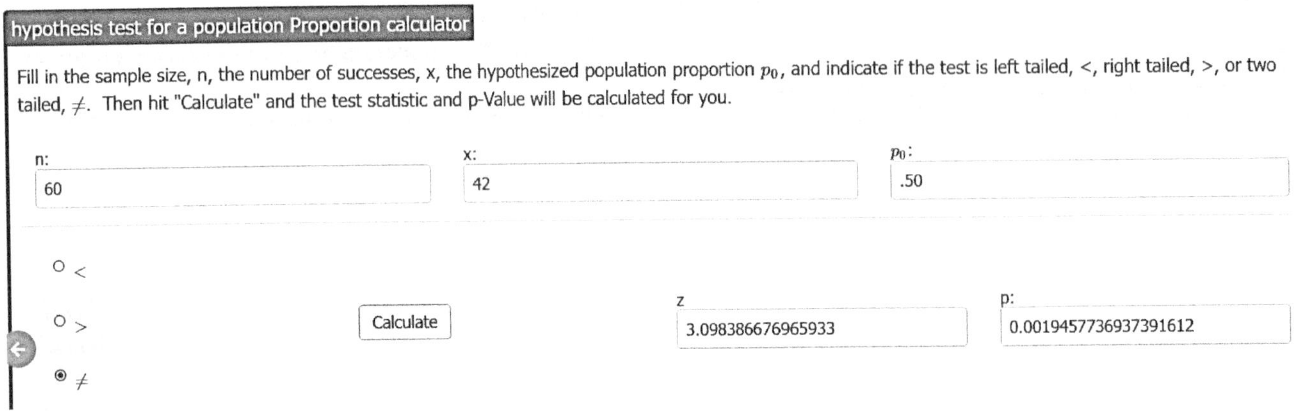

Chapter 10, Single-Sample Inference
Chapter Exercises

Some questions are purely factual and have only one correct answer. Others require greater thought on your part and may have more than one possible answer.

1. Fill in the missing term(s) that makes the statement true.

    a. In the four steps in hypothesis testing, the third step involves consulting the _____ of the statistic in question.
    b. The issue that was glossed over in Chapter 9 was that we didn't know _____ in testing hypotheses about μ.
    c. When $\sigma_{\bar{X}}$ is estimated from $\hat{s}$ and n is small, $(\bar{X} - \mu_{Ho})/s_{\bar{X}}$ is not _____.
    d. Compared to the _____, the t distribution is _____ in shape.
    e. This is because _____ is a _____ across repeated random samples.
    f. In estimating μ from a sample, df = _____.
    g. In estimating σ from a sample df = _____ because _____.
    h. The requirement for the t-test procedure that $\bar{X}$ and $\hat{s}$ be independent across repeated random samples is only met if the population is _____.
    i. For a 2-tail t test of a null hypothesis, $t_{\bar{X}}$ = 2.06. With α = .05 and df = 7, the result _____.
    j. In a binominal test, p = _____ and q = _____.
    k. The two assumptions of the binominal test are that _____ and _____.
    l. The major advantage of the large sample test of a proportion over the binominal test is that _____.

2. Among the many helpful pamphlets available at the College Health Center is one entitled, *Sleep or Fail!* In the very first paragraph on page one of the pamphlet is the statement that college students average only 5.2 hr. of sleep a night. Being a big sleeper yourself, you wonder whether your fellow students in general are different from the national sample. You select a random sample of 16 students and have them monitor their number of hours of sleep a night for a week. From this you compute the average hours of sleep a day. You obtain the following statistics for your sample: $\bar{X}$ = 6.6 hr. per day and $\hat{s}$ = 2.4 hr. per day.

    a. Test the H_0 that μ_{COL} = 5.2 hr. using α =.05.
    b. Compute the 90% CI for the estimated value of μ.

3. Thirty-four students in an introductory psychology class are asked to write down the number of books that they have read "for pleasure" in the past month. Here are the data:

0	2	5	1	2	3	0	1	0	1
4	2	0	2	4	2	1	3	2	1
0	3	4	1	2	1	0	3	0	2
2	0	1	5						

    Use *JASP* to test H_0 that μ_{COL} = 3.0 bks/Mo. Set α =.05. Summarize your findings and write a brief report of the analysis. What can you conclude about the students' reading habits?

4. Probably everyone has read the disclaimer on cereal boxes that the product is packed by weight and not by volume, and that "...some settling may occur during shipping." Okay, but did you ever wonder how accurately the box is filled "by weight?" Haven't you wanted to check out how much cereal there really is in the box? For some products (cereal is not one of them), underfilling the package by 5% or so would have a significant positive impact on a manufacturer's profits. Suppose you had the leisure and resources to indulge your morbid curiosity. You select for review canned cashews, because nuts are relatively expensive, and a consistent underfilling would have a noticeable effect on a company's profits. Besides, you like cashews. You obtain a random sample of 30 1-pound cans of a particular brand of salted cashews from the stores in your hometown of Madison, Wisconsin. Each can is carefully opened, and the contents are weighed on a sensitive electronic scale, accurate to 1/100 of an ounce. Below are the weights in ounces of the contents for the 30 cans comprising your sample.

15.21	16.21	16.12	15.73	16.05	15.68
15.79	15.84	16.09	16.12	15.79	16.02
16.03	16.21	15.87	16.04	15.89	15.85
15.93	15.75	15.24	15.93	15.97	15.81
15.71	15.97	15.96	15.82	15.92	15.94

Provide a complete univariate description of the data. Based on the data, what is your conclusion concerning the claim that the company is selling a "1-pound can" of cashews? State your Null and Alternative Hypotheses and the alpha level used in your tests. Then present and interpret the relevant statistics.

5. Based on casual observation during the 2020 Presidential Election, you believe that people who own large, obnoxious, polluting SUVs were much more likely to have "Trump" stickers on their vehicles than "Biden" stickers. To find out whether this is true, you systematically observe SUVs in the parking areas at several local malls on a particular Saturday. You count only those vehicles that meet your definition of large and obnoxious and have a "vote for" sticker displayed. By 5:00 p.m., you have identified 96 large, obnoxious SUVs with stickers. Of the 96 vehicles, 64 had "Trump" stickers and 32 had "Biden" stickers.

   a. You are going to test the Null Hypothesis that owners of large, obnoxious SUVs are equally likely to have either type of sticker. Formally state the <u>non-directional</u> Null and Alternative Hypotheses that you will test using the large sample test of a proportion. Do your data meet the assumptions of the test?
   b. Carry out a large-sample test of a proportion to evaluate the Null Hypothesis and state the results of the test.
   c. Compute the 90% CI for your estimate of the population p.

Chapter 11
Two-Sample Inference

What's In This Chapter?

When we test whether the values of a statistic are **significantly different** for two samples, we are testing the H_0 that the samples are drawn from populations having the same value for the parameter in question. Inferential procedures of this type are particularly important in research because they permit the analysis of data from experiments. This chapter discusses procedures for testing the significance of the difference between **pairs of sample means and sample proportions.**

Testing of the significance of the difference between two sample means almost always involves estimating the population standard deviations from sample data. As a result, the **t distribution** must be used to evaluate the difference between the means. When the two samples are independent of one another, the procedure is called an **independent t test**.

Sometimes it is advantageous to collect the sample data in such a way that each score in one sample is paired with a score in the other sample (usually by collecting the data on the same cases), and the pairs of scores are positively correlated. The test of the difference between correlated sample means is known as a **correlated t test**. An alternative to the correlated t test is the nonparametric **Wilcoxon Matched-Pairs Signed-Ranks** test.

If the t test reveals a significant difference between the two sample means, a statistic called the **point biserial correlation coefficient squared (r_{pb}^2)** can be used to quantify the magnitude of the difference in terms of explained variability in much the same way as r^2. Another effect size statistic called **Cohen's d** quantifies the size of the obtained difference as a percentage of the average variability of the scores.

Confidence-interval estimates of the size of the difference between two population means can be made for both independent and correlated samples.

Testing the significance of the difference between two sample proportions is made using the **large sample test of the difference between proportions**. This procedure uses the normal distribution to approximate the discrete sampling distribution of the difference between two sample proportions. The normal curve also permits the construction of confidence interval estimates of the size of the difference between the two population proportions.

Introduction

This chapter teaches you how to use the data from two samples to test hypotheses about and make estimates of the *difference* between two population means and proportions. The procedures in this chapter represent a straightforward extension of the logic developed in Chapter 10 for single-sample inference.

To decide whether two populations differ in value for some parameter, the researcher begins by formulating a Null and Alternative Hypothesis concerning the differences between the two parameters. Although one can test any hypothetical difference, the logic of research almost always leads to the H_0 that the two parameters are equal in value, i.e.:

$$H_0: Parameter_1 = Parameter_2$$
$$H_A: Parameter_1 \neq Parameter_2$$

Directional hypotheses of equality/inequality between the two parameters are also possible, e.g.:

$$H_0: Parameter_1 \geq Parameter_2$$
$$H_A: Parameter_1 < Parameter_2$$

If the sample data lead us to reject the H_0 of equality between two populations, we say that the two samples are **significantly different** with respect to the statistic in question, as in "there was a significant difference between the two sample means." Be clear that, technically, a "significant difference" means that the samples represent two populations that have different values for the parameter in question, **not necessarily that the difference is practically or theoretically important.**

Statistical inference based on two samples is perhaps the most widely used of all inferential procedures. This is because the comparison of results between two samples forms the basis of the experiment—the premier method in science for discovering causality. As we indicated in Chapter 2, an experiment manipulates the presence or amount of some variable called the independent variable and determines whether this variable causally influences another variable called the dependent variable. The essential feature of an experiment is that values of the dependent variable are measured under two conditions. In the experimental condition, the cases of the sample are exposed to the independent variable and the values of the dependent variable measured. In the control or reference condition, the cases are treated in the same manner as the experimental cases, except that they are not exposed to the independent variable. If there is a significant difference between the two samples in the value of the dependent variable at the end of the experiment, one may conclude that independent variable influenced the dependent variable. Most often, the statistic used to quantify the dependent variable is the sample mean, although other statistics such as the median and sample proportion are used as well.

One can also test the significance of the difference between statistics collected on non-experimental samples. For example, one could compare the mean number of hours spent studying each day by samples of male and female freshman college students. If a significant difference were found between the means for the two samples, one could conclude that the populations of males and females in question have different values of the population mean, μ, for the variable of study time. Unlike experiments, studies of this type yield *correlational relationships*, in which the nature of causality is uncertain. Many things are correlated with gender, and we have no idea which of these variables is/are the cause of the average difference in studying.

Basic Logic of Two-Sample Inference

Here is the basic logic underlying two-sample inference. Once the H_0 and H_A are formulated and an α level decided upon, the researcher ideally selects a random sample from each population and computes the value of the relevant statistic for each sample. The *difference* in the value of the statistic for the two samples is used to test the H_0. The key idea to be grasped in this process is that **the difference in the value of a statistic for two samples is itself a statistic and can be evaluated using** the **sampling distribution of differences**. The sampling distribution of the difference in the value of a statistic between two samples can tell us the likelihood of obtaining the sample difference by chance if H_0 were true. If the likelihood is small, we will reject H_0 and accept H_A, otherwise, we will fail to reject H_0.

Here is a "generic" definition for a sampling distribution of differences for any statistic: The sampling distribution of differences between pairs of statistics is **a continuous probability distribution of values of the differences between pairs of statistics for an infinitely large number of pairs of samples of size n_1 and n_2, drawn with replacement from two defined populations**.

In the three sections that follow, we will explain the details of hypothesis testing and estimation for pairs of independent and correlated sample means and independent proportions. Don't become lost in the details; keep focusing on the fact that the same logical process is used each time.

Inference With Pairs of Independent Means

In developing the procedures for testing whether two sample means are significantly different, we will distinguish between **independent** (**uncorrelated**) samples and **correlated** samples. Two samples are independent if the observations in one sample are not linked or associated with the observations in the other sample. Independence is the usual consequence of random sampling.

In this section, we will develop the procedures for testing the significance of the difference between two independent sample means. The test is commonly called an **independent *t* test** because *t* is used as the test statistic. The hypothetical study titled, "Rewrite It Again, Sam" will serve as the basis for developing the test. **Read the description of the study carefully before continuing.**

The independent variable for this little experiment might be called *Opportunity to Rewrite* and the dependent variable *Quality of In-Class Essay*. Presumably, the *No Rewrite* condition serves as the control group. The experimental group did five originals and five rewrites, for a total of 10 essays. The control group did 10 original essays. Ideally, the total amount of time spent writing was the same for both groups. The question of interest was whether there was a significant difference between the two groups in the mean score for the dependent variable.

Because the professor wanted to be able to detect a difference in either direction between the two groups, he elected to use a nondirectional formulation of H_0 and H_A:

$$H_0: \mu_R = \mu_{NR}$$
$$H_A: \mu_R \neq \mu_{NR}$$

Rewrite It Again, Sam

Did a teacher ever make you rewrite a paper? The rationale for rewriting is that it forces students to learn how to correct and thereby avoid future lapses in style, organization, and grammar. The problem with rewriting is that it takes time—time that might be used in producing other papers. The following hypothetical experiment was undertaken to see whether rewriting is an effective instructional strategy. Each of 40 students was randomly assigned to one of two sections of Composition 101, both taught by Professor Finicky. The 20 students in Section A (*Rewrite Condition*, abbreviated "R") were required to write five 5-page essays during the semester. After each essay was graded, the students were required to rewrite their essays with instructions to "correct all mistakes and enhance the quality of your writing." These rewrites were carefully graded and were given equal weight in determining the students' final grades. Students in Section B (*No Rewrite Condition*, abbreviated "NR") were assigned 10 5-page papers to write during the course of the semester. They were graded in the same manner as those in the other class section, but students were not given the opportunity to rewrite them. At the end of the semester, all 40 students were required to write an in-class, 3-hour essay on a common topic. The percentage grade on this writing assignment constituted the measure of writing skill attained by the class members and is the **dependent variable** of the experiment. Below are the percentage grades on the in-class paper for the two classes.

Rewrite Condition (*R, n*= 20)

```
82  70  95  83  85
75  60  88  70  78
80  82  71  79  82
84  77  69  69  76
```

No Rewrite Condition (*NR, n* = 20)

```
81  94  80  79  82
72  92  76  94  89
89  79  81  97  87
73  88  82  90  86
```

To minimize the probability of making a Type I Error the professor set $\alpha = .01$.

Assumptions of the Independent *t* Test

The procedure described below uses the *t* distribution to test hypotheses about the difference between two population means and to make confidence interval estimates of the size of the difference. The *t* distribution will be appropriate only if the following assumptions are met for the two populations.

1. **The data are numerical**. The reason for this is the same as in the single-sample test of a population mean.

2. **The samples are random samples from the two populations of interest**. No mention is made of the sampling strategy, so we don't know how random or representative they are of any particular population.

3. **The two populations of interest are normally distributed**. This ensures that the sampling distribution of the differences between means will be normal in shape and that the sample estimates of the population mean and variance will be independent of one another. Given that the CLT applies to differences between sample means and the sample sizes are fairly large, the sampling distribution of the difference between means will be close to normal in shape.

4. **The two populations have the same variance, i.e., $\sigma_R^2 = \sigma_{NR}^2$**. This is referred to as **homogeneity of variance** assumption.

What Happens If You Violate These Assumptions?

The procedures in this chapter and the four chapters that follow on analysis of variance make fairly restrictive assumptions about the populations from which the samples are drawn. The assumption of normality and homogeneity of variance are called **parametric assumptions** because they concern parameters of populations. A good deal of research has shown that parametric procedures such as the *t* test and the analysis of variance are relatively unaffected by even fairly large violations of these two assumptions **when two conditions are met: (a) The samples are reasonable large, i.e., *n* = 10 or more and (b) the samples are very close to the same size.** Under these conditions, these procedures are said to be **robust**, which means they yield correct inferential conclusions and accurate confidence interval estimates.

Computational Procedures for Testing H_0

Figure 1 shows the descriptive analysis of these data using the *JASP* DESCRIPTIVES procedure. We'll use this information to show how the *t* test is computed. Note that there is a moderate difference between the sample means

Figure 1

Table of Statistics Produced by the JASP DESCRIPTIVES Procedure.

Descriptive Statistics

	R Condition	NR Condition
Valid	20	20
Median	78.500	84.000
Mean	77.750	84.550
Std. Error of Mean	1.797	1.592
Std. Deviation	8.039	7.119
Minimum	60.000	72.000
Maximum	95.000	97.000

in favor of the NR class. The standard deviations for the two conditions are approximately the same (8.04% and 7.12%) and indicate only a modest variation in the scores. The median and mean for both conditions are similar in magnitude, suggesting that the distributions of scores for both samples are symmetrical.

The statistic of interest for testing H_0 is the **difference between the sample means**. Because H_0 is nondirectional, the subtraction can be made in either direction, i.e., $\bar{X}_{NR} - \bar{X}_R$ or $\bar{X}_R - \bar{X}_{NR}$. Subtracting the NR mean from the R mean gives: 77.8% - 84.6% = -6.8%. The question is whether a difference of this magnitude is likely to occur by chance if H_0 is true.

To answer this question, we need to know what the sampling distribution of differences between independent means looks like:

1. **Shape of the Distribution.** The sampling distribution of differences between means behaves similarly to the sampling distribution of a single mean. It is exactly normal in shape if both populations are normally distributed or, if they are not, the sampling distribution tends toward normality as the size of the two samples increases.

2. **Distribution Mean.** The mean of the sampling distribution of differences between means is designated $\mu_{\bar{X}_1-\bar{X}_2}$ and is equal to the difference between the population means, i.e.,

$$\mu_{\bar{X}_1-\bar{X}_2} = \mu_1 - \mu_2$$

If the H_0 tested is that $\mu_1 = \mu_2$ then the mean of the sampling distribution will equal 0, i.e.,

$$\mu_{\bar{X}_1-\bar{X}_2} = 0$$

3. **Distribution Standard Deviation.** The standard deviation of the sampling distribution of differences between means (the "standard error" of the difference between means) is abbreviated $\sigma_{\bar{X}_1-\bar{X}_2}$ and is given by the following equation:

$$\sigma_{\bar{X}_1-\bar{X}_2} = \sqrt{\frac{\sigma_1^2}{n_1} + \frac{\sigma_2^2}{n_2}}$$

where: σ_1^2 and σ_2^2 = the variances of the two populations and
n_1 and n_1 = the sizes of the two samples.

Using this knowledge of the sampling distribution of differences between sample means, we could test H_0 that $\mu_1 = \mu_2$ using z as the test statistic, i.e.,

$$z_{\bar{X}_1-\bar{X}_2} = \frac{(\bar{X}_1 - \bar{X}_2) - (\mu_1 - \mu_2)}{\sigma_{\bar{X}_1-\bar{X}_2}}$$

$$z_{\bar{X}_1-\bar{X}_2} = \frac{(\bar{X}_1 - \bar{X}_2) - 0}{\sigma_{\bar{X}_1-\bar{X}_2}}$$

Figure 2 shows what the sampling distribution of differences between means would look like if the two populations of interest are normally distributed and $\mu_1 = \mu_2$.

Figure 2

Normal Sampling Distribution of Differences Between Sample Means when $\mu_1 = \mu_2$ and σ_1^2 and σ_2^2 are Known.

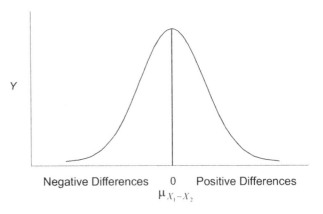

Unfortunately, as was the case with tests concerning a single population mean, in testing hypotheses about the difference between population means, we almost never know the values of the population variances and, thus, the exact value of $\sigma_{\bar{X}_1-\bar{X}_2}$. Instead, it is necessary to estimate the population variances from the sample data. Historically, what was done was to compute the **estimated standard deviation for the sampling distribution of**

differences between means according to the following formula:

$$s_{\bar{X}_1 - \bar{X}_2} = \sqrt{\frac{\hat{s}_1^2}{n_1} + \frac{\hat{s}_2^2}{n_2}}$$

where: $\hat{s}_1^2 = \sum(X_1 - \bar{X}_1)^2 / n_1 - 1$, and

$\hat{s}_2^2 = \sum(X_2 - \bar{X}_2)^2 / n_2 - 1$

For large samples (i.e., $n_1 > 25$ and $n_2 > 25$), the quotient

$$\frac{(\bar{X}_1 - \bar{X}_2) - (\mu_1 - \mu_2)}{s_{\bar{X}_1 - \bar{X}_2}}$$

is almost perfectly normally distributed and can be evaluated using the test statistic z. For smaller samples, however, sample estimates of $\sigma_{\bar{X}_1 - \bar{X}_2}$ become too unstable and the resulting quotient is no longer adequately approximated by the unit normal curve.

Fortunately, William Gossett (brewery guy) was able to demonstrate that if the normality and homogeneity of variance assumptions are met, and a slightly different formula is used to compute $s_{\bar{X}_1 - \bar{X}_2}$, then the quotient is distributed as t, with $df = (n_1 - 1) + (n_2 - 1)$. The new way of computing $s_{\bar{X}_1 - \bar{X}_2}$ uses what is called the "pooled estimate" (averaged estimate) of the population variance, which is computed as:

$$s_P^2 = \frac{\sum(X_1 - \bar{X}_1)^2 + \sum(X_2 - \bar{X}_2)^2}{(n_1 - 1) + (n_2 - 1)}$$

This pooled estimate is then used to compute $s_{\bar{X}_1 - \bar{X}_2}$ according to the following formula:

$$s_{\bar{X}_1 - \bar{X}_2} = \sqrt{\frac{s_P^2}{n_1} + \frac{s_P^2}{n_2}}$$

When $s_{\bar{X}_1 - \bar{X}_2}$ is computed in this manner, hypotheses about the differences between the two population means can be tested using Student's t distribution, i.e.

$$t_{\bar{X}_1 - \bar{X}_2} = \frac{(\bar{X}_1 - \bar{X}_2) - (\mu_1 - \mu_2)}{s_{\bar{X}_1 - \bar{X}_2}}$$

Let's apply this procedure to the data in our sample problem. To give you a feel for the procedure, we will do the computations by hand. The computation of $s_{\bar{X}_1 - \bar{X}_2}$ directly from the raw data can be quite tedious, so we'll take a shortcut that computes the pooled estimate from already-available statistics.

$$s_P^2 = \frac{(n_1 - 1)\hat{s}_1^2 + (n_2 - 1)\hat{s}_2^2}{(n_1 - 1) + (n_2 - 1)}$$

Extracting the required values from the *JASP* statistics table in **Figure 1** yields the following:

$$s_P^2 = \frac{(19)8.04^2 + (19)7.12^2}{(19) + (19)}$$

$$= \frac{(19)64.64 + (19)50.69}{(19) + (19)}$$

$$= \frac{1228.19 + 963.19}{38} = \frac{2191.38}{38}$$

$$= 57.68$$

Substituting this pooled estimate of the assumed common population variance into Gossett's formula for computing $s_{\bar{X}_1 - \bar{X}_2}$ yields

$$s_{\bar{X}_1 - \bar{X}_2} = \sqrt{\frac{s_P^2}{n_1} + \frac{s_P^2}{n_2}} = \sqrt{\frac{57.68}{20} + \frac{57.68}{20}}$$

$$= \sqrt{2.88 + 2.88} = \sqrt{5.76} = 2.40$$

We can now compute the test statistic t for evaluating H_0 that $\mu_R = \mu_{NR}$.

$$t_{\bar{X}_1 - \bar{X}_2} = \frac{(\bar{X}_1 - \bar{X}_2) - (\mu_1 - \mu_2)}{s_{\bar{X}_1 - \bar{X}_2}}$$

$$= \frac{(77.8\%) - (84.6\%) - 0}{2.40\%} = \frac{-6.80\%}{2.40\%} = -2.83$$

This obtained value of t must be compared with the critical value for the t distribution having $(n_1 - 1) + (n_2 - 1) = 38\ df$ for a 2-tail test with $\alpha = .01$. **Table B** in the Appendix gives the value as 2.712. Because our obtained value of t exceeds this critical value, we know that the difference between sample means of -6.8% falls among the most unlikely 1% of outcomes that would occur if H_0 were true. We will therefore reject H_0 and conclude that $\mu_R \neq \mu_{NR}$. Based on this study, it seems that it is better to write more new essays than to spend the time rewriting old ones. **Figure 3** shows graphically the results of the H_0 test.

Figure 3

Sampling Distribution of t Showing the Result of the H_0 Test that $\mu_R \neq \mu_{NR}$.

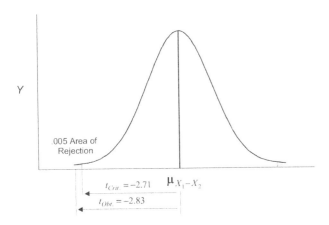

Computational Procedures for Making Confidence Interval Estimates

What do we estimate the difference between the population means to be? The point estimate of the difference between the population means is the difference between the sample means, -6.8%. A confidence interval estimate of the differences is obtained using the following general formula:

$$CI = (\bar{X}_1 - \bar{X}_2) \pm (t_{1-CI})(s_{\bar{X}_1 - \bar{X}_2})$$

Substituting in the problem values and looking up t_{1-CI} in **Table B** in the Appendix, we'll compute the 90% *CI* for the estimated difference:

$$90\% \; CI = (77.8\% - 84.6\%) \pm 1.69(2.40\%)$$

$$= -6.80\% \pm 4.06\%$$

$$= -10.86\% \; to -2.74\%$$

JASP Analysis of the Data

In case you hadn't noticed, carrying out an independent *t* test by hand is a lot of work. Happily, you will never need to do so again. We can get the same results for the "Rewrite It Again, Sam" study much more easily by using the *JASP* T-TESTS, Independent-Samples T-Test.

For this procedure and for most of the statistical tests covered in the next chapters, *JASP* assumes that the data have been entered into the data set in a particular manner. Each case needs to include the condition of the experiment he or she was in (R or NR) and the value of the dependent variable (% Grade). This allows the program to use the condition variable to split the cases into two groups for the purposes of the *t* test. **Figure 4** shows what the data set would look like in the *Excel* Spreadsheet.

Figure 4

Excel Data Editor Showing the Format of the Data for the Independent T Test.

	A	B
	A1	
	A	B
16	Rewrite	79
17	Rewrite	69
18	Rewrite	85
19	Rewrite	78
20	Rewrite	82
21	Rewrite	76
22	No-Rewrit	81
23	No-Rewrit	72
24	No-Rewrit	89

Figure 5 shows the data analysis table produced by this procedure. The value of *t* of 2.83 for the Independent Samples T-Test is the same value that we obtained by hand with a lot of work. The *p* value of .007 makes it clear that the difference of 6.8% between the Rewrite and No-Rewrite conditions is very unlikely to have occurred by chance. *JASP* subtracted the smaller mean from the larger mean yielding a positive value of *t*. This produces positive values for the 90% confidence interval which also are the same as those we obtained by hand. The Cohen's d statistic will be discussed below.

Below the *t*-test table in **Figure 5** are the results of two other tests which were selected from the Assumption Checks option in the procedure. The first is the **Shapiro-Wilk** test of the assumption of normality of the populations represented by the samples. Basically, what the test does is to compare each sample distribution to a fitted normal curve and measure the magnitude of the departures. The H_0 tested is that the fit is perfect and the *p* value is the probability that any departures are due to chance. A significant *p* value (e.g., < .05) means that the departure is large and improbable. For both samples, *p* is nowhere near significant. Thus, for our data the assumption of normality is extremely likely to be met.

The second test is **Levene's test** of the assumption of homogeneity of variance for the two populations. The test statistic is the *F* ratio, which we will consider in the next chapter. At the moment, notice that the *p* value for the test

Figure 5

JASP Independent T Test of the Rewrite Experiment and Shapiro-Wilk test of Normality and Levene's Test of the Equality of Variances Assumption.

Independent Samples T-Test

	t	df	p	Cohen's d	SE Cohen's d
GRADE	-2.832	38	0.007	-0.896	0.346

Note. Student's t-test.

Test of Normality (Shapiro-Wilk)

		W	p
GRADE	Rewrite	0.977	0.885
	No Rewrite	0.968	0.720

Significant results suggest a deviation from normality.

Test of Equality of Variances (Levene's)

	F	df$_1$	df$_2$	p
GRADE	0.029	1	38	0.866

is .866, which indicates that you cannot reject the hypothesis of *equal* variances, i.e., the homogeneity of variance assumption appears to be met.

Effect Size Statistics—Cohen's *d*

The last entry in the T-Test table in **Figure 5** is a statistic called **Cohen's *d*** used to quantify the size or importance of the significant difference between the two sample means. It was added to the T-Test table by selecting the Effect size option where Cohen's is the default. Cohen's d is defined as:

$$d = \frac{\bar{X}_1 - \bar{X}_2}{s_P}$$

Where: s_P = the weighted average of the two sample standard deviations.

For the rewrite experiment, d would equal:

$$d = \frac{\bar{X}_1 - \bar{X}_2}{s_P} = \frac{-6.800\%}{7.759\%} \approx -.896$$

Notice that the units of measurement (percentage grade) cancel out, leaving a unitless statistic. The value of -.896 is considered a very large effect. Frequently the decimal fraction is converted into a percentage as in, "the difference between the means was almost 90% of the size of the average standard deviation."

Effect Size Statistics—r_{pb}^2

Another effect size statistic is the point biserial correlation coefficient squared or r_{pb}^2 which was introduced in Chapter 5. Rather than using the formula for Pearson *r* to compute it, it can be computed more directly from the obtained value of *t* using the following formula:

$$r_{pb}^2 = \frac{t^2}{t^2 + n + n_2 - 2} = \frac{t^2}{t^2 + df}$$

In the present example, r_{pb}^2 is computed as:

$$\frac{-2.83^2}{-2.83^2 + 38} = \frac{8.01}{8.01 + 38} = \frac{8.01}{46.01} = .174$$

This statistic, r_{pb}^2, is the proportion of the variability in the dependent variable accounted for by the independent variable. For descriptive purposes, the proportion is usually converted to a percentage. Expressed in percentage terms, the rewrite vs. no-rewrite variable accounts for about 17% of the variability in final grades—a modest, but not inconsequential effect. At this time, *JASP* does not compute r_{pb}^2 for the independent *t* test procedure, so you will need to do it by hand.

Independent *t* Test Graphs

We'll consider two graphs that are commonly used to display *t* test results. The first, **Figure 6**, is what *JASP* calls a **Bar Plot**, but which we have called a "bar-like function graph" to distinguish it from a bar graph. The 90% *CI* brackets are used to quantify how different the two sample means are.

Figure 6

Mean Percentage Grade for the Two Experimental Conditions. Brackets Indicate the 90% Confidence Interval for Each Mean.

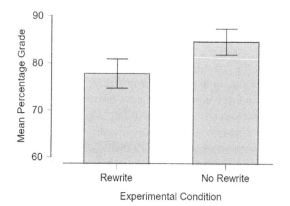

The second, **Figure 7**, shows a **Raincloud Plot** of the data for the two experimental conditions. The raincloud plot produced by *JASP* conveys a huge amount of information by combining the raincloud portion with a box plot. The cloud portion is a smoothed representation of the histogram distribution of scores for each condition and the dots (the "rain") are the values of the individual scores. Recall from Chapter 2 that the box plot at the bottom of each cloud conveys the following information: (a) the width of the box is the IQR, i.e., C_{25} to C_{75}, (b) the line in the middle of the box is the value of the median, i.e., C_{50}, and (c) the ticks to the left and right of the box are the Minimum and Maximum scores. Needless to say, raincloud plots are cool.

Figure 7

Raincloud Plot of the Sample Data Using a Horizontal Orientation.

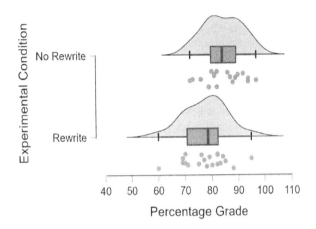

It might be useful at this point to summarize all that we have learned about the "Rewrite It Again, Sam" experiment. Here is a sample report of the descriptive and inferential analysis of the data:

The 20 students in the Rewrite Condition of the experiment earned a mean score of 77.8% on the final paper, while the 20 students in the No Rewrite Condition had a mean score of 84.6%. Both groups showed relatively little variation in their scores (SD's = 8.1% and 7.1%, respectively), and the distributions for both groups were unimodal and nearly symmetrical. To decide whether the difference of 6.8% between the two sample means was statistically significant, an independent *t* test between means was carried out with α set at .01. The *t* test was statistically significant, *t* (38) = 2.83, *p* = .007, indicating that students in the No Rewrite Condition, on average, did a better job on the final paper than those in the Rewrite Condition. The mean difference of 6.8% was associated with a point biserial squared (r_{pb}^2) value of .174, meaning that about 17% of the variability in scores on the final paper could be attributed to the experimental variable. The 90% confidence interval estimate of the size of the difference produced by the two experimental conditions was -2.7% to -10.9%.

Inference with Pairs of Correlated Means

Two sample means are correlated when the researcher has structured the sample in such a way that the observations in one sample are positively correlated with the observations in the second sample. The reason for using correlated samples is that the resulting *t* test is usually **more powerful than an independent *t* test**, i.e., it makes it more likely that one will detect real differences in population means. There are three general ways to produce correlated samples.

1. Select a random sample of cases from some population and measure the cases under two different conditions of a study. Studies of this sort are called "test-retest" or "repeated-measures" studies.

2. Select a random sample of cases from some population and match pairs of cases on their scores on one or more variables that correlate substantially with the dependent variable. Then randomly assign one member of each pair to one condition and the other member to the other condition. This is sometimes called a "matched pairs" type of study. For example, if you were studying the efficacy of two different methods of teaching elementary physics, it would be desirable to create correlated samples by matching pairs of students on their SAT math scores.

3. Select a random sample of naturally occurring pairs of cases that are similar with respect to variables that correlate with the dependent variable (e.g., identical twins), and randomly assign one member of each pair to the two conditions.

The effect of each of these procedures is to "build in" a pair-wise correlation between the **scores for the dependent variable** for the two samples. These samples are then assumed to represent two **correlated populations**. Two populations are correlated when each element in one population is linked to a corresponding element in the other population, and the two elements tend to be similar in value. A test of the significance of the difference between two correlated sample means is called a **correlated *t* test**. We will develop the procedures for a correlated *t* test using the example titled "Living with Pigs" **Read the description of the study before proceeding.**

Living with Pigs

Research suggests that "date rape" is a sadly common phenomenon on college campuses. The term is used to describe situations in which the victim is coerced by a friend or acquaintance to engage in unwanted sex during some type of social activity. The victim is almost always female and alcohol abuse by both parties is a common element in many incidents. Women have long decried the tendency for victims, perpetrators, and authorities to view date rape as an "unfortunate event" or a "misunderstanding," rather than a crime. To see whether *reported* incidence of date rape is changing on college campuses, two sociology students obtained incident reports for sexual assaults classified as date rapes for 12 colleges and universities in their region for the years 2018 and 2022. The institutions ranged from small, private schools to a large urban university. To adjust for differences in enrollment, the data were converted to number of reported assaults per 100 students, per academic year.

Institution	2018	2022	Institution	2018	2022
A	4.41	3.44	G	2.09	4.06
B	1.42	2.93	H	1.22	2.23
C	2.17	4.65	I	2.04	.68
D	4.46	7.64	J	1.86	.99
E	3.00	4.47	K	1.52	2.45
F	1.42	2.60	L	3.18	4.44

The research question is whether there is a significant difference between the reported mean rate of date rape in 2018 and the mean rate in 2022. The populations of interest presumably are the incident rates for sexual assault for all American colleges and universities for 2018 and for 2022. Whether these samples are representative of the populations is open to question. Notice that the dependency between the samples (and the populations) has been 'built in' by virtue of measuring the same cases (colleges and universities) under two different conditions—in this case, different times.

Because the investigators wanted to be able to detect a significant difference between the sample means in either direction, they employed a nondirectional H_0 and H_A

$$H_0: \mu_{2018} = \mu_{2022}$$
$$H_A: \mu_{2018} \neq \mu_{2022}$$

Alpha was set at .05 to maximize the power of the statistical procedure.

Assumptions of the Procedure

The H_0 of no difference between the correlated population means may be tested using the t distribution if the following assumptions are met:

1. **The data are numerical**. The data from the date-rape study clearly meet this assumption.

2. **The samples are random samples from the two correlated populations of interest**. The schools were not selected on a random basis and their representativeness is unknown. As a result, even if significant differences are found, it will not be clear to which colleges and universities the results apply.

3. **The two populations of interest are normally distributed**. This assumption is almost certainly not met. The 2018 sample data are highly positively skewed and it is likely that the same thing is true of the populations. With small sample sizes, severe violations of this assumption result in increases in both Type I and Type II Errors. Were we doing a study of this type, we might want to consider the use of a "nonparametric" comparison procedure as an alternative to the correlated t test that doesn't have such restrictive assumptions. **JASP provides such an alternative called Wilcoxon Matched-Pairs Signed-Ranks test** which is described below.

Computational Procedures for Testing H_0

Figure 8 shows the *JASP* Descriptives statistics for the two samples. The mean number of assaults per 100 increases modestly across the 4-year period. For both years, the standard deviations are very large compared with the means, and the 2018 data show marked positive skewing.

The sampling distribution of the difference between correlated means is similar in two respects to the sampling distribution of differences between independent means. First, the distribution will be exactly normal in shape if the two populations are normally distributed, or it will tend

Figure 8

JASP Descriptive Statistics for the Living with Pigs Study.

Descriptive Statistics

	2018 Data	2022 Data
Valid	12	12
Median	2.065	3.185
Mean	2.399	3.382
Std. Error of Mean	0.325	0.540
Std. Deviation	1.125	1.872
Minimum	1.220	0.680
Maximum	4.460	7.640

toward normality with increasing sample size if they are not. Second, the mean of the sampling distribution of differences between means will equal the difference between the two population means, i.e.,

$$\mu_{\bar{X}_1-\bar{X}_2} = \mu_1 - \mu_2$$

Under the H_0 tested that $\mu_{2018} = \mu_{2022}$, the mean of the sampling distribution, $\mu_{\bar{X}_1-\bar{X}_2}$, will equal 0, i.e.,

$$\mu_{\bar{X}_1-\bar{X}_2} = 0$$

The difference between the sampling distribution for independent and correlated means lies in how the standard error of the sampling distribution is computed. Recall that for independent means, $\sigma_{\bar{X}_1-\bar{X}_2}$ was given as:

$$\sigma_{\bar{X}_1-\bar{X}_2} = \sqrt{\frac{\sigma_1^2}{n_1} + \frac{\sigma_2^2}{n_2}}$$

For correlated means, the $\sigma_{\bar{X}_1-\bar{X}_2}$ of the sampling distribution is given by the following formula:

$$\sigma_{\bar{X}_1-\bar{X}_2} = \sqrt{\frac{\sigma_1^2}{n_1} + \frac{\sigma_2^2}{n_2} - 2\rho\sigma_{\bar{X}_1}\sigma_{\bar{X}_2}}$$

The new part of the equation, $-2\rho\sigma_{\bar{X}_1}\sigma_{\bar{X}_2}$, reduces the size of $\sigma_{\bar{X}_1-\bar{X}_2}$ by an amount that is proportional to the correlation (ρ – rho) that exists between the cases in the two populations. This makes sense because when samples are drawn from correlated populations, the cases comprising the samples tend to be similar in value, thus making the two sample means more alike in value. Because the sample means tend to be closer in value, the differences between them will be smaller, i.e., will show less variability.

As usual, the problem with the formula for $\sigma_{\bar{X}_1-\bar{X}_2}$ is that the population variances are unknown and must be estimated from sample data. A similar situation exists with the parameter ρ. It is also unknown and must be estimated from the sample value of r. The appropriate estimation formula thus becomes:

$$s_{\bar{X}_1-\bar{X}_2} = \sqrt{\frac{\hat{s}_1^2}{n_1} + \frac{\hat{s}_2^2}{n_2} - 2rs_{\bar{X}_1}s_{\bar{X}_2}}$$

If the assumptions of the procedure are met, then the quotient

$$\frac{(\bar{X}_1 - \bar{X}_2) - (\mu_1 - \mu_2)}{s_{\bar{X}_1-\bar{X}_2}}$$

will follow the t distribution. The df for the t distribution for evaluating correlated means differs from that for independent means. Recall that for independent means, $df = (n_1 - 1) + (n_2 - 1)$. For correlated means, however, the $df = (n_{Pairs} - 1)$. The reason for this is that in correlated samples, the estimate of the population variance provided by the first sample is not independent of the estimate generated by the second sample (the samples are correlated, remember). Recall the definition of df—the number of elements that *independently* contribute to estimating a population parameter. Only $(n_{Pairs} - 1)$ of the total set of observations are independent in estimating $\sigma_{\bar{X}_1-\bar{X}_2}$.

Applying these procedures to our sample data, we first calculate the difference between the sample means. Because H_0 is nondirectional, the direction of subtraction is arbitrary; we'll subtract the mean for 2018 from the mean for 2022:

$$\bar{X}_{2022} - \bar{X}_{2018} = 3.38 - 2.40 = +.98$$

The estimated standard deviation of the sampling distribution of differences between means is equal to:

$$s_{\bar{X}_1-\bar{X}_2} = \sqrt{\frac{\hat{s}_1^2}{n_1} + \frac{\hat{s}_2^2}{n_2} - 2rs_{\bar{X}_1}s_{\bar{X}_2}}$$

$$= \sqrt{\frac{1.13^2}{12} + \frac{1.87^2}{12} - 2(.67)\left(\frac{1.13}{\sqrt{12}}\right)\left(\frac{1.87}{\sqrt{12}}\right)}$$

$$= \sqrt{\frac{1.28}{12} + \frac{3.50}{12} - (1.34)(.33)(.54)}$$

$$= \sqrt{.11 + .29 - .24}$$

$$= \sqrt{.16} = .40$$

The value of the test statistic t is then computed as:

$$t_{\bar{X}_1 - \bar{X}_2} = \frac{(\bar{X}_1 - \bar{X}_2) - (\mu_1 - \mu_2)}{s_{\bar{X}_1 - \bar{X}_2}}$$

$$= \frac{.98 - 0}{.40} = 2.45$$

The t distribution appropriate to the test is the one for $df = (n_{Pairs} - 1) = 12 - 1 = 11$. According to **Table B** in the Appendix, the critical value of t for this distribution for a 2-tail test with $\alpha = .05$ is 2.201. Because our obtained t exceeds this value, we may reject H_0 that the two population means are equal.

Seemingly, there has been an increase in the reported incidence of date rape on the college and university campuses between 2018 and 2022. This conclusion must be tempered by the fact that the representativeness of the samples is unknown, and the normality assumption probably violated. A consequence of the latter is that the probability of a Type I Error in the present situation could be greater than the .05 α level adopted for the comparison.

Computational Procedures for Making Confidence Interval Estimates

A confidence-interval estimate of the size of the difference between the two population means uses the same formula as that for independent samples. For our sample data, the 90% *CI* would be calculated as follows:

$$90\% \ CI = (\bar{X}_2 - \bar{X}_1) \pm (t_{1-CI})(s_{\bar{X}_1 - \bar{X}_2})$$

$$= (3.38 - 2.40) \pm 1.80(.40)$$

$$= .98 \pm .72$$

$$= .26 \text{ to } 1.70$$

JASP Analysis of the Data

With a lot less effort, we can analyze these data using *JASP* T-TESTS, Paired Samples T-Test. **Figure 9** is a screen capture of the *JASP* Data Editor showing how the data need to be entered to carry out the correlated t test.

Figure 9

JASP Data Editor Showing the Format of the Data for the Correlated Pairs T Test.

	Institution	2018 Data	2022 Data
1	A	4.41	3.44
2	B	1.42	2.93
3	C	2.17	4.65
4	D	4.46	7.64
5	E	3	4.47

Figure 10 on the next page shows the table giving the results of the H_0 hypothesis test of no difference between the population means. Notice first that the correlated samples t-test value of -2.44 is significant at $p = .033$, which is more precise than that obtained using hand calculations.

The lower table in **Figure 10** contains the Shapiro-Wilk test of normality, which is actually the test that the **difference scores between pairs** are normally distributed. The test did not reject the hypothesis of normality of the difference scores, but the p value of .277 suggests that we are headed in that direction. For that reason, most investigators would add the appropriate non-parametric test of the difference between means to verify the t-test result.

For correlated samples the appropriate non-parametric test is called the **Wilcoxon Matched-Pairs Signed-Ranks** test. This is an exceedingly handy test that makes no assumptions about the shape of the distributions involved. Instead, one first finds the signed differences between the paired scores, i.e., subtract each 2018 score from the 2022 score, maintaining the sign of the difference. The signed differences are then rank ordered by absolute (unsigned) size. The positive and negative ranks then are added up and the smaller of the two sums is called T. Under the H_0 of no difference in the scores for the two populations of interest, the sampling distribution of T will be approximately normally distributed with:

$$\mu_T = \frac{n(n+1)}{4}$$

and

$$\sigma_T = \sqrt{\frac{n(n+1)(2n+1)}{24}}$$

Where: n = sample size.

Figure 10

JASP Paired Samples T Test of the Means for the Living with Pigs Study and the Shapiro-Wilk Test for Normality. Also Included is the Wilcoxon Signed-Ranks Non-Parametric Test.

Paired Samples T-Test

Measure 1	Measure 2	Test	Statistic	z	df	p	Effect Size	SE Effect Size
2018 Data	- 2022 Data	Student	-2.439		11	0.033	-0.704	0.262
		Wilcoxon	11.000	-2.197		0.027	-0.718	0.316

Note. For the Student t-test, effect size is given by Cohen's *d*. For the Wilcoxon test, effect size is given by the matched rank biserial correlation.

Test of Normality (Shapiro-Wilk)

		W	p
2018 Data	- 2022 Data	0.919	0.277

The results of the Wilcoxon test are shown in the second line of the T-Test table. The value of *T* for the sample data is 11 and the associated *z* score is -2.197. The *p* value is .027, which is even smaller than that obtained with the *t* test. Clearly, our significant difference in rape incidence between 2018 and 2022 is not an artifact of assumption violations.

The last bit of information in the table that we'll consider is the Effect Size statistic for the *t* test. It is Cohen's *d* computed as the ratio of the mean of the **difference scores** (-.983) divided by the standard deviation of the difference scores (1.395). Again, it is a fairly large value, indicating that the mean difference is approximately 70% of the size of the standard deviation of differences.

Pearson r_{pb}^2 also can be computed as an effect size statistic for these data using the following formula:

$$r_{pb}^2 = \frac{t^2}{t^2 + (n_{Pairs} - 1)}$$

$$= \frac{2.44^2}{2.44^2 + 11} = \frac{5.95}{16.95} = .35$$

Thus, year of reporting accounts for about 35% of the variability in incidences of date rape.

Figures 11 and **12** are the two *JASP* graphs showing how individual colleges and universities changed from 2018 to 2022.

Here is how one might report the results of the analysis of the "Living with Pigs" study in narrative form:

The data are the number of sexual assaults reported per 100 students for the same 12

Figure 11

Reported Mean Number of Rapes per 100 for Years 2018 and 2022. Brackets Indicate the 90% Confidence Interval for Each Mean

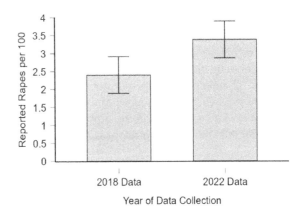

Figure 12

Raincloud Plot of Reported Rapes per 100 for Years 2018 and 2022.

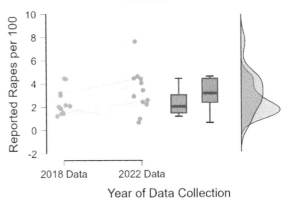

colleges and universities for the academic years 2018 and 2022. The number of assaults per 100 students increased by .98 across the four-year period. A correlated t test with α set at .05 revealed that the increase in mean assault rates over the three-year period was statistically significant, $t(11) = 2.43$, $p = .033$. A parallel analysis using the non-parametric Wilcoxon Matched-Pairs Signed-Ranks test yielded comparable results. The 90% CI estimate of the size of the increase in assault rates between the two periods was .26% to 1.70%. The r_{Pb}^2 value of .35 indicates that the four-year period accounts for 35 % of the variability in assault rates.

Inference with Pairs of Sample Proportions

Recall from Chapter 10 that the sample proportion, P, is the proportion of cases in the sample falling into one of two dichotomous categories. Also recall that the proportion for a population is abbreviated p (not ρ, which is rho). Now, just as you can test the significance of the difference between two sample means, you can also test whether two sample proportions (P_1 and P_2) differ significantly. The procedure described below is called the **large sample test of the difference between independent proportions**. It is appropriate for testing *only* the specific H_0 that two independent population proportions (p_1 and p_2) are equal. We will also describe a procedure for estimating the size of the difference between two population proportions. **We will develop these procedures in the context of the example below titled, "It's the Pits." Read it carefully before proceeding.**

Assuming that our samples of male and female students could be considered representative of all male and female students at the institution, we'll use the large sample test of proportions to test the nondirectional H_0 that the proportion of males at the college using scheduling calendars was equal to the proportion of females who used scheduling calendars, i.e.,

$$H_{0:}\ p_M = p_F$$
$$H_{A:}\ p_M \neq p_F$$

Assumptions of the Procedure

The assumptions of the large sample test of the difference between independent proportions are as follows.

1. **The data are categorical in nature and can assume only two possible values**. The present data meet this assumption.

2. **The samples are independently and randomly selected from the populations of interest**. The two samples are independent of one another but were not randomly selected. Possibly, however, the samples are substantially representative of the student body of Carol's college.

3. **The sample is sufficiently large that the normal curve will adequately approximate the discrete sampling distribution followed by the difference between two sample proportions**. There is a rule of thumb discussed below for deciding whether this assumption is met. The data meet this assumption.

Computational Procedures for Testing H_0

The large sample procedure described here tests only the H_0 that the two population proportions are equal. As a result, H_0 does not specify what the values of p and q are.

It's the Pits

Recall grad student Carol from the practical intelligence study described in "The Right Stuff" problem for correlation and linear regression. The subtest of the "Practical Intelligence Test" (PIT) that measured Intrapersonal Intelligence contained the following item:

True False   I use a scheduling calendar to keep track of assignments and appointments.

While analyzing the test items, Carol noticed that female students seemed much more likely to answer "True" to the question than the male students. Here are the data for the entire sample of students to whom she administered the PIT:

	Males	Females
Proportion answering "True"	.072	.168
Number of Cases	157	175

Because these values are needed to compute the standard deviation of the sampling distribution of differences between proportions (the standard error of the sampling distribution), we must estimate p and q from the sample data. On the tentative assumption that H_0 is true, the best estimate of p and q is obtained by pooling the sample data together to yield what is called p_{est} and q_{est}. The formulas for doing so are as follows:

$$p_{est} = \frac{n_1 P_1 + n_2 P_2}{n_1 + n_2}$$

and,

$$q_{est} = 1 - p_{est}$$

Using this "pooled estimate" of the assumed values of p and q for the two populations, the **estimated standard error of the sampling distribution of differences between proportions**, $s_{P_1-P_2}$, is computed as follows:

$$s_{P_1-P_2} = \sqrt{\frac{p_{est}\, q_{est}}{n_1} + \frac{p_{est}\, q_{est}}{n_2}}$$

If $n \times P$ and $n \times Q$ are larger than 10.0 for both samples, the sampling distribution of **differences between sample P's** will be adequately approximated by a normal distribution and the test statistic z can be used to evaluate the probability of obtaining the sample value of P_1-P_2 by chance, if H_0 were true. For the data for our example problem, for the male sample, $n \times P = 11.3$ and $n \times Q = 145.7$; for the female sample, $n \times P = 29.4$ and $n \times Q = 145.6$. In all cases, the quotient is larger than 10.0, so we are justified in using the large sample procedure to test H_0.

The formula for the test is:

$$z_{P_1-P_2} = \frac{(P_1 - P_2) - 0}{s_{P_1-P_2}}$$

Because H_0 is nondirectional, we will arbitrarily subtract P_{Male} from P_{Fem}, yielding:

$$P_{Fem} - P_{Male} = .168 - .072 = .096$$

The manual computation of $s_{P_1-P_2}$ is a bit laborious. We'll first find p_{est} and q_{est}.

$$p_{est} = \frac{n_1 P_1 + n_2 P_2}{n_1 + n_2}$$

$$= \frac{(157)(.072) + (175)(.168)}{157 + 175}$$

$$= \frac{11.304 + 29.400}{332} = \frac{40.704}{332} = .123$$

and,

$$q_{est} = 1 - p_{est} = 1 - .123 = .877$$

The next step is to use the estimated values of p and q to compute the estimated standard error of the difference in proportions, $s_{P_1-P_2}$:

$$s_{P_1-P_2} = \sqrt{\frac{p_{est}\, q_{est}}{n_1} + \frac{p_{est}\, q_{est}}{n_2}}$$

$$= \sqrt{\frac{(.123)(.877)}{157} + \frac{(.123).887)}{175}}$$

$$= \sqrt{\frac{.108}{157} + \frac{.108}{175}} = \sqrt{.00069 + .00062}$$

$$= \sqrt{.00069 + .00062} = \sqrt{.00131} = .036$$

The test statistic z is then computed as:

$$z_{P_1-P_2} = \frac{(P_1 - P_2)}{s_{P_1-P_2}}$$

$$= \frac{(.168 - .072)}{.036} = \frac{.096}{.036} = 2.67$$

For the unit normal curve, the critical value of z needed to reject H_0 for a 2-tail test with $\alpha = .05$ is ± 1.96. The obtained value of z exceeds the critical value, so we can reject H_0 that the two population proportions are equal. A higher proportion of the female students use scheduling calendars than do the male students.

Computational Procedures for Making Confidence Interval Estimates

The confidence interval estimate of p_1-p_2 is computed using a formula which should now be quite familiar:

$$CI = (P_1 - P_2) \pm (z_{1-CI})(s_{P_1-P_2})$$

The only alteration in the procedure is that when estimating the difference between two population proportions, a somewhat different formula is used to compute $s_{P_1-P_2}$, i.e.,

$$s_{P_1-P_2} = \sqrt{\frac{P_1 Q_1}{n_1} + \frac{P_2 Q_2}{n_2}}$$

For our example, $s_{P_1-P_2}$ is computed as:

$$= \sqrt{\frac{(.072)(.928)}{157} + \frac{(.168)(.823)}{175}}$$

$$= \sqrt{.000426 + .000799}$$

$$= \sqrt{.00122} \approx .035$$

The 90% *CI* for our estimate of the difference between the population proportions is:

$$90\%\ CI = .096 \pm (1.65)(.035)$$

$$= .096 \pm .058$$

$$= .038\ to\ .154$$

We are 90% confident that the difference in the population proportions of male and female students who use personal calendars is between .038 and .154, i.e., between about 4% and 15%.

Analysis of the Data Using INTERACTIVE STATISTICS

It's obvious that analyzing differences between sample proportions and finding confidence intervals by hand is way too much work to be fun. Again, in its current version *JASP* doesn't provide a simple way to carry out such analyses, so we'll fall back on **Interactive Statistics** to do the work. Remember, its location is Libretexts.org → Platforms → Libraries → Statistics → Learning Objects → Interactive Statistics.

The procedure for two-sample proportion analysis is Module 33, *Hypothesis Test and Confidence Interval Calculator- Difference Between Population Proportions*.

Figure 13 is a screen shot of the data entry window and results. Notice that raw numbers rather than proportions are entered. The "Pooled" option has to be selected to correspond to the calculational formulas we used and the .90 *CI* entered. The program computes the proportions from the raw data. The resulting values of $z_{P_1-P_2}$ and the 90% *CI* are the same as those computed by hand (the subtraction of sample *P*s was done in reverse of that which we did by hand).

Final Comment

As was the case with single-sample inference, the procedures introduced in this chapter are only a portion of the many tests available for testing whether two samples differ in the value of some statistic. But, if you really understand the logic of two-sample inference developed in this chapter (as opposed to memorizing steps of a technique), you will be able to learn any of the other procedures on your own, as they all follow the same logical process.

Figure 13

Screen Shot of Module 33 From Interactive Statistics with Data from It's the Pits Entered and Analyzed.

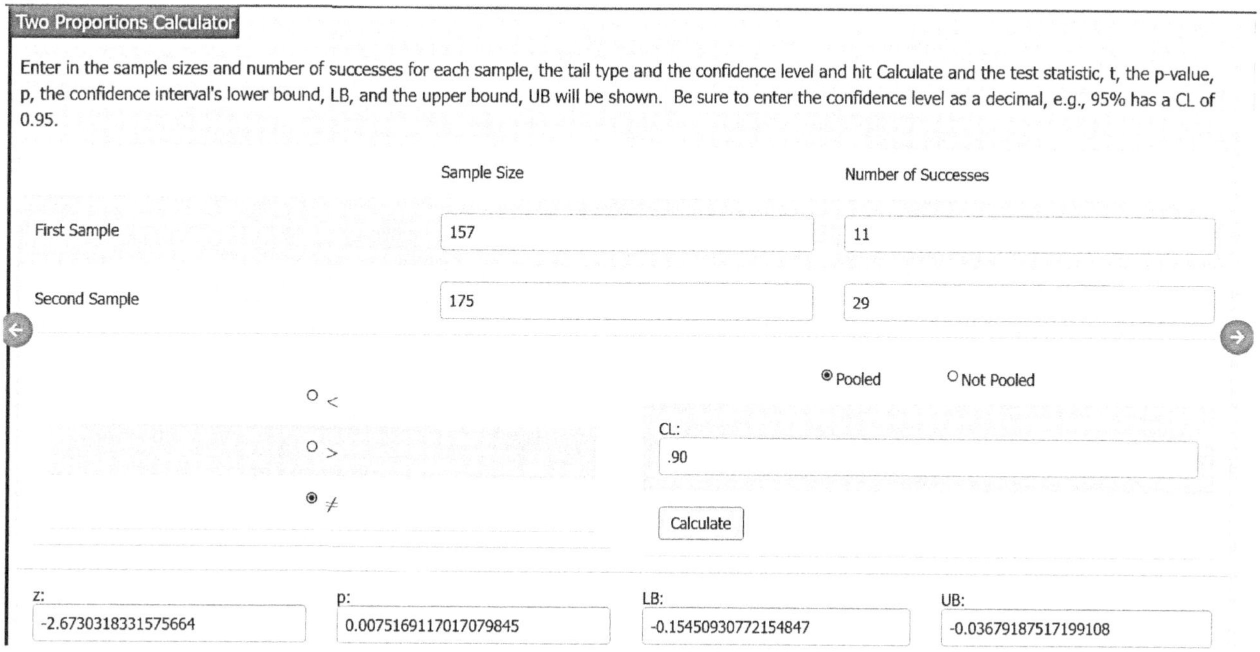

Two Proportions Calculator

Enter in the sample sizes and number of successes for each sample, the tail type and the confidence level and hit Calculate and the test statistic, t, the p-value, p, the confidence interval's lower bound, LB, and the upper bound, UB will be shown. Be sure to enter the confidence level as a decimal, e.g., 95% has a CL of 0.95.

	Sample Size	Number of Successes
First Sample	157	11
Second Sample	175	29

○ <
○ >
● ≠

● Pooled   ○ Not Pooled

CL: .90

[Calculate]

z: -2.6730318331575664
p: 0.0075169117017079845
LB: -0.15450930772154847
UB: -0.03679187517199108

"If someone is able to show me that what I think or do is not right, I will happily change, for I seek the truth, by which no one was ever truly harmed. It is the person who continues in his self-deception and ignorance who is harmed."

Marcus Aurelius

Chapter 11, Two-Sample Inference
Chapter Exercises

Some questions are purely factual and have only one correct answer. Others require greater thought on your part and may have more than one possible answer.

1. Fill in the missing term(s) that makes the statement true.

    a. In most two-sample situations, the H_0 tested is that _____.
    b. The two samples usually compared represent _____.
    c. The difference between two sample means is itself a _____ and has a _____.
    d. Another name for independent samples is _____ samples.
    e. The Rewrite study is technically an _____.
    f. Requiring that two populations have the same variance is called the _____ assumption.
    g. $\mu_{\bar{X}_2-\bar{X}_2}$ is the _____ of the sampling distribution of the _____.
    h. As was the case with single samples of the mean, we almost never know _____.
    i. s_P^2 is called the _____ estimate of σ^2
    j. In a 2-sample independent t test, the df = _____ + _____.
    k. The measures of "effect size" in a 2-sample t test are _____ and _____.
    l. Levene's test is used to assess the assumption of _____.
    m. Correlated t tests are almost always more _____ than independent groups tests.
    n. Testing the same cases under two different conditions is called a _____ or _____ study.
    o. In psychology, the most common form of "naturally occurring" correlated pairs is _____.
    p. As ρ (rho) approaches + 1.00, $\sigma_{\bar{X}_2-\bar{X}_2}$ approaches _____.
    q. The large sample test of two sample proportions only tests H_0 that _____.

2. Recall Professor Zandar from the data analysis problems for Chapter 4. She had the students in her *Introduction to Woman's Studies* class complete an anonymous questionnaire regarding a variety of issues of particular importance to young adults. One question asked, "Assuming you have the financial means, how many children would you like to have?" Below are the responses of the 32 women and the 12 men in the class. At the time all you had to do was produce a Descriptive Statistics table of the data. Now, we'll conduct an independent t test to determine whether the men and women differed significantly in the number of children desired.

Women					Men				
5	3	0	2	2	3	0	1	2	3
0	1	0	0	1	2	3	4	3	2
4	6	0	1	3	3	2			
2	1	2	3	0					
0	4	2	4	0					
2	2	5	2	3					
0	1								

    What's your conclusion?

3. Does implied "social status" affect helping behavior? A social psychologist conducts the following study in a large mall. A 25-year-old female research assistant holding a stack of 20 paperback books stations herself near an elevator door. When randomly selected, solitary male shoppers approach within five feet of the elevator door, the research assistant appears to accidentally drop her books. She says, "Oh gosh!" and starts to bend down and slowly pick them up. Another research assistant concealed some distance away records the number of books that the shopper helps to retrieve. On some randomly selected days, the female research assistant is dressed in faded blue jeans and an old shirt (low status condition); on other days, she is wearing an expensive tailored suit (high status condition). Her makeup and hairdo are the same for both conditions. Below are the data (number of books retrieved) by each shopper in the two conditions.

High Status Condition	Low Status Condition
19 17 18 16	19 0 0 0
17 0 11 14	10 6 17 0
13 0 13 0	17 0 0 13
15 0 12 17	0 14 0 17
14 13	0 16 0 0

Use *JASP* to do a complete analysis of the data, beginning with the statement of the Null and Alternative Hypotheses. Summarize your findings in a short narrative report. There's a clever wrinkle to the data.

4. The chief engineer in a large concrete products manufacturing plant has developed a formula for a new curing agent that she believes will substantially accelerate the rate at which pre-stressed concrete beams reach maximum strength. Concrete, as you may know, gradually becomes stronger after it "sets." The strengthening process is a negatively accelerated function, with large initial gains, followed by progressively smaller gains over a very long period of time. Even a modest speedup of the early part of the strengthening process would be commercially valuable. To test whether the new agent is really effective, she has 12 60' beams cast using the standard procedure, and 12 beams cast with the new curing agent added to the mix. All beams are allowed to cure for a 20-day period and then subjected to a stress test to determine the cracking point at the middle of each beam. The data given below are the imposed forces (pounds per square inch) required to crack each of the 24 beams.

Standard Beams		Beams w/ New Agent	
2617 lbs.	2370 lbs.	2738 lbs.	2866 lbs.
2143	2455	2464	2938
2840	2096	2852	2771
2936	2417	2860	2581
3184	2368	3387	2831
2580	2470	2517	2500

Present a univariate description of these data using the most appropriate statistics. Include an appropriate graph. Then use *JASP* to test the Null Hypothesis that the curing agent is ineffective.

5. As a class project, a student in experimental psychology compares the reaction times of 10 other students in the class to a visual signal and an auditory signal. Participants are instructed to press the space bar of a computer keyboard as quickly as possible when either signal is given. The time to press the key is recorded by the computer in milliseconds (msec). The test sequence consists of 20 trials, with 10 visual and 10 auditory signals presented in a random order. The dependent variable is the mean reaction time for the 10 trials in each of the two sensory modalities. Here are the data.

	Auditory	Visual		Auditory	Visual
Student 1	674.2	698.0	Student 6	604.6	614.0
Student 2	549.2	567.9	Student 7	565.5	578.3
Student 3	589.5	583.2	Student 8	394.2	392.1
Student 4	423.7	439.3	Student 9	546.8	571.0
Student 5	782.0	802.3	Student 10	579.1	590.3

Use *JASP* to do a complete analysis of the data, beginning with the statement of the Null and Alternative Hypotheses. Summarize your findings in a short narrative report.

6. A random sample of students at Southeastern Wesleyan College are asked toward whom they are leaning in the upcoming election for student body president. The finalist candidates are Rob Merkle, an outspoken advocate for LGBTIQA+ rights and Norma Sellers, President of the Young Republicans chapter on campus. Here are the findings broken down by respondent's gender.

Females	87 for Merkel	81 for Sellers	12 Undecided
Males	22 for Merkle	52 for Sellers	9 Undecided

Do a complete analysis of the data, beginning with the statement of the Null and Alternative Hypotheses. Summarize your findings in a short narrative report.

Chapter 12
One-Way ANOVA

What's In This Chapter?

The **one-way analysis of variance** (**ANOVA**) is a statistical technique for determining whether two or more sample means differ significantly, i.e., for testing H_0 that two or more population means are equal. The one-way ANOVA is used in preference to carrying out multiple *t* tests because the probability of making a Type I Error is not influenced by the number of groups involved in the analysis.

The one-way ANOVA employs the **F ratio** as the test statistic. The *F* ratio is defined as **the ratio of two independent, unbiased estimates of a population variance**. The precise shape of the sampling distribution of *F* is determined by the *df* in the numerator and denominator variance estimates. Sampling distributions of *F* are highly positively skewed for smaller *df*. For all combinations of *df*, however, the mean of the sampling distribution of *F* is always equal to 1.00.

In applying the *F* ratio to the one-way ANOVA, two variance estimates are derived from the data. The first, called the **within-groups variance estimate**, estimates just the population variance, σ^2, regardless of whether H_0 is true or false. The second independent estimate, called the **between-groups variance estimate**, is derived from the variability of the group means about the overall mean. If H_0 is true, this estimate is a second estimate of the population variance. But, if H_0 is false, the between-groups estimates will reflect not only the population variance, but also differences among the population means. The between-groups estimate is placed in the numerator of the *F* ratio. If the obtained value of *F* is unusually large, it is taken to mean the between-groups estimate is being influenced by differences in the population means, i.e., H_0 is false.

If the *F* ratio is significant, the statistics η^2 ("**eta squared**") and ω^2 ("**omega squared**") can be used to quantify the "size" of the effect of the independent variable in the same way as r^2 or r_{pb}^2.

Following rejection of H_0 by a significant *F* ratio, comparisons are made among the group means to determine which of the means are significantly different. One test, called Tukey's **Honestly Significant Difference** (**HSD**) test, is frequently used to make these comparisons. Rejection of H_0 does not imply that the differences among the group means are large or important—only that the differences are unlikely to have occurred by chance.

Function graphs are used to visualize the results of a significant one-way ANOVA. Continuous function graphs are used when the independent variable is ordered and continuous in nature. Categorical function graphs are used to represent the relationship between a dependent variable and a categorical independent variable.

Introduction

The analysis of variance (ANOVA, pronounced an-noh-vah) technique was invented in the early part of the 20th century by a very versatile mathematician, Sir Ronald Fisher. It is a highly general procedure that permits researchers to efficiently analyze the results of experiments having one or more independent variables and two or more conditions for an independent variable. In this chapter we will consider the simplest form of the technique, which is called a **one-way ANOVA**.

The one-way ANOVA is used to analyze the results of an experiment in which there is one independent variable, and there are two or more **levels** or **categories** of the independent variable. For example, suppose that you wanted to investigate whether taking large doses of Vitamin C reduced the frequency of colds in older adults. You might have four experimental conditions: (a) Control Condition in which the volunteers received no additional vitamin C beyond that available in their ordinary diet, (b) a condition in which the volunteers received a supplement of 500 mg of vitamin C per day, (c) a condition involving a 1000 mg supplement, and (d) a condition involving a 2000 mg supplement. The independent variable would be *Amount of Supplementary Vitamin C* and there would be four "levels" of the independent variable (0, 500, 1000, and 2000 mg conditions). Your dependent variable presumably would be the *Number of Colds* developed by each volunteer over some specified period.

The term "levels" implies that the independent variable is ordered along some dimension such as quantity or amount. Some independent variables, however, are represented by unordered categories. Suppose that you wanted to compare three different methods for teaching statistics: (a) the traditional textbook approach with a teacher (Textbook), (b) an on-line graphics and text approach (On-Line Text), and (c) an on-line lecture approach (Talking Head). The independent variable would be *Methods of Teaching Statistics*, which would be represented by three categorical conditions.

Regardless of whether the independent variable is of the level or category type, the researcher is faced with the task of deciding whether any of the two or more sample means representing the conditions of the experiment differ significantly. More technically, the researcher wants to test the H_0 that the samples representing the conditions of the experiment are drawn from populations having the same value of μ for the dependent variable, i.e.,

$$H_0: \mu_1 = \mu_2 = \mu_3 = +\ldots+\mu_K$$

where: μ_k = any number of population means.

For our vitamin C experiment, the H_0 would be stated:

$$H_0: \mu_{0mg} = \mu_{500mg} = \mu_{1000mg} = \mu_{2000mg}$$

The alternative hypothesis, H_A could be stated in several different ways in this situation. The most reasonable approach, however, is to state that at least two of the population μ's are not equal, i.e.,

$$H_A: \text{At least two } \mu\text{'s are not equal}$$

One might suppose that the simplest way to test the H_0 would be to carry out a series of comparisons between pairs of sample means using the *t* test for independent sample means. If any pair of means were found to differ significantly, the H_0 could be rejected, and the H_A accepted. In fact, that's the approach that researchers took prior to Fisher's invention of ANOVA. There is a serious problem with this approach, however. It dramatically increases your risk of making a Type I Error. Let's see why.

Imagine four populations of scores, each having the same mean and standard deviation, i.e., H_0 is true. We select one random sample from each population. We can designate these samples A, B, C, and D. We are going to compare every possible pair of sample means using the *t* test for independent means. The number of possible comparisons that we will have to make is given by the formula:

$$\text{\# Comparisons} = K(K-1)/2$$

where: K = Number of samples to be compared.

For four samples, the number of comparisons is

$$\text{\# Comparisons} = 4(4-1)/2 = 6$$

We can verify the formula by listing all the possible comparisons, i.e.,

A vs. B	B vs. C
A vs. C	B vs. D
A vs. D	C vs. D

If we conduct each comparison with α = .05, the probability of making a Type I Error on *any one comparison* is .05. But, because we are making *six* comparisons, the probability that *at least one* of the comparisons will result in a Type I Error goes up quite rapidly.

A simple formula for approximating this probability is:

$$p(\geq 1 \text{ Type } I \text{ Error}) = 1 - (1-\alpha)^C$$

where: C = Number of comparisons made.

For C = 6 comparisons, the probability of making at least one Type I Error, given that H_0 is always true is:

$$p(\geq 1 \text{ Type } I \text{ Error}) = 1 - (1-\alpha)^C$$
$$= 1 - (1-.05)^6$$
$$= 1 - (.95)^6$$
$$= 1 - .74$$
$$= .26$$

One can, of course, reduce the probability of making Type I Errors in this situation by using a more rigorous α level. If all six comparisons are carried out with α = .01, the probability of at least one spurious rejection is much lower, i.e.,

$$p(\geq 1 \text{ Type } I \text{ Error}) = 1 - (1-\alpha)^C$$
$$= 1 - (1-.01)^6$$
$$= 1 - (.99)^6$$
$$= 1 - .94$$
$$= .06$$

On the surface, this might seem like a simple solution to the problem and, indeed, it is. One can hold the overall risk of falsely rejecting H_0 to any desired α level. There is, however, a severe cost to this approach. If H_0 is false, the probability of making a Type II Error increases rapidly as α

is made more rigorous (recall the discussion of Power in Chapter 9). Fisher's ingenious solution was to come up with an altogether different approach for testing H_0—one that holds the overall probability of making a Type I Error at any desired level *without* increasing the probability of making a Type II Error.

The ANOVA procedure uses a new test statistic called *F*, in honor of its inventor, Fisher. To understand how ANOVA tests the H_0 about two or more population μ's, it is first necessary to study the *F* statistic in some detail.

The *F* Distribution

The *F* statistic is often referred to as the **F ratio**. This is because *F* is defined as **the ratio of two independent, unbiased estimates of a population variance, σ^2**. We can best understand this definition by conducting a "thought experiment." Imagine a normally distributed population of scores having known μ and σ. From the population we draw two random samples of size n_1 and n_2. The samples do not need to be of the same size. Because they are randomly selected, the two samples are independent of one another. For each sample, we compute the unbiased estimate of the population variance, $\hat{s}^2$, according to the following formula:

$$\hat{s}^2 = \frac{\sum(X - \bar{X})^2}{n - 1}$$

The estimate from the first sample we will call $\hat{s}_1^2$ and the estimate from the second sample, $\hat{s}_2^2$. We form the *F* ratio by arbitrarily placing the estimate derived from Sample 1 in the numerator and the estimate from Sample 2 in the denominator, i.e.,

$$F = \frac{\hat{s}_1^2}{\hat{s}_2^2}$$

What values can the *F* ratio statistic assume? First, it should be clear that *F* must always be positive, because both $\hat{s}^2$'s are positive numbers. Second, the *F* ratio *could be* a very small value close to 0 if, by chance, the numerator estimate happened to be much smaller in value than the denominator estimate. Third, the *F* ratio *could be* quite large if, by chance, the numerator estimate happened to be much larger than the denominator estimate.

Let's use a concrete example to make this clear. Suppose that the population σ = 15. The population variance would then be $\sigma^2 = 15^2 = 225$. We arbitrarily set sample sizes at $n_1 = 5$ and $n_2 = 20$. For our first pair of random samples, $\hat{s}_1^2 = 122.4$ and $\hat{s}_2^2 = 246.8$. The *F* ratio would be:

$$F = \frac{122.4}{246.8} = .496$$

We draw a second pair of random samples and this time $\hat{s}_1^2 = 486.0$ and $\hat{s}_2^2 = 185.2$. The *F* ratio becomes:

$$F = \frac{486.0}{185.2} = 2.624$$

F seems to "bounce around" quite a bit in value in our example. Why is that? The reason is that neither sample is terribly large, and Sample 1 is particularly small. Small samples yield unstable estimates of σ^2. If both of our sample sizes had been larger – say 25 or 30 – then the numerator and denominator estimates for any given pair of random samples would tend to be close to the population value of 225. Because both estimates would be close in value to 225 and thus to each other, the ratio of the two estimates would tend to approximate a value of 1.00.

This leads us to our fourth and fifth generalizations about the *F* ratio. The fourth is that as sample size increases, the *F* ratio tends not to vary as much in value as when sample sizes are smaller. Extremely low or high values are possible, but unlikely. The fifth generalization is that, because both the numerator and denominator of the ratio are estimating the same thing—the population variance, σ^2—the mean or "expected" value of *F* across an infinite number of pairs of samples is 1.00, i.e.,

$$E(F) = 1.00$$

Based on these five points, we are now able to discuss the **sampling distribution of the *F* ratio**. Formally, the sampling distribution of *F* is defined as a **continuous probability distribution of values of *F* arising from an infinite number of pairs of samples of size n_1 and n_2 drawn with replacement from a normally distributed population**.

From this definition, it should be obvious that, like *t*, there are an infinite number of sampling distributions of *F*, with a different distribution for every possible combination of sample sizes. Moreover, like *t*, the different sampling distributions of *F* technically are based on the *df* for the two sample estimates, rather than sample size *per se*. Recall that the *df* for $\hat{s}$ (and thus $\hat{s}^2$) is computed as $n - 1$. For sample sizes of $n_1 = 5$ and $n_2 = 20$, the sampling distribution of *F* would be based upon 4 *df* for the numerator estimate of σ^2 and 19 *df* for the denominator estimate of σ^2.

Figure 1 shows what the sampling distribution of F looks like for df = 4 and 19. For comparison, the sampling distribution of F for df = 3 and 19 is also included.

Figure 1

Sampling Distributions of the F Ratio for df = 3, 19 and df = 4, 19.

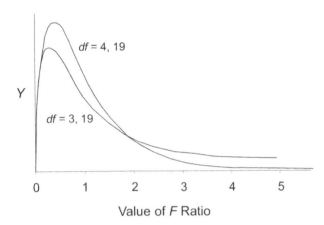

Note that both sampling distributions are very positively skewed, with that for df = 3 and 19 slightly more so than the distribution with 4 and 19 df. As df increases, the sampling distribution becomes more symmetrical. For very large df in the numerator and denominator estimates of the population variance, the sampling distribution of F approaches a normal distribution approximately centered on μ_F = 1.00. In fact, as noted above, regardless of the df involved, the mean for each distribution is always 1.00.

Using the F Ratio in ANOVA

Having described the F ratio statistic and its sampling distribution, we are now able to show how this statistic is used to test the H_0 that two or more population means are equal. We will develop the logic of one-way ANOVA by doing an analysis "by hand" with a very simplified data set and then we'll show how JASP can be used to analyze a somewhat more realistic set of data.

Assumptions of the Procedure

The one-way ANOVA procedure makes the same assumptions as those for the independent t test.

1. **The data are numerical.** This assumption is required to perform the arithmetic operations required by the procedure.

2. **The samples are randomly selected from the populations of interest and are independent of one another (uncorrelated samples).** This assumption is met if cases in an experiment are randomly selected from some definable population and then randomly and independently assigned to the conditions representing the levels or categories of the experiment. The procedures in this chapter are not appropriate when the observations in two or more samples show a pairwise correlation. ANOVA with correlated samples is covered in Chapter 15.

3. **The scores for the populations represented by the samples are normally distributed.** Normality of the populations ensures that the two variance estimates derived from the sample data (see below) will be independent of one another, and that the resulting F ratio will follow its known sampling distribution. The validity of this assumption is usually unknown. If the populations are unimodal and not too badly skewed, violation of this assumption has little effect on the results of the ANOVA—particularly when sample sizes are larger (e.g., n's > 20).

4. **The populations of interest all have the same variance.** This is referred to as the homogeneity of variance assumption. An "eyeball" check of the homogeneity of variance assumption is done by finding the ratio of the largest sample variance to the smallest sample variance. If the ratio is no more than 4:1, many investigators assume that the assumption is held to be adequately approximated. Of course, Levene's test of the assumption is also available (see below).

Logic of the ANOVA Procedure

We will show how the one-way ANOVA works by analyzing the results of our hypothetical vitamin C experiment described in the introduction. Assume that we conducted the experiment with middle-aged adult volunteers. We selected a sample of 20 volunteers from a pool of volunteers and randomly assigned five to each of the four conditions of the experiment. Each person received the amount of vitamin C appropriate to his or her condition for a period of one year. The dependent variable was the number of colds that each person developed over this period. The data are shown in **Table 1**. The mean for each condition ($\bar{X}$) is given, along with the grand mean ($\bar{\bar{X}}$), which is the mean of all 20 cases.

This is how the F ratio is applied to evaluate these hypothetical data. Recall that the F ratio statistic requires two independent, unbiased estimates of a population variance. It turns out that we can obtain two such estimates from the sample data. The first estimate is called the **within-groups estimate** and is abbreviated s^2_{WG}. Because all four populations are assumed to have the same variance, σ^2, we can get our best estimate of σ^2 by "pooling" the sample data together to yield one estimate based on a single large sample instead of four separate estimates based on smaller samples. Recall that this is what

Table 1

Number of Colds Developed by Each Participant in the Vitamin C Experiment for Each Level of the Independent Variable.

	Daily Amount of Vitamin C (mg)			
	0	500	1000	2000
	4	4	4	2
	5	3	2	1
	3	3	2	2
	3	4	3	3
	5	2	2	3
$\bar{X}$ =	4.0	3.2	2.6	2.2
$\bar{\bar{X}}$ = 3.0				

was done in the computation of the standard error of the sampling distribution for the independent *t* test.

The formula for pooling the sample data to yield the within-groups estimate of the population variance is:

$$s^2_{WG} = \frac{\sum(X_1 - \bar{X}_1)^2 + \ldots + \sum(X_K - \bar{X}_K)^2}{(n_1 - 1) + \ldots + (n_K - 1)}$$

where: *K* = number of samples and
+...+ = expand expression to include all *K* number of samples.

The numerator of the expression is called the pooled **sums of squares** (sum of the squared deviations) of the within-group estimate. It is abbreviated SS_{WG}. The denominator is the *df* that contribute to the pooled estimate and is abbreviated df_{WG}. In the formula, the squared deviations for each sample are taken about that sample's mean. As a result, even if the samples are drawn from populations having different means, the value of the within-group variance reflects **only random variation of the scores—not any differences between the population means themselves.**

For our sample data the within-groups variance estimate is computed as:

$$s^2_{WG} = \frac{\sum(X_1 - \bar{X}_1)^2 + \ldots + \sum(X_K - \bar{X}_K)^2}{(n_1 - 1) + \ldots + (n_K - 1)}$$

$$= \frac{4.0 + 2.8 + 3.2 + 2.8}{4 + 4 + 4 + 4} = \frac{12.8}{16} = .80$$

Our within-groups estimate of σ^2 is .80. It is based on 16 *df*.

The second estimate of σ^2 required for the *F* ratio is obtained *indirectly* from the variability among the sample means themselves. This is a key point that many students fail to understand. The idea is this. **Variability among sample means may be converted back into an estimate of variability among scores.** This relationship between score variability and variability among sample means is already familiar to us. Recall that the third part of the Central Limit Theorem concerning the sampling distribution of the mean states that:

$$\sigma_{\bar{X}} = \frac{\sigma}{\sqrt{n}}$$

The formula simply says that the amount of variability among sample means is a function of the amount of variability among the scores themselves. Of course, because σ is almost never known, we are usually forced to use the estimation version of this formula in which the standard deviation of the sampling distribution is estimated from the sample data, i.e.,

$$s_{\bar{X}} = \frac{\hat{s}}{\sqrt{n}}$$

This formula says that if we have an estimate of the population standard deviation, we can use it to estimate the standard deviation of the sampling distribution of the mean. *But*—and here is the key point—the reverse is also true. If you had a value for $s_{\bar{X}}$ you could rearrange the formula to generate the value of $\hat{s}$, or more precisely, for $\hat{s}^2$. To do so, we first square both sides of the equation:

$$(s_{\bar{X}})^2 = \left(\frac{\hat{s}}{\sqrt{n}}\right)^2$$

$$s_{\bar{X}}^2 = \frac{\hat{s}^2}{n}$$

Then, multiple both sides by *n*:

$$n\, s_{\bar{X}}^2 = \hat{s}^2$$

According to this formula, if you have an estimate of the variance of the sampling distribution, you can multiply it by sample size to get an estimate of the population variance.

In the ANOVA situation, the unbiased estimate of the variance of the sampling distribution of the mean is obtained by:

$$s_{\bar{X}}^2 = \frac{\sum(\bar{X} - \bar{\bar{X}})^2}{K - 1}$$

where: $\bar{X}$ = means of the groups,
$\bar{\bar{X}}$ = overall or grand mean, and
K = number of groups.

This formula treats the sample means as though they were scores. The discrepancy between each sample mean and the mean of the means is squared, the squared discrepancies are summed, and then divided by the number of means minus one—the *df* for the estimate. Because the discrepancies are taken around the overall mean, one of the discrepancies is not independent, so the estimate of the variance of the sampling distribution is based on the number of means minus one, i.e., $df = K-1$. By dividing by *df*, we have computed the estimated variance of the sampling distribution of the mean, i.e., $s_{\bar{X}}^2$.

For our Vitamin Experiment data, the estimated variance of the sampling distribution of the mean, $s_{\bar{X}}^2$ is computed as follows:

$$s_{\bar{X}}^2 = \frac{\sum(\bar{X} - \bar{\bar{X}})^2}{K - 1}$$

$$= \frac{1.00^2 + .20^2 + -.40^2 + -.80^2}{4 - 1}$$

$$= \frac{1.00 + .04 + .16 + .64}{3} = \frac{1.84}{3} = .613$$

If the variance of the sampling distribution of the mean is multiplied by the sample size for the groups, it yields an estimate of the population variance. This estimate is called the **between-groups estimate** and is abbreviated s_{BG}^2. When the sample size is the same for all groups in the ANOVA, the simplest formula for s_{BG}^2 is:

$$s_{BG}^2 = n\, s_{\bar{X}}^2$$

where: n = the sample size of each group.

For our sample, the between-groups estimate is computed as:

$$s_{BG}^2 = n\, s_{\bar{X}}^2$$
$$= 5 \times .613$$
$$= 3.065$$

Out between-groups estimate of σ^2 is 3.065. It is based on 3 *df*.

If the samples are independent of one another (Assumption 1) and are normally distributed (Assumption 2), the two estimates, s_{WG}^2 and s_{BG}^2, will be independent of one another. This derives from the fact that **in samples drawn from normally distributed populations, the sample $\bar{X}$ and $\hat{s}^2$ will be uncorrelated across an infinitely large number of samples.** Thus, if our assumptions are met, we have been able to derive two independent, unbiased estimates of the common population variance, σ^2.

We form the *F* ratio by always dividing the between-groups estimate by the within-groups estimate, i.e.,

$$F = \frac{s_{BG}^2}{s_{WG}^2} = \frac{3.065}{.801} = 3.826$$

The *F* ratio for our sample data is equal to 3.826. It is based on 3 and 16 *df*.

How is this *F* ratio used to evaluate the H_0 that the population means are equal? The answer lies in the values that the *F* ratio will assume when the H_0 is true and when the H_0 is false. Let's begin with the within-groups estimate, s_{WG}^2. **Regardless of whether H_0 is true or false, if the assumptions of the procedure are met, s_{WG}^2 is a pure estimate of σ^2.**

Now, what about the between-groups estimate, s_{BG}^2? If H_0 is true, the differences among the sample means are entirely due to chance and their variability about their own mean yields a pure estimate of σ^2. On the other hand, if H_0 is false, then part of the differences between the sample means is because two or more means were drawn from populations in which the μ's were not equal. The variability of the sample means about $\bar{\bar{X}}$ would then be due to both chance factors and a systematic tendency for the means to be different in value. **In this case, s_{BG}^2 will be an estimate not only of σ^2, but also of the magnitude of the differences that exist between the population μ's.**

Table 2 summarizes this discussion of what s_{WG}^2 and s_{BG}^2 estimate when H_0 is either true or false. If H_0 is true, the numerator and denominator of the *F* ratio are both estimates of σ^2 and the *F* ratio will follow its known sampling distribution for the relevant *df* for the numerator and denominator estimates, with $E(F) = 1.00$, i.e.,

$$F = \frac{s_{BG}^2}{s_{WG}^2} = \frac{\text{an estimate of } \sigma^2}{\text{an estimate of } \sigma^2}$$

On the other hand, if H_0 is false, the between-groups estimate will tend to be larger than the within-groups estimate, and the *F* ratio will be larger than would be

Table 2

The Nature of the Within- and Between-Group Estimates When H_0 is True and False.

If H_0 Is True	If H_0 Is False
$s_{WG}^2 \to \sigma^2$	$s_{WG}^2 \to \sigma^2$
$s_{BG}^2 \to \sigma^2$	$s_{BG}^2 \to \sigma^2 + \text{Diff among } \mu's$

expected through random sampling effects, with $E(F) > 1$, i.e.,

$$F = \frac{s_{BG}^2}{s_{WG}^2} = \frac{\text{an estimate of } \sigma^2 + \text{diff among } \mu's}{\text{an estimate of } \sigma^2}$$

How large does an obtained value of F have to be before we will conclude that the H_0 is false and that at least two population means differ in value? Here is where the α level criterion comes into play. If we decide to use a .05 α level, we will reject H_0 if the obtained value of F is among the *highest* 5% of F ratios that would be obtained by chance if H_0 were true. The entire area of rejection is placed at the upper end of the sampling distribution of F because a false H_0 always leads to unusually large values of F. **Figure 2** shows the sampling distribution of F for $df = 3$ and 16, with the critical values corresponding to the upper 5% ($\alpha = .05$) and 1% ($\alpha = .01$) areas of rejection under the curve indicated.

The F ratio for our sample data (3.826) falls in the area of rejection at the .05 α level. We would thus reject the H_0 that all of the population means are equal.

Partitioning the Variance in ANOVA

The forgoing introduction to ANOVA focused on finding the two unbiased, independent estimates of the population variance in order to use the F ratio to test the H_0 that the population means were equal. That's a good strategy for teaching the basic concepts of the F ratio and its use in hypothesis testing. But in a more general sense, "analysis of variance" literally means breaking down the total variance in a set of scores into its component parts.

To understand how, you need to recall from the end of Chapter 6 on multiple linear regression that variances are additive, i.e., the variance (s^2) of a set of scores is the sum of the variances of the components of those scores. And

Figure 2

Sampling Distributions of the F Ratio for df = 3, 16, Showing Critical Values of F for .05 and .01 α Levels.

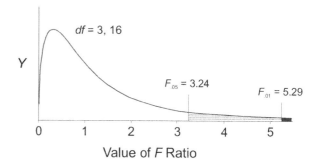

just as variances are additive, the numerator for the variances (the sums of squares) are also additive.

Let's apply this idea to ANOVA. We'll begin finding what is called the **total sums of squares (SS_{Total})**. This is the variability of all of the scores in the data set around the grand mean, i.e., the mean of all of the scores. The formula for finding the SS_{Total} for our small data set is:

$$SS_{Total} = \sum (X - \bar{\bar{X}})^2 = 22.0$$

We've already found the sums of squares for the within group variance estimate as:

$$s_{WG}^2 = \frac{\sum(X_1 - \bar{X}_1)^2 + \ldots + \sum(X_K - \bar{X}_K)^2}{(n_1 - 1) + \ldots + (n_K - 1)}$$

$$= \frac{SS_{WG}}{df_{WG}} = \frac{12.8}{16} = .80$$

The numerator for the variance estimate is the **within groups sums of squares (SS_{WG})**, which you can see in the formula is equal to 12.8.

The **between groups sums of squares (SS_{BG})** is again the numerator in the between groups variance estimate and is equal to:

$$n \sum (\bar{X} - \bar{\bar{X}})^2 = 5(1.84) = 9.2$$

Here's the neat part. The SS_{Total} is equal to the sum of the SS_{BG} and SS_{BG}, i.e.,

$$SS_{Total} = SS_{BG} + SS_{WG}$$

which is:

$$22.0 = 9.2 + 12.8$$

This is an extremely general concept. For all ANOVA situations, the total sums of squares can always be partitioned into all of its component parts—the more complex the ANOVA, the more parts there are! Moreover, the various variance estimates for use in the *F* ratio are always computed as a sums of squares divided by the relevant degrees of freedom.

The results of an ANOVA are always placed in what is called a **source table**, which is an organized presentation of all of the various components (see below).

One-Way ANOVA With *JASP*

To see how *JASP* is used to carry out a one-way ANOVA, we will analyze a data set for a hypothetical experiment that investigates the influence of hourly wages on worker performance. **The hypothetical experiment is called, "Your Order, Please." Carefully read through the description of this experiment before continuing.**

How well do these data meet the four assumptions of the one-way ANOVA? First, the rating data are numerical. Second, the cases of the experiment were randomly selected from the population of company stores and were randomly and independently assigned to the conditions of the experiment. Third, we have no information concerning the normality assumption. Because sample sizes are so small, it is impossible to use the sample distributions to evaluate the validity of this assumption. Fourth, as you will see below, the sample standard deviations (and, thus, the sample variances) are very similar in value, suggesting that the homogeneity of variance assumption is likely to be met.

The H_0 and H_A to be tested are stated as follows:

H_0: $\mu_{\$12} = \mu_{\$15} = \mu_{\$18} = \mu_{\$21}$
H_A: At least two μ's are not equal

We will adopt α = .05 for purposes of testing the H_0.

ANOVA Source Table

The *JASP* one-way ANOVA is done using the ANOVA procedure, with the sub procedure ANOVA. In running the analysis, we have instructed *JASP* to (a) generate the ANOVA source table, including two effect size statistics (b) include Levine's test of the assumption of homogeneity of variance, and (c) use Tukey's Honestly Significant Difference (*HSD*) procedure to carry out what are called "post-hoc" comparisons among the sample means.

Figure 3 is the ANOVA source table. As noted above, this table is the traditional way that ANOVA results are reported. The first column of the table labeled "Cases" identifies the "**Between Groups**" (Salary) and the "**Within Groups**" (Residual) variance estimates. The second column lists the value of the sum of squares for the two variance estimates.

The third column lists the *df* for the between groups, and within groups variance estimates. There are 3 *df* for the between-groups estimate (4 means – 1), 20 *df* for the

---

Your Order, Please

Service in "fast food" restaurants is often slow, marginally discourteous, and inaccurate. Part of the reason probably lies in the generally poor wages paid to part-time fast food restaurant employees. It's tough to hire bright, friendly, high-energy people to do a stressful job for $12 an hour. What would happen to service if fast food workers made significantly better wages?

A large fast food restaurant chain decides to find out. The company plans to try out four different starting salary levels— $12, $15, $18, and $21 per hour. Twenty-four company owned restaurants in the Midwest are randomly selected for the experiment. Six restaurants are randomly assigned to each wage level. Restaurant managers are given instructions to increase salaries to the indicated level and to use the incentive provided by the higher wages to increase worker performance. After three months, each restaurant is visited several times by quality control inspectors from the company's home office and given an overall customer service evaluation score (CSE) that could range from 0 to 100. Below are the service quality scores for the 24 restaurants.

$12 / hr.		$15 / hr.		$18 / hr.		$21 / hr.	
87	81	86	89	93	94	96	93
89	86	93	89	92	88	87	93
84	87	85	92	94	93	93	94

Figure 3

JASP ANOVA Source Table For the One-Way Analysis of the "Your Order, Please" Experiment.

ANOVA - CSE Is the Dependent Variable

Cases	Sum of Squares	df	Mean Square	F	p	η^2	ω^2
Salary (BG)	193.833	3	64.611	8.076	0.001	0.548	0.469
Residuals (WG)	160.000	20	8.000				

Notes. Type III Sum of Squares. I have annotated the table for expository purposes.

within-groups estimate (6 scores – 1) for each sample, pooled across the four samples.

The fourth column is labeled "**Mean Square**," which is an old-fashioned (but still used) way of referring to the between-groups and within-groups population variance estimates. It is the sum of squares divided by the *df*.

The fifth column is the value of the *F* ratio which is computed by dividing s_{BG}^2 by s_{WG}^2. In our example, *F* = 64.611 / 8.000 = 8.076.

The sixth column is the *p* value for the obtained *F* ratio. The value of .001 listed in the table is the probability of obtaining an *F* ratio of this magnitude or larger by chance *if* H_0 were true. As usual, we compare the *p* value to our α level criterion to decide whether to reject H_0. Because .001 is lower than .05, we will reject H_0 that the samples were drawn from populations having the same means. According to our results, at least two of the population means differ. In terms of our independent variable, the salary paid to hourly workers does influence the quality of the restaurant service. Exactly which sample means differ significantly remains to be determined. We will describe how this is done in the section on post-hoc comparisons.

The last two columns contain the values of the effect size statistics η^2 (eta squared) and ω^2 (omega squared), which are discussed below.

Levene's Test of Homogeneity of Variance

Figure 4 contains the results of Levene's test of the homogeneity of variance assumption. The formula for Levene's test is sufficiently complex that it is not given here and virtually all researchers are happy to let the data analysis program do the somewhat complex calculations.

The completely non-significant *p* value for the test is .892, which is consistent with the fact that the sample standard deviations were all very close in value.

Figure 4

JASP Levene's Test for the Homogeneity of Variance Assumption.

Test for Equality of Variances (Levene's)

F	df1	df2	p
0.204	3.000	20.000	0.892

Descriptive Statistics

Because the *F* ratio for the independent variable was significant, we'll compute the basic descriptive statistics for the experiment using the JASP DESCRIPTIVES procedure. These are displayed in **Figure 5**.

Figure 5

Descriptive Statistics for the Four Conditions of the "Your Order, Please" Experiment.

Descriptive Statistics

	CSE			
	$12	$15	$18	$21
Valid	6	6	6	6
Median	86.500	89.000	93.000	93.000
Mean	85.667	89.000	92.333	92.667
Std. Error of Mean	1.145	1.291	0.919	1.229
Std. Deviation	2.805	3.162	2.251	3.011
Minimum	81.000	85.000	88.000	87.000
Maximum	89.000	93.000	94.000	96.000

Inspection of the sample means indicates that there is a clear tendency for the sample means to increase as the starting hourly wage increases. Also note that the sample standard deviations are all small and approximately the same magnitude. The means and medians are very close in value, suggesting that the parent populations are symmetrical.

Post-Hoc ANOVA Comparisons

Rejection of H_0 tells us that at least two population means differ in value, which is to say, at least two sample means differ significantly (necessarily, the largest difference will be significant). But it is possible that more than just two means differ significantly. Determining which pairs of means are significantly different is referred to as making **post-hoc** ("after the fact") **comparisons** among the sample means.

Statisticians have developed several different approaches to making these comparisons. These approaches differ in the extent to which they seek to "protect" the researcher from making Type I Errors, i.e., falsely claiming that two means differ significantly. Approaches that are very protective in this regard are called "conservative" post-hoc procedures.

In the opposite extreme are the "liberal" approaches that seek to increase the power of the contrast, i.e., the ability to detect a real difference. *JASP* provides many procedures to choose from; we have selected **Tukey's Honestly Significant Difference (HSD) test**. This test is moderately powerful and allows you to test the significance of the difference between all possible combinations of sample means. The *HSD* test assumes that the overall H_0 has been rejected.

JASP computes the value of *t* for every possible pair of sample means, using Tukey's correction to determine which comparisons are statistically significant. **Figure 6** shows the results of this analysis.

The first two columns indicate which comparisons are being made, the third the difference in the two means involved, the fourth the standard error of the test, and the fifth and sixth columns the value of *t* and the *p* value for the comparison. Notice the first three rows contain the comparison between the mean for the $12 condition with each of the other three conditions. Examination of the *p* values indicates that the $12 hr. condition mean differs significantly from the means for the $18 hr. and $21 hr. conditions, but not from the $15 hr. condition mean. Inspection of the information in the fourth and fifth rows indicates that the $15 hr. condition mean did not differ significantly from any of the other means.

Overall, the comparisons among the sample means indicate that the optimal starting salary probably is $18 hr. Workers earning $18 hr. perform significantly better than workers earning $12 hr. There appears to be no further advantage to increasing starting salaries to $21 hr., because the $18 hr. and $21 hr. groups did not differ significantly. This interpretation needs to be qualified by the fact that the samples used in the experiment are small and the failure to find differences between group means could very well reflect the operation of Type II Errors.

Assessing the Strength of the Independent Variable with η^2 and ω^2

Recall in Chapter 11 when we discussed the distinction between a statistically significant difference and a theoretically or practically important difference? In the case of the independent samples *t* test, we introduced a statistic called r_{pb}^2, which quantified how much of the variance in the dependent variable was due to the independent variable. A similar statistic is available for the one-way ANOVA and is included in the second to the last column of the ANOVA source table above. It is called eta

Figure 6

JASP One-Way ANOVA Post Hoc Comparisons Among All Sample Means for the "Your Order, Please" Experiment.

Post Hoc Comparisons - Salary

		Mean Difference	SE	t	p_{tukey}
($12/Hr)	($15/Hr)	-3.333	1.633	-2.041	0.207
	($18/Hr)	-6.667	1.633	-4.082	0.003
	($21/Hr)	-7.000	1.633	-4.287	0.002
($15/Hr)	($18/Hr)	-3.333	1.633	-2.041	0.207
	($21/Hr)	-3.667	1.633	-2.245	0.145
($18/Hr)	($21/Hr)	-0.333	1.633	-0.204	0.997

Note. P-value adjusted for comparing a family of 4

squared and is abbreviated η^2. This statistic was introduced briefly at the end of Chapter 5. This is another one of those odd cases where a Greek letter is used to symbolize a statistic. The statistic η^2 is defined and computed as:

$$\eta^2 = \frac{SS_{BG}}{SS_{WG} + SS_{BG}} = \frac{193.8}{353.8} = .548$$

This statistic can be interpreted similarly to r^2 and r_{pb}^2. In words, it is the **proportion of the total variation in the dependent variable associated with changes in the value of the independent variable**. As usual, for ease of expression, the proportion is translated into a percentage. A large value of η^2 means that much of the total variation among the scores is because they are in groups that have substantially different means.

For our study of fast-food restaurant salaries, the value of η^2 is .548. This is an unusually large value to find in social science research and indicates that starting salary exerts a powerful effect on the index of service quality. In words, we would say that about 55% of the variability in the service quality scores was attributable to differences in wages paid to the workers.

The second effect size statistic listed in the source table is called omega squared (ω^2) and measures much the same thing that η^2 measures. Many investigators prefer it because η^2 has been shown to overestimate the portion of variance that actually would be accounted for by the independent variable in the population. In other words, η^2 is a **biased estimator**, and the extent of the bias is greatest when sample sizes are small. The statistics ω^2 corrects for that bias. For the present experiment ω^2 equals .469, which is to say that starting salary more conservatively accounts for 47% of the variability in service quality scores. The difference of .079 between the two statistics reflects the extent of the bias in η^2. The prudent thing is probably to report both effect size statistics, particularly because η^2 is widely used in the research literature.

The formula for manually computing ω^2 is

$$\omega^2 = \frac{SS_{BG} - (df_{BG})(\hat{s}_{WG}^2)}{SS_{Total} + \hat{s}_{WG}^2}$$

Graphing ANOVA Results

The results of our hypothetical experiment can be displayed graphically using a **function graph**. Recall that a function graph displays in pictorial fashion how values of a dependent variable change as a function of changes in an independent variable. Obviously, we would construct such a graph only if the ANOVA indicated that some of the group means differed significantly. A non-significant F Ratio implies (but does not prove) that the dependent variable is not a function of the independent variable.

In a function graph, the X axis displays the levels or categories of the independent variable, while the Y axis displays the value of some statistic computed for the dependent variable for each level or category. Usually, we are interested in how the average value of the dependent variable changes across levels or categories of the independent variable, so the Y axis would contain either group means or medians (usually the former). Sometimes, however, the independent variable may influence other aspects of the dependent variable, such as the extent of variation in the scores. In that case, you might show how the value of the standard deviation changes for different values of the independent variable.

Figure 7 shows a *JASP* generated function graph (under the Plots submenu) showing how the mean quality score for our fast-food restaurants changed as the hourly wage was increased. In this graph, the value of the dependent variable mean for each level of the independent variable is indicated with a point. These points are connected with straight lines. The resulting "curve" is an estimate of what the function might look like for the four population μ's. 90% CI bands have been included as an option of the procedure.

The use of lines to connect the points indicates the

Figure 7

JASP Function Graph Showing Mean CSE Scores with 90% CI Brackets for All Four Conditions of the "Your Order, Please" Experiment.

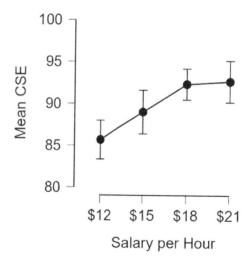

continuous nature of the independent variable, i.e., that the levels of the independent variable used in the experiment are points on a continuum of salary values ranging from low to high. Because the independent variable is continuous in nature, this type of graph is sometimes called a **continuous function graph**. Theoretically, from the function graph, one could estimate the value of the dependent variable for salary levels not actually included in the experiment. For example, for a salary of $13.50 / hr. the height of the function line is at a mean quality score a little over 87.

A somewhat different form of the function graph is used when the independent variable is categorical in nature. Such graphs are referred to as **categorical function graphs**. Recall our example at the beginning of the chapter in which the independent variable was different methods of teaching statistics—the Text approach, the On-Line Text approach, and the Talking Heads approach.

Suppose that the dependent variable had been performance on a comprehensive multiple-choice test of knowledge of statistics using three different teaching methods. If a one-way ANOVA had revealed significant differences among the three condition means, we would want to show how the mean score on the test varied as a function of type of teaching method.

Currently, *JASP* doesn't provide an option to produce categorical function graphs in the ANOVA procedure, although it does so under T-TESTS as Box Plots. **Figure 8** shows a graph of hypothetical means for the teaching method experiment generated by *PowerPoint*. As noted in Chapter 2, the graph looks suspiciously like a bar graph, but there is one important difference. In a bar graph, the *Y* axis reports the frequency with which cases fell into two or more unordered categories. In a categorical function graph, the *Y* axis gives the value of some statistic summarizing the dependent variable for each category of the independent variable. The bars are used to indicate the categorical, unordered nature of the independent variable.

We've covered a lot of different topics in this chapter, so it may be useful to try to pull it together by putting all the results of the hourly wage experiment into a single narrative summary:

The descriptive data for the experiment are summarized in Table 3. A one-way ANOVA of these rating data yielded a significant *F* ratio, $F(3, 20) = 8.08$, $p = .001$, indicating that the means of at least two experimental conditions differed significantly. Post-hoc comparisons among the four conditions using Tukey's *HSD* test with $\alpha = .05$ revealed that the mean for the $12 condition was significantly lower than the $18 and $21 conditions. The $15 and $18 conditions were intermediate in value and not significantly different from the $21 condition. The value of η^2 was .548, indicating that about 55% of the variability in quality scores could be accounted for by the independent variable. The ω^2 statistic yielded a more conservative estimate of 47%.

Figure 8

PowerPoint Produced Categorical Function Graph Showing Mean Class Performance as a Function of Teaching Method.

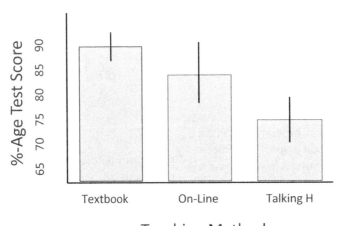

Chapter 12, One-Way ANOVA
Chapter Exercises

Some questions are purely factual and have only one correct answer. Others require greater thought on your part and may have more than one possible answer.

1. Fill in the missing term(s) that makes the statement true.

   a. The principles underlying ANOVA were developed by _____.
   b. Unlike the two-sample t test, ANOVA allows comparisons among _____.
   c. The problem with doing multiple t tests to compare three or more conditions is that _____.
   d. For an experiment with five conditions, you would have to do _____ t tests.
   e. With five conditions and H_0 true, with α = .05 the probability of at least one Type I Error is _____.
   f. The F distribution is named after _____ and is defined as _____.
   g. F can never be _____ because _____.
   h. For df_1 = 4 and df_2 = 5, E(F) = _____ and the sampling distribution is _____.
   i. The assumption of normality in the populations of interest is required because _____.
   j. In ANOVA, the H_0 is that _____.
   k. The term SS stands for _____.
   l. If H_0 is true or false, s^2_{WG} estimates only _____.
   m. If H_0 is false, s^2_{BG} estimates _____.
   n. The F test is always "one tailed" because _____.
   o. In a one-way ANOVA, SS_{Total} is equal to _____ and _____.
   p. Levene's test is used to _____.
   q. If F yields a p value smaller than the α criterion, _____ are done to determine which means differ.
   r. Tukey's HSD test is used to control for _____.
   s. η^2 measures the _____ associated with (caused by) the independent variable.
   t. ω^2 is needed because _____.
   u. A function graph shows how a(n) _____ changes as a function of a(n) _____ variable.

2. Do pop quizzes work? In theory, pop quizzes increase student learning by making it more likely that students will be diligent in studying if there is some chance that they will be tested every time they go to class. Researchers secure a substantial grant to find out. Thirty-three teachers of introductory psychology classes at five universities agree to participate in such an experiment. There are four conditions: (a) No pop quizzes during the semester, (b) Five pop quizzes during the semester, (c) 10 pop quizzes during the semester, and (d) 15 pop quizzes during the semester. All teachers giving quizzes agree to have them count for 25% of the course grade. Most classes met twice a week for the semester, although a few met three times. Fifty-four sections were available for the study and were assigned to condition entirely at random. The dependent variable was the final class average at the end of the semester (% grade). Given different teachers, different teaching times, different testing, and different institutions, there is a lot of slop (randomness) in the experiment—but that's the nature of educational research. Here are the data:

No Pop Quizzes	5 Pop Quizzes	10 Pop Quizzes	15 Pop Quizzes
76 78 83 71	80 77 67 75	74 82 89 78	84 85 79 88
83 71 70 75	78 83 74 76	69 79 74 83	86 79 84 77
71 78 83 69	77 79 82 77	79 75 79 80	86 90 78 85
70 68	72 76	79	88

Use JASP to do a complete analysis of these data, including an appropriate graph, and write a summary of your results.

3. Can a short video change attitudes about a controversial subject? Currently there is a good deal of emotional debate about the topic of gender reassignment in children. Many people argue that it is a non-problem that has been manufactured by adherents of "woke" ideology. Others say that cross-gender identification is fairly common in children and sometimes needs to be dealt with medically. To see whether people's beliefs and attitudes about the subject can be influenced by a brief educational video, two psychologists construct three 30-minute video podcasts on the topic. The first is a medically oriented presentation of the issues involved and statistics concerning treatments and outcomes (Medical

Podcast). The second is an emotional interview with a family of parents and a young girl portrayed by professional actors who discuss the nasty conflicts that occurred with peers and teachers and the somewhat negative outcome of hormone replacement therapy (Negative Family). The third is the same actor family discussing how peers and teachers mostly accepted the change and the positive outcome of hormone replacement therapy (Positive Family). The participants in the experiment were 90 volunteers taking various psychology and sociology night classes at a large junior college. The participants were chosen because they were older adults working toward an associate degree. All were told that they would be watching a podcast on a controversial issue and would be asked to evaluate how they felt about the issue at the end of the podcast. The dependent variable was a 10-item questionnaire that used an 11-point rating scale to measure participant's agreement with various statements (1 ="Strongly Disagree", 6 = "Not Sure", 11= "Strongly Agree".) concerning how informative, engaging, and interesting the presentation was. The question of particular interest was:

"I think that it's a real problem that the government needs to regulate."

Here are the data for that question:

Medical Podcast					Negative Family					Positive Family				
9	6	8	4	7	11	8	6	7	9	7	2	6	6	2
7	8	4	1	5	5	10	7	4	5	2	3	4	6	1
10	8	6	7	4	10	9	4	11	6	8	2	6	3	2
8	5	7	6	7	5	8	8	9	6	4	7	2	6	7
3	7	5	8	4	9	8	11	9	7	1	1	4	6	2
5	7	4	6	4	8	5	9	10	6	5	2	2	6	1

Use *JASP* to do a complete analysis of these data, including an appropriate graph, and write a summary of your results.

4. Are women with tattoos viewed as less competent and/or intelligent by male executives? To find out, a psychologist prepares three different versions of a 15-minute videotaped "lecture" on workplace sexual harassment delivered by "Dr. Susan Reynolds." In fact, in each case the lecture is delivered by a well-rehearsed professional actress using a standard script. The only difference between versions is the presence or absence of conspicuous tattoos on the speaker's arm(s). In the control condition, she has no tattoos. In the second version, she has a tattoo on the left arm. In the third, on both arms. (Fake of course) Dress and hair style are constant. Middle level managers from several large corporations serve as subjects for the experiment. Each subject (randomly assigned) views one of the tapes and afterwards rates the speaker on a variety of dimensions. Below are the data for the rating of the speaker's "level of expertise" (scores range from 1 to 9, with 1 being very low).

No Tattoos					Tattoo on Left Arm					Tattoo on Both Arms				
8	7	7	7	8	7	9	7	8	5	4	1	6	7	4
8	6	6	8	7	8	5	8	7	6	6	5	2	6	5
7	8	7	9	8	7	5	6	9	6	1	6	2	8	1
7	8				4					3	5	4		

Do a complete descriptive and inferential analysis of these data and write a thoughtful interpretation of the results.

Chapter 13
Two-Way ANOVA

What's In This Chapter?

The **two-way ANOVA** is used to evaluate the data from experiments in which there are two independent variables. Three H_0 are tested in a two-way ANOVA. Two of these are concerned with what are called the **main effects** of the two variables. These are the differences among the means for the levels of each variable averaged across the levels or categories of the other variable. The test of the main effects in a two-way ANOVA is conceptually the same as the test carried out in a one-way ANOVA, except that in a two-way ANOVA, the **within-cells estimate** replaces the within-groups estimate in the denominator of the F ratio. The third H_0 tested concerns the potential **interaction** of the two variables. An interaction occurs when the effect of one of the independent variables is different for different levels or categories of the other independent variable.

If an interaction is present, interpretation of the main effects of the two variables must be qualified because they do not fully represent the influence of the independent variables on the dependent variable. In this case, the **simple effects** need to be evaluated using post-hoc comparisons. An **interaction plot** shows the effects of an independent variable on the dependent variable, separately for each level or category of the other independent variable.

As was the case for the one-way ANOVA, η^2 and ω^2 should be used to quantify the effect size for significant main effects and for a significant interaction. In evaluating the effect size for significant effects, **partial η^2** is sometimes used.

Introduction

In many experiments, investigators simultaneously manipulate more than one independent variable. This is done because most of the interesting dependent variables in the world are complexly determined, i.e., they are causally influenced by several different independent variables. To assess fully how two or more variables influence some dependent variable, it is necessary to incorporate all the variables into a single experiment. In this chapter we will restrict our discussion to the analysis of experiments involving two independent variables. In such an experiment, each level or category of the first independent variable is combined with every level or category of the second independent variable. Each case is independently and randomly assigned to one combination of the independent variables. The statistical procedure for analyzing the results of such experiments is called a **two-way ANOVA**. Chapter 14 covers the situation where there are more than two independent variables.

The logic underlying the two-way ANOVA represents, for the most part, a relatively straightforward extension of the logic involved in the one-way ANOVA. The test statistic is still the F ratio, and you are still comparing within-group and between-group variance estimates. The difference lies in the number of between-group variance estimates. In the one-way ANOVA, there was one between-group estimate derived from the single independent variable. In the two-way ANOVA, there are *three* different between-group variance estimates—one for each of the two independent variables and one based upon what is called the **interaction** of the independent variables. To understand how these variance estimates are obtained, we need to visualize the structure of an experiment having two independent variables.

Figure 1 on the next page shows in abstract form the results of an experiment having three levels for the first independent variable (Variable A) and two levels for the second independent variable (Variable B).

Diagrams of this type show the **design of the experiment**. The levels of the two independent variables are called the "experimental treatments," and the combinations of treatments are called the "**cells**" of the design. The experiment would be called a "3 x 2 design" because there are three levels for the A variable and two levels for the B variable. In this example, there are five observations per cell, indicated by the subscripted X's. These represent the values of the dependent variable obtained under each treatment combination.

The mean of the five observations in each cell has been indicated by a subscripted mean, e.g., $\bar{X}_{a1,b1}$ is the mean of the cell representing the combination of Level 1 of the A variable and Level 1 of the B variable. The means computed

Figure 1

Diagrammatic Representation of a 3 X 2 ANOVA Design with Five Cases Per Cell.

Variable B	Variable A: a_1	a_2	a_3	
b_1	$X_1\ X_2\ X_3\ X_4\ X_5$ $\bar{X}_{a1b1}$	$X_6\ X_7\ X_8\ X_9\ X_{10}$ $\bar{X}_{a2b1}$	$X_{11}\ X_{12}\ X_{13}\ X_{14}\ X_{15}$ $\bar{X}_{a3b1}$	$\bar{X}_{b1}$
b_2	$X_{16}\ X_{17}\ X_{18}\ X_{19}\ X_{20}$ $\bar{X}_{a1b2}$	$X_{21}\ X_{22}\ X_{23}\ X_{24}\ X_{25}$ $\bar{X}_{a2b2}$	$X_{26}\ X_{27}\ X_{28}\ X_{29}\ X_{30}$ $\bar{X}_{a3b2}$	$\bar{X}_{b2}$
	$\bar{X}_{a1}$	$\bar{X}_{a2}$	$\bar{X}_{a3}$	$\bar{\bar{X}}$

for the cells represent the **simple effects** of the two variables. There are also means indicated for the columns and rows of the design. The three column means corresponding to the three levels of the *A* variable are labeled $\bar{X}_{a1}$, $\bar{X}_{a2}$, and$\bar{X}_{a3}$. These values are computed by averaging *across* the 10 observations in the two levels of the *B* variable and represent what are called the **main effects** of the *A* variable. The two means labeled $\bar{X}_{b1}$ and $\bar{X}_{b2}$ are the main effect means for the *B* variable. These are computed by averaging across the 15 observations in the three *A* levels. The mean indicated by $\bar{\bar{X}}$ is the overall mean and is the average of all 30 cases in the design.

Nature of the H_0's in a Two-Way ANOVA

The two-way ANOVA permits the independent test of three H_0's concerning the data. The first two H_0's concern the main effects of the two independent variables. For the main effect of the *A* variable, the H_0 and H_A would be stated as:

H_0: $\mu_{a1} = \mu_{a2} = \mu_{a3}$
H_A: At least two μ's are not equal

The H_0 asserts that, averaging across the *B* variable, the three populations corresponding to the three categories of Variable *A* have the same μ's.

Similarly, the H_0 and H_A for the main effect of the *B* variable would be stated as follows:

H_0: $\mu_{b1} = \mu_{b2}$
H_A: The two μ's are not equal

The third H_0 tested concerns what is called the **interaction** of the two independent variables. The concept of an interaction between variables is a bit complex, so we will have to spend a little time developing the idea. Basically, an interaction between two independent variables means that **the effect of one independent variable is different for different levels or categories of the second independent variable.**

The classical example of an interaction concerns the effects of alcohol and barbiturates on central nervous system functioning. Let one independent variable be *Amount of Alcohol Consumed*, represented by the levels of 0 oz. and 2 oz. The second independent variable is *Amount of Barbiturate Ingested*, represented by the levels of 0 mg and 50 mg. The dependent variable is *%-Age Central Nervous System* (*CNS*) *Depression.*

Figure 2 diagrams the result of this hypothetical 2 x 2 experiment. The numbers in the cells represent the mean percentage reduction of CNS activity. Let's examine the **simple** or **cell effects** of each variable. When no barbiturate is present, ingestion of 2 oz. of alcohol produces a 25% reduction of CNS activity. When no alcohol is consumed, ingestion of 50 mg of barbiturate produces a 25% reduction in CNS activity. From these means, we might predict that the combination of alcohol and barbiturate will produce about a 50% reduction in activity. If that turned out to be

Figure 2

Cell Means for a 2 X 2 ANOVA Design Showing the Amount of CNS Depression for Combinations of Alcohol and Barbiturates.

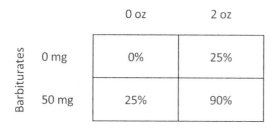

Figure 3

Interaction Plot of Alcohol and Barbiturate Data Shown in Figure 2.

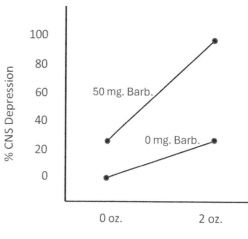

the case, we would say that the two variables operate in an **additive** fashion, i.e., the combined effect of the two variables is the sum of their separate effects. In fact, however, the combination of 2 oz. of alcohol and 50 mg of barbiturate produces a 90% reduction in activity. Here, the effect of either variable on CNS activity varies as a function of the level of the other variable. We would say that alcohol and barbiturate "interact" in their effect on CNS activity. When the nature of the interaction is to produce a combined effect which is greater than the sum of the effects of each variable taken separately, we say that the interaction is **synergistic**.

The best way to make sense of an interaction is to graph it. A graph of an interaction is called an **interaction plot**. An interaction plot is a function graph showing the relationship between one of the independent variables and the dependent variable separately for each of the levels or categories of the second independent variable. **Figure 3** depicts an interaction plot of the alcohol and barbiturate data. The X axis is scaled for the amount of alcohol and the Y axis for the percentage depression in CNS activity. Separate function lines are shown for the 0 mg. and the 50 mg. levels of the barbiturate variable. (We could just as well have put the amount of barbiturate on the X axis and plotted separate function lines for the different levels of the alcohol variable.)

Notice that the two function lines are not parallel. **Non-parallel function lines indicate that an interaction between the variables may be present.** Of course, even if an interaction were not present, random sampling and assignment effects would make it unlikely that the lines would be exactly parallel. The determination of whether an apparent interaction is statistically significant is made using the two-way ANOVA.

This leads us to the third H_0 tested in the two-way ANOVA. Basically, the H_0 states that the two variables do not interact—that the effects of the two variables (if any) are additive in nature. Sometimes it is difficult to state this H_0 in a concise manner. Here's how the null and alternative hypotheses concerning the possible interaction component could be stated for the 3 x 2 design displayed in **Figure 1**.

H_0: $(\mu_{a1,b1} - \mu_{a1,b2}) = (\mu_{a2,b1} - \mu_{a2,b2}) = (\mu_{a3,b1} - \mu_{a3,b2})$
H_A: At least two mean differences are not equal

The H_0 says that the simple effects of the B variable (should they exist) are the same for all levels of the A variable. Had there been three levels of the B variable, the symbolic statement of H_0 would be complex indeed. To avoid that, we frequently fall back on the following shortcut statement of H_0 and H_A concerning the possibility of an interaction.

H_0: The variables are additive in their effects
H_A: The variables interact

Assumptions of the Procedure

The assumptions for the two-way ANOVA are substantially the same as those for the one-way ANOVA.

1. **The data are numerical in character.** This is assumed as usual. The equations given below to develop the logic of the procedure assume equal numbers of cases per cell. When cell sizes are unequal, a more sophisticated computational approach is required. The *JASP* ANOVA procedure which we will use to carry out such analyses provides the appropriate corrections for unequal cell sizes.

2. **The samples representing the cells of the design are randomly selected from the populations of interest and are independent of one another (uncorrelated samples).** This assumption is met if the subjects in an experiment are randomly selected from some identifiable population and are randomly and independently assigned to the treatment combinations representing the cells of the design.

3. **The scores for the populations represented by the row, column, and cell means are normally distributed.** The validity of this assumption is usually unknown, generally because cell sample sizes are too small to permit estimates of the shape of the population. Fortunately, if the populations are unimodal and not badly skewed, violation of this assumption will have little effect on the results of the ANOVA, particularly when the sample size for each of the cells is relatively large ($n \geq 10$) *and* cell n's are approximately equal in size.

4. **The populations represented by the row, column, and cell means all have the same variance.** The homogeneity of variance assumption is usually tested by comparing the size of the within-cell sample variances. If the ratio of the largest to the smallest variance is no more than 4:1, the assumption is held to be adequately approximated.

Testing the H_0's With the F Ratio

Within-Cells Variance Estimate

The following discussion is built around the abstract 3 x 2 analysis presented in **Figure 1**. Refer to this figure as you read about how each of the H_0's is tested in a two-way ANOVA.

As you recall from the discussion of the one-way ANOVA, the denominator variance estimate in the F ratio was called the within-groups estimate (s^2_{WG}) because it was obtained from the variability of the scores in each group (condition) around the group mean. Regardless of whether the H_0 was true or false, s^2_{WG} estimated only the random variation in the populations, i.e., σ^2. The same principle applies in a two-way ANOVA. Here we obtain what is called the **within-cells variance estimate** (s^2_{WC}) from the variability of the scores within cells about the cell means. The formula for this is:

$$s^2_{WC} = \frac{SS_{WC}}{df_{WC}} = \frac{\sum_{a=1}^{A} \sum_{b=1}^{B} \sum_{i=1}^{n} (X_{a,b,i} - \bar{X}_{a,b})^2}{AB(n-1)}$$

where: SS_{WC} = sums of squares for the within-cells estimate,

df_{WC} = df for the within-cells estimate,

A = number of columns in the Design,

B = number of rows in the design, and

n = number of observations per cell.

The numerator of the above formula uses embedded summation signs to indicate that the sum of the squared discrepancies of each score about the cell mean is computed for each cell and then summed for all cells, starting with the first cell ($a1,b1$) and ending with the last cell ($a3,b2$). This is a bit cumbersome, but in Chapter 14 we'll see there's an easier way to compute within cells variance estimate.

The *df* is computed by multiplying the number of cells (A times B) times $n - 1$ for each cell. **For the 3 X 2 design shown in Figure 1, the within-cells estimate is based on 24 df**, i.e., 4 df for each cell times six cells.

Regardless of whether any of the three H_0's to be tested are true or false, if the assumptions of the two-way ANOVA are met, s^2_{WC} is an estimate only of the population variance, σ^2.

The Main Effect of the A Variable

The H_0 tests of the main effects of the A and B variables are conceptually identical to the H_0 test in a one-way ANOVA. Variability of the main effect means about the overall mean is used to derive an estimate of the population variance (σ^2). For Variable A, the estimate is designated s^2_A and is obtained as follows:

$$s^2_A = \frac{SS_A}{df_A} = \frac{n_a \sum_{a=1}^{A} (\bar{X}_a - \bar{\bar{X}})^2}{A - 1}$$

where: SS_A = sums of squares for the A estimate,

df_A = df for the A estimate,

A = number of levels of the A Variable,

$\bar{X}_a$ = main effect mean for each level of the A variable,

$\bar{\bar{X}}$ = overall mean, and

n_a = sample size for each level of the A variable.

The formula says to take each of the main effect means for the A Variable, subtract the overall mean, square the difference, and then sum the result across all main effect means. The sum is multiplied by the number of observations for each level of the A variable. The sums of squares is then divided by the number of main effect means minus one. **For the 3 X 2 design shown in Figure 1, the main effect estimate for the A variable is based on 2 df**, i.e., 3 main effect means minus 1.

If the H_0 for Variable A is true (i.e., $\mu_{a1} = \mu_{a2} = \mu_{a3}$), then s_A^2 will estimate only the random variation of the population. If the H_0 is false and at least two population μ's differ, then the size of s_A^2 will reflect not only the random variation of the population (σ^2), but also the differences among the population μ's. The F ratio test of the H_0 is computed as:

$$F_A = \frac{s_A^2}{s_{WC}^2}, \quad \text{with 2 and 24 } df$$

If the H_0 is true, both the numerator and denominator estimates are estimating the same quantity, σ^2, then the obtained value of F will follow the sampling distribution of F for df_A and df_{WC} degrees of freedom, with an expected value of 1.00. If the obtained value of F is improbably large (i.e., p value is less than α), we will conclude that s_A^2 reflects the presence of differences among population means.

The Main Effect of the B Variable

The test of the H_0 for the main effect of the B variable is carried out in the same manner. You begin by obtaining an estimate of σ^2 from the variation among the Variable B main effect means. This estimate is abbreviated s_B^2 and is computed as:

$$s_B^2 = \frac{SS_B}{df_B} = \frac{n_b \sum_{b=1}^{B}(\bar{X}_b - \bar{\bar{X}})^2}{B-1}$$

where: SS_B = sums of squares for the B estimate,
df_B = df for the B estimate,
B = number of levels of the B Variable,
$\bar{X}_b$ = main effect mean for each level of the B variable,
$\bar{\bar{X}}$ = Grand mean, and
n_b = sample size for each level of the B variable.

For the 3 X 2 design shown in Figure 1, the main effect estimate for the B variable is based on 1 df, i.e., 2 main effect means minus 1.

If H_0 concerning the main effect of Variable B is true, the numerator estimate, s_B^2, will estimate σ^2. If H_0 is false, s_B^2 will estimate σ^2 *plus* systematic differences due to differences among population means. The F ratio test is computed as:

$$F_B = \frac{s_B^2}{s_{WC}^2}, \quad \text{with 1 and 24 } df$$

As with the test of the A variable, an improbably large value of F implies that the H_0 is false and that the two main effect population means are different.

The Interaction of A and B

The F-ratio test of the H_0 that the two variables are additive in their effects (do not interact) is somewhat more complicated. The logic runs like this. If the two variables A and B are additive in their effects on the dependent variable, then any differences among the cell means will be accounted for by the main effects of the A and B variables (if any), plus the operation of random error. We can see this by considering the hypothetical cell, column, and row means for 2 x 2 design shown in **Figure 4** on the next page.

There are three bits of information for each cell of the 2 X 2 design. The first line is the cell mean, the second line is the difference between the cell mean and the overall mean, and the third line is the amount of that difference not predicted or accounted for by the additive effects of the two independent variables. To compute this third quantity, we have expressed the main effects of each variable in terms of the distance that a main effect mean lies from the overall mean. These differences are indicated by the numbers adjacent to the lines linking each main effect mean to the overall mean.

Consider first the *difference* between the cell mean ($\bar{X}_{1,1} = 10$) and the overall mean ($\bar{\bar{X}} = 25$) for the a1,b1 cell. Its value is -15. Can this difference be accounted for by the fact that the cell is at the a1 level of Variable A and the b1 level of Variable B? The answer is yes. The difference of -15 is the additive effect of being in the a1 condition (-5) and in the b1 condition (-10). The amount of the difference that is not accounted for by the additive effects of the two variables is 0, which is what the third line in the cell indicates ("Unpredicted = 0"). Similarly, the difference of -5 for the a2,b1 cell is the purely additive effect of a2 (+5) and b1 (-10), i.e., +5 + (-10) = -5. You should satisfy yourself that the differences for the other two cell means are also completely "accounted for" by the additive effects of the two variables.

In contrast, consider the means for the 2 x 2 design shown in **Figure 5**. The *difference* between the cell mean ($\bar{X}_{1,1} = 10$) and the overall mean ($\bar{\bar{X}} = 22.5$) for the a1,b1 cell is -12.5. Is this difference entirely accounted for by the additive effects of the two variables? The answer is no. The main effect for a1 = -10 and for b1 = -7.5, which adds up to

Figure 4

Representation of a 2 X 2 Design in Which Cell Means Are an Additive Function of the Main Effects.

Variable A

	a1	a2	
b1	$\bar{X}_{1,1} = 10$ $\bar{X}_{1,1} - \bar{\bar{X}} = -15.0$ Unpredicted = 0	$\bar{X}_{2,1} = 20$ $\bar{X}_{2,1} - \bar{\bar{X}} = -5.0$ Unpredicted = 0	$\bar{X}_{b1} = 15$
b2	$\bar{X}_{1,2} = 30$ $\bar{X}_{1,2} - \bar{\bar{X}} = +5.0$ Unpredicted = 0	$\bar{X}_{2,2} = 40$ $\bar{X}_{2,2} - \bar{\bar{X}} = +15.0$ Unpredicted = 0	$\bar{X}_{b2} = 35$
	$\bar{X}_{a1} = 20$	$\bar{X}_{a2} = 30$	$\bar{\bar{X}} = 25$

Column effect: +5 / -5. Row effect: -10 / +10.

-17.5 units. The a1,b1 mean is not as far away from the overall mean as would be suggested by the main effects of the two variables. The unpredicted part of the distance is +5. A similar situation exists for the other cell means. The main effects of a2 and b1 are +10.0 and -7.5, respectively. Their additive effect is thus +2.5 units, yet the difference between the cell mean and the overall mean for cell a2,b1 is -2.5 units, an unpredicted amount of -5. A similar situation exists for the other two cells of the design.

Clearly then, the values of the cell means in this situation are not perfectly predictable from the additive main effects of the two independent variables. Does this mean that an interaction is present? The answer is "perhaps." It is also possible that the apparent non-additivity is just a result of error (random effects). This is what the F- ratio test is designed to determine.

To test the H_0 that the two variables exert only additive effects, we will derive a variance estimate from that portion of the variability of the cell means about the overall mean that is not accounted for by the main effects of the two variables. This is done by "adjusting" the differences between the cells means and the overall mean by "subtracting out" the additive influences. The formula for adjusting the differences for each cell mean is:

$$adj.(\bar{X}_{a,b} - \bar{\bar{X}}) = (\bar{X}_{a,b} - \bar{\bar{X}}) - (\bar{X}_a - \bar{\bar{X}}) - (\bar{X}_b - \bar{\bar{X}})$$

where: $\bar{X}_{a,b}$ = each cell mean,
$\bar{X}_a$ = main effect mean for the *a* level of the cell,
$\bar{X}_b$ = main effect mean for the *b* level of the cell, and
$\bar{\bar{X}}$ = overall or grand mean.

What this equation says in words is that you can adjust the difference between a cell mean and the overall mean by subtracting from it the amount that the row and column main effect means for that cell differ from the overall mean. This expression can be simplified a bit if we first get rid of the parentheses by multiplying through by the appropriate signs, yielding:

$$adj.(\bar{X}_{a,b} - \bar{\bar{X}}) = \bar{X}_{a,b} - \bar{\bar{X}} - \bar{X}_a + \bar{\bar{X}} - \bar{X}_b + \bar{\bar{X}}$$

If we then add up the negative and positive values of $\bar{\bar{X}}$ and rearrange the terms, we get the following form of the equation:

$$adj.(\bar{X}_{a,b} - \bar{\bar{X}}) = \bar{X}_{a,b} - \bar{X}_a - \bar{X}_b + \bar{\bar{X}}$$

Figure 5

Representation of a 2 X 2 Design In Which Cell Means Are an Interactive Function of the Main Effects.

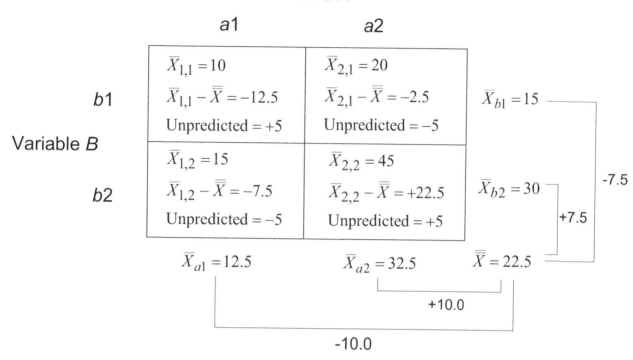

We can then use this adjusted difference for each cell mean to derive the population variance estimate according to the following formula:

$$s_{AB}^2 = \frac{SS_{AB}}{df_{AB}} = \frac{n_{a,b} \sum_{a=1}^{A} \sum_{b=1}^{B} (\bar{X}_{a,b} - \bar{X}_a - \bar{X}_b + \bar{\bar{X}})^2}{(A-1)(B-1)}$$

where: SS_{AB} = sums of squares for the interaction estimate,
df_{AB} = degrees of freedom for interaction estimate,
$n_{a,b}$ = number of scores in each cell,
A = number of levels or categories Of the A variable, and
B = number of levels or categories of the B variable.

In words, what the formula says to take each cell mean, subtract from it the main effect means for the row and column means corresponding to the cell, and then add to it the overall mean. Square this quantity and add up the values for all the cells. Multiply this sum by the number of observations per cell and then divide this sum by the number of levels of the A variable minus one times the number of levels of the B variable minus one.

It may seem odd to you that the formula requires adding $\bar{\bar{X}}$ to the cell mean after subtracting out the two main effect means. The reason for doing so is that each time you subtract out a main effect mean, you also subtract out the overall mean. The overall mean gets subtracted out twice, when it should only be subtracted from the cell mean one time. The addition of the overall mean compensates for the double subtraction.

The df for this variance estimate is the number of cell means that are "free to vary" once you consider the main effects of the two variables. The row and column main effect means constrain the values of the cell means. For example, in our 2 x 2 design, once any one cell mean is given a value, the remaining three cell means are constrained in value by virtue of the fact that the row and column means all have particular values. There is, thus, only 1 df for the adjusted cell estimate. **The general formula for computing the df for the interaction variance estimate is $(A - 1) \times (B - 1)$.**

If H_0 is true and the two variables are purely additive in their effects in the population, then s_{AB}^2 derived from the adjusted differences between the cell means and the overall mean will estimate only random error, i.e., σ^2. If, however, the variables interact in their influence on the dependent variable, then s_{AB}^2 will reflect both random error and non-additive differences among population means. The F test of the H_0 for the 3 X 2 design is carried out as

$$F_{AB} = \frac{s_{AB}^2}{s_{WC}^2}, \text{ with 2 and 24 } df$$

If H_0 concerning additivity is true, both the numerator and denominator estimates will be estimating the same quantity, σ^2, and the obtained value of F will follow the sampling distribution of F for df_{AB} and df_{WC} degrees of freedom. An improbably large value of F is taken to mean that s_{AB}^2 has been influenced by the presence of interaction effects.

Interpreting the Results of a Two-Way ANOVA

The simplest outcome of a two-way ANOVA is that none of the F ratios will reach statistical significance, i.e., their p values all will be greater than the α level criterion that you decided to use. If this occurs, your work is ended—albeit in a not very satisfactory manner. The interpretation is simple; neither independent variable exerts an influence on the dependent variable.

A happier outcome is to find that at least one of the F ratios is significant. If one or both F ratios for the main effects of the two variables is significant and the interaction F ratio is not, the interpretation of the results is straightforward. You will carry out post-hoc contrasts among the *main effect means* for each of the significant variables to decide which means are significantly different (assuming there are more than two means). There is no point in comparing the cell means (simple effects) because the lack of a significant interaction means that the effect of the significant variable will be the same for all levels or categories of the other variable

A more complex situation arises when the test for the interaction is significant. In this case, our interpretation of the main effects of the two variables is complicated by the presence of the interaction. This is because the main effect means do not completely reflect the nature of the effects of the independent variables. As a result, we need to consider the simple effect means to understand how the two variables combine to affect the dependent variable. Post-hoc contrasts are conducted among these means and as indicated above, we will need to construct an interaction plot of the simple effect means to fully interpret the nature of the interaction. Part of this interpretative process is to examine the value of η^2 for each significant effect to compare their relative strengths (see below).

Partitioning the Sums of Squares

As we learned in the previous chapter on one-way ANOVA, the SS_{Total} can be broken into its component parts. In a two-way ANOVA, SS_{Total} is given by the following:

$$SS_{Total} = SS_{WC} + SS_A + SS_B + SS_{AB}$$

Two-Way ANOVA With *JASP*

The following example problem ("Two Possible Encounters") will be analyzed using the same *JASP* ANOVA procedure used in the one-way ANOVA. **Read the description of the problem on the next page completely before continuing.**

Are the Assumptions Met?

The precise sampling strategy used in the experiment is not described, but it is clearly a convenience sample. As such, there is no clear-cut population to which the results may be generalized. Given the nature of the dependent variable and the fact that college students are used as subjects we would probably want to restrict our conclusions to young, well-educated Americans. The subjects were, however, randomly and independently assigned to the conditions of the experiment. No information is presented here concerning the shapes of the populations represented by the samples for each treatment combination, but inspection of the rating scores suggest that they are probably at least unimodally distributed. The cell variances are not terribly homogeneous thanks to the low ceiling of the rating scale, making it questionable whether the homogeneity of variance assumption is met.

Format for the Data Set

Figure 6 shows a screen shot of the *JASP* Data Editor for five cases showing how the data need to be formatted for a two-way ANOVA.

Figure 6

Screen Shot of JASP Data Editor Showing Data Format for the Experiment.

	Type E	Gender	Rating
12	Emot.	Male	5
13	Emot.	Male	5
14	Emot.	Male	4
15	Emot.	Female	7

Two Possible Encounters

Several students and I carried out a simple experiment to investigate sources of jealousy in college students. Volunteers from several psychology classes read a short fictional story about an encounter that one's boyfriend (in the case of female students) or girlfriend (in the case of male students) reported back to them. One story was about an emotional encounter and the other about a sexual encounter. Our interest was in which story would make the person most jealous and whether males and females differed in their response to the story. Students read **one** of the two stories and then used a 9-point scale to indicate how jealous they would be. Anger was measured by a different question. The scale looked like this, and the volunteers circled their response:

How jealous would this be if it happened to you?

1	2	3	4	5	6	7	8	9
Not At all Jealous				Quite a Bit Jealous				Intensely Jealous

Here are the two stories. Remember, the volunteers read only one of the two stories and the gender was changed to be appropriate for male and female volunteers:

Emotional Encounter
"I just met this girl in my class and somehow, we really connected emotionally. We went to Starbucks after class and talked for hours until it closed. I shared things with her that I've never shared with anyone else. I'm not sexually attracted to her at all, but it was an awesome experience."

Sexual Encounter
"I just met this girl in my class and one thing led to another and we wound up having amazing sex at her apartment. I spent the whole night and we did things together I'd never done with anyone else. She's not the type of girl I'd ever get really serious about, but it was an awesome experience."

Here are the jealousy rating scale data for the male and female volunteers for type of encounter (*Type-E*):

Emotional Encounter Story				Sexual Encounter Story			
Males		Females		Males		Females	
4	3	7	9	9	8	9	9
3	8	6	8	7	9	8	7
5	5	7	5	8	6	9	7
7	3	8	6	7	6	6	9
4	5	7	7	8	8	7	8
6	5	9	9	7	4	9	9
2	4	7	6	7	9	8	10*
		8	6	7		9	8
		7	7			8	9
		8					

* One volunteer added 10 and circled it.

Running the Analysis

The analysis of the data uses the same *JASP* ANOVA sub procedure that we used for a one-way analysis. The only difference is that the two experimental variables (*Type E* and *Gender*) are moved over to the active "Fixed Factors" box. In running the procedure, the following options are selected: Descriptive Statistics and Estimates of Effect Size (η^2 and partial η^2 and ω^2) and Homogeneity Tests (Levine's test). Also selected are Post-Hoc comparisons of the main effect and interaction (simple effects) means using **Tukey's HSD procedure** and an interaction plot under the Descriptive Plots option.

Homogeneity of Variance Test

Figure 7 shows the results of Levene's test for homogeneity of variance for the populations represented by the four samples. The *p* value of .359 suggests that there is a modest risk that the assumption may be violated, probably with a very limited effect.

Figure 7

Levene's Test For Homogeneity of Variance.

Test for Equality of Variances (Levene's)

F	df1	df2	p
1.092	3.000	62.000	0.359

ANOVA Source Table

We'll begin with an examination of the ANOVA source table, considering the *F* ratios for each treatment. We'll consider the effect size statistics in a separate section below. The ANOVA source table is shown in **Figure 8**.

The *F* ratio for the main effect of type of encounter (*Type E*) is obtained by dividing the variance estimate for *Type E* (the "mean square" for *Type E*) by the within-cells ("Residuals") variance estimate. The *F* ratio of 36.355 is associated with a *p* value of < .001. From this we know that, overall, type of encounter has a highly significant effect on jealousy.

The *F* ratio for the main effect of *Gender* is also very large, *F* = 31.841, and highly significant (*p* value < .001), indicating that, overall, the gender of the participant has a large effect on jealousy.

Finally, the *F* ratio for the interaction of *Type E* and *Gender* is also significant (*F* = 7.121, *p* value = .01). This tells us that the effect of type of encounter on jealousy is different, depending upon the gender of the person.

Interpreting the Effect Size Statistics

In the previous chapter on one-way ANOVA, we encountered the statistics η^2 and ω^2 and indicated that many investigators preferred the latter as a more accurate estimate of the percentage of variance accounted for by the independent variable. In the present data, the two are much closer in value than was the case for the "Your Order, Please" experiment. The reason for this is that the sample size in the present experiment (*n* = 66) is much larger than that in the salary experiment (*n* = 24), and that the degree of bias with η^2 diminishes with increasing sample size.

JASP also computes another effect size statistic called partial η^2 (η^2_p), which is used when there is more than one independent variable.

Recall that Eta squared was defined in the last chapter as:

$$\eta^2 = \frac{SS_{BG}}{SS_{Total}}$$

which also can be expressed as:

$$\eta^2 = \frac{SS_{BG}}{SS_{Total}} = \frac{SS_{BG}}{SS_{WG} + SS_{BG}}$$

because the total sums of squares is simply the sum of between group and withing group sums of squares.

But when there are two independent variables (call them *A* and *B*), η^2 for the *A* variable looks like this:

$$\eta^2 = \frac{SS_A}{SS_{Total}} = \frac{SS_A}{SS_{WC} + SS_A + SS_B + SS_{A \times B}}$$

where: SS_{WC} = sums of squares for the within cell error variance estimate.

Figure 8

JASP ANOVA Source Table for the "Type of Encounter" Experiment.

ANOVA - Rating of jealousy for type of encounter (*Type-E*) and Participant's gender (*Gender*)

Cases	Sum of Squares	df	Mean Square	F	p	η^2	η^2_p	ω^2
Type E	59.535	1	59.535	36.355	<.001	0.276	0.370	0.266
Gender	52.142	1	52.142	31.841	<.001	0.241	0.339	0.232
Type E ✶ Gender	11.661	1	11.661	7.121	0.010	0.054	0.103	0.046
Residuals (WC)	101.531	62	1.638					

Note. Type III Sum of Squares Note: I heavily edited the title of the table and inserted (WC).

Now the total sums of squares is made up of the within cell (error) sums of squares plus the sums of squares for the two main effects and the interaction. As a result, the values of η^2 and ω^2 now reflect the **portion of the total sums of squares contributed by each component of the analysis**. For the analysis shown in **Figure 8**, ω^2 for the main effect of type of encounter equals .266, which means that *Type E* accounts for 27% of the total variability in the experiment. Moreover, the two main effects and the interaction combined account for 54% of the total variability (.266 + .232 + .046). This is a very large value for psychology experiments and clearly indicates that jealousy and gender are powerful motivators!

Some statisticians have argued that η^2 (and ω^2) understate the magnitude of the effect size when computed in this manner. They argue that including all components of the analysis in the denominator for η^2 reduces the size of the ratio for each component, i.e., the estimate of the magnitude of its effect.

Instead, they suggest that the effect size for each component of the analysis can be assessed relative to just the error sums of squares, i.e., the random variation among participants. For example, for Variable A, the corrected formula would be:

$$\eta^2_p = \frac{SS_A}{SS_{WC} + SS_A}$$

The resulting quotient is called a "partial η^2" and is symbolized as η^2_p. The "partial" part of the name means that the formula "partials out" the influence of the other components of the analysis from the denominator. The same principle applies to the partial eta squared for the *B* variable and the *A* X *B* interaction.

As you can see in the source table for the experiment, the values for η^2_p for both variables and the interaction are appreciably larger than those for η^2. The difficulty lies in interpreting exactly what the numbers mean. Although η^2_p is clearly a ratio of sums of squares, it is not a measure of the percentage of the total variability present in the data, so it makes no sense to talk about the relative importance of the various components *within the experiment*. Moreover, unlike η^2, η^2_p can exceed 1.00 when summed across all of the components. Finally, unlike η^2, η^2_p is not considered as an estimator of a corresponding parameter in the populations of interest. It exists only for the sample.

So, what do you use η^2_p for in the ANOVA? Basically, consider it another index that allows you to make comparative statements about the "size" of one component compared with another component, e.g., η^2_p for *Type E* is 9% larger than that for *Gender* and 3.7 times bigger than for the interaction of *Type E* and *Gender*.

More importantly, it is also used to compare the magnitude of the effect of a particular variable across multiple experiments that may not include the same combination of variables.

Descriptive Analysis of the Data

Figure 9 shows a table of the basic descriptive statistics for the experiment included as an option under the ANOVA sub-procedure. The four cell means show substantial

Figure 9

JASP Descriptive Statistics for the Four Conditions of the Two Possible Encounters Experiment.

Descriptives – Jealousy Ratings for both main effects.

Type E	Gender	N	Mean	SD	SE
Emot.	Female	19	7.211	1.134	0.260
	Male	14	4.571	1.651	0.441
Sexual	Female	18	8.278	1.018	0.240
	Male	15	7.333	1.345	0.347

Note: SE stands for the standard error of the mean.

differences, with female ratings tending to be higher than those of the males. For both genders sexual infidelity produces greater jealousy ratings than emotional infidelity. The unusually low ratings for males for emotional infidelity suggests a potential interaction of gender and type of infidelity. The standard deviations are all similar in value, mostly consistent with the results of Levene's test.

Post-Hoc Comparisons

Post-hoc comparisons among the sample means show that for both men and women jealousy is significantly greater for sexual v. emotional encounters. For emotional encounters, women are significantly more jealous than men, but the sexes don't differ in jealousy for sexual encounters.

Interaction Plot

The nature of the interaction between T*ype E* and *Gender* is best seen in pictorial form. **Figure 10** shows a function graph with *Type E* on the X axis and separate function lines for men and women. Given that both independent variables were categorical in nature, a comparative bar-like function graph would also have been appropriate, but as we noted in Chapter 12, the current version of *JASP* doesn't produce them for ANOVA.

The presence of the Gender X Type of Encounter interaction means that a portion of the main effect for Gender is coming from the tendency for females to be markedly more jealous than males at a partner's emotional infidelity in addition to their greater general tendency toward jealousy. (This isn't some lame sexist comment by the way. There are extremely good evolutionary reasons for females to be more prone than males toward partner jealousy.)

We will end this chapter we an example narrative description of the results of the analysis of the experiment:

The data for the Two Possible Encounters experiment were subjected to a 2 X 2 ANOVA. Both main effects and the interaction were statistically significant. The Sexual Encounter scenario produced significantly greater jealousy than the Emotional Encounter in both men and women. Women were more jealous than men in both scenarios, significantly so in the Emotional Encounter scenario. Collectively, the main effects and interaction accounted for 54% of the variance in jealousy ratings.

Figure 9

JASP Interaction Plot of the Four Conditions of the Two Possible Encounters Experiment (Female Data in Open Circles).

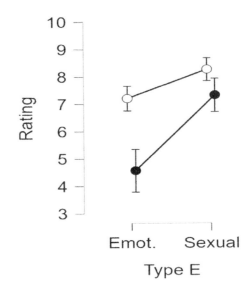

Chapter 13, Two-Way ANOVA
Chapter Exercises

Some questions are purely factual and have only one correct answer. Others require greater thought on your part and may have more than one possible answer.

1. Fill in the missing term(s) that makes the statement true.

   a. In experiments, researchers manipulate more than one independent variable because _____.
   b. If the "design of an experiment" is a 3 X 3, there are _____ combinations of independent variables.
   c. Two-way ANOVAs are divided into _____ and _____ effects.
   d. $\bar{\bar{X}}$ is the _____, which is _____.
   e. In a 3 X 3 "design", $\bar{X}_{a2,b3}$ is the cell mean for the _____ level of the A variable and the _____ level of the B variable.
   f. When two variables interact in a "synergistic" fashion, their combined effect is _____.
   g. s_A^2 / s_{WC}^2 tests _____.
   h. In the 3 X 2 design described in **Figure 1** in the chapter, the F test for A is carried out with _____ df.
   i. The assumption of normality in the populations represented by the cells is required because _____.
   j. $(\bar{X}_{a,b} - \bar{X}_a - \bar{X}_b + \bar{\bar{X}})$ is the _____ for a given _____.
   k. For a 3 X 3 design, the interaction F test has _____ df in the _____.
   l. For the Two Possible Encounters experiment, $\eta^2 = .24$ for gender, which means that _____.
   m. In the Two Possible Encounters experiment, the significant interaction reflects the fact that _____.

2. Why do we help others? Three students undertake a casual study of variables that influence altruistic behavior in college students. The experiment manipulates the target of the altruistic act (Cousin, Brother, Sister) and the degree of social closeness to the target (Close growing Up, Fairly Distant Growing Up), making a 3 X 2 ANOVA design. Sixty-five volunteers read one of six stories about a person needing a loan to get him or her through a difficult situation. The volunteer is told that he or she has $10,000 saved and is asked how much of that amount he/she would lend to the target with only a small chance of getting it back. Here is the story:

   Your older _____ has gotten himself (herself) into a financial bind and comes to you for a loan, saying, "Anything would help." You have managed to save $10,000 by working two jobs and have plans for the money. You have always been close to your _____ growing up (Or, You were never close to your _____, although you got along well enough). Understanding that you're not likely to get it back, how much would you be willing to loan your _____ ?

Ok, their advisor says that this was not the most well-designed study, but maybe the results would be interesting. Here are the data, rounded to the nearest $1,000:

Close Growing Up						Never Close Growing Up					
Cousin		Brother		Sister		Cousin		Brother		Sister	
5	10	8	10	10	10	2	3	6	7	10	9
7	5	10	7	9	10	3	3	7	4	8	10
6	4	8	10	10	8	1	0	5	6	9	8
5	6	6	9	8	10	3	5	5	6	10	7
5	5	10	8	10	9	2	0	4	5	9	8
9		8		10		6		5			

Use *JASP* to do a complete analysis of the data and write an insightful summary and interpretation of the findings.

3. Recall the problem in Chapter 1 describing an altruism study in children. Here's the description of the study again. Using your new knowledge of statistics, you're going to do a complete analysis of the data.

Altruism refers to the willingness to help others without receiving any immediate benefit in return. In a study of altruism in children, 85 children of different ages were given the opportunity to share a cup containing 10 wrapped hard candies with another child from your school "who is in the next room who doesn't have any candy." Here are the data for the study. The dependent variable was the number of pieces of candy that the child placed in an empty cup to be given to the other child.

	Number of Children Sharing Various Amounts of Candy					
	0 Pieces	1 Pieces	2 Pieces	3 Pieces	4 Pieces	5 Pieces
3 – 4 year old Boys (*n* = 12)	5	5	2	0	0	0
3 – 4 year old Girls (*n* = 15)	4	7	2	1	1	0
5 – 6 year old Boys (*n* = 13)	2	5	5	1	0	0
5 – 6 year old Girls (*n* = 15)	1	4	4	3	2	1
7 – 8 year old Boys (*n* = 14)	2	4	6	1	1	0
7 – 8 Year old Girls (*n* = 16)	0	3	5	5	2	1

Begin by creating the appropriately formatted data file and then carry out a complete descriptive and inferential analysis of the results. Code *Gender* as 1 = Boys and 2 = Girls. For the independent variable of *Age*, you can use either the lower or upper value; it makes no difference in the analysis.

If you are feeling particularly adventurous, this is one of those situations where the data could also be analyzed using multiple linear regression. If you change both *Gender* and *Age* to numerical variables ("Scale") you can enter them as Covariates (predictors) in the REGRESSION – Linear regression procedure to predict number of pieces of candy that each child shares.

(Normally in this case the investigator would also create an "interaction variable" to be included as a predictor along with *Gender* and *Age*. Such variables are created as the product of the two other variables to be included in the analysis and adjusting the values of the two variables to minimize multicollinearity. For these data, the ANOVA shows no hint of an interaction between *Gender* and *Age* so you don't need to create an interaction variable.)

4. As part of a senior thesis, a student investigated how "bogus feedback" to a person about his/her level of physiological arousal influenced that person's altruistic behavior. The experiment derived from the theory that people act in an altruistic manner to reduce the distress they experience when they see others experiencing misfortune. The more distress that a person experiences, the greater will be his/her willingness to act altruistically. After being fitted with a heart-rate monitor, individual student volunteers seated in a small room watched a video depicting the plight of children in a small African country experiencing severe draught. The student volunteers were led to believe that the purpose of the study was to assess how viewing the distressing video influenced their heart rate and were told that they could listen to their heartbeat over a small speaker while they watched the video. In fact, depending upon condition, some volunteers heard sounds indicating their actual heart rate or a rate either much slower or faster than their actual rates. This was "bogus feedback" and its purpose was to mislead the volunteers about their level of arousal induced by viewing the distressing video. There were three feedback conditions to which volunteers were randomly assigned: (a) Actual Rate (a control condition), (b) Artificially Slow Rate, and (c) Artificially Fast Rate. In addition to the actively manipulated variable, the experimenter used a self-report questionnaire to divide her volunteers into dispositional "low empathic" vs. "high empathic" groups. Level of dispositional empathy constituted a second variable of the experiment. At the conclusion of the video, volunteers were told that the experimenter had agreed with the relief agency that produced the video to collect donations to relieve the situation. Each subject was given an official donation envelope and asked to make whatever donation he/she could. The amount donated was the dependent variable of the experiment. (The money *was* actually donated to the agency in question.) Below are the amounts donated by each volunteer for the different conditions of the experiment.

	Actual Rate			Artificially Slow Rate			Artificially Fast Rate		
Dispositional Low Empathic	1.00	.50	1.00	.25	0	1.00	.50	0	1.00
	0	0	1.25	1.00	.50	.50	.50	1.00	.50
	1.25	.75	1.00	1.25	1.00	0	0	.75	1.00

dispositional	2.00	1.75	3.00	1.75	2.15	3.25	10.00	2.95	5.00
High Empathic	3.25	1.00	2.25	1.50	5.00	1.00	5.75	8.59	7.50
	1.50	2.00	2.75	1.00	2.50	2.25	3.75	5.00	12.00

Use *JASP* to do a complete analysis of the data and write an insightful summary and interpretation of the findings.

5. There's a famous article in psychology titled "What is Beautiful is Good," published in 1972 and replicated literally dozens of times. In essence, throughout history people who are particularly physically attractive are seen as kinder, more intelligent, happier, etc. Two psychology majors speculate that the times have changed, and modern people are no longer taken in by a pretty face. They do a simple study in which classmates are shown a photograph either an average looking young woman OR a stylistically beautiful young woman. Both women are dressed identically, are of the same apparent age, and both are smiling. Under the guise of having participants judge personality qualities from photographs, they get 11-point ratings for several qualities—one of which is "Having a Good Heart." The scale ranged from 1 (Not at All Nice) to 6 (Average Niceness) to 11 (Incredibly Nice). Here are the data:

Male Participants				Female Participants			
Average Face		Beautiful Face		Average Face		Beautiful Face	
6	4	8	10	7	8	4	2
8	5	10	9	6	9	6	3
4	5	7	8	8	6	1	2
6	7	11	10	3	7	3	4
7	5			5	5	2	3
				6	5	1	8
				5		2	3
						1	

Use *JASP* to completely analyze the data. The results are quite interesting.

Chapter 14
Factorial ANOVA

What's In This Chapter?

Factorial ANOVA simply a more general case of two-way ANOVA except for the complexity of having multiple interactions. This chapter explores factorial ANOVA using two three-way ANOVA examples. A three-way ANOVA has three independent variables, which means that there are three main effect *F* tests, three **first order** (two-way) interaction *F* tests, and one **second order** (three-way) interaction *F* test.

A first-order interaction is the interaction of two of the independent variables summed across the third variable not involved in the interaction. In a three way ANOVA having independent variables *A*, *B*, and *C*, the first order interactions are *A* X *B*, *A* X *C*, and *B* X *C*. The second order interaction is *A* X *B* X *C*.

As was the case for the two-way ANOVA, post-hoc comparisons are made to identify the source of significant main effects or interactions. The statistics ω^2 and η^2 should be used to quantify the effect size for significant main effects and interactions.

Introduction

This chapter expands on the topic of analysis of variance discussed in the previous chapter on two-way ANOVA. It does so by expanding the general concept to three-way ANOVA, i.e., the situation where there are three independent variables. A four-way ANOVA example could be used, but the analysis and interpretation of such designs is hopelessly tedious and complex. Designs beyond three independent variables are almost never used in practice due to the difficulty in securing enough cases to maintain adequate and equal sample sizes for all cells of the design. Where there are more than three independent variables, investigators frequently will adopt a regression solution in place of factorial ANOVA.

This is a good spot to introduce some new terms and concepts. In ANOVA parlance, the independent variables are often called **factors** and when there is more than one independent variable the term **factorial ANOVA** is used. In all factorial ANOVAs, every level of an independent variable is paired with every level of all other independent variables. When there are no missing or "empty" cells, the ANOVA is said to be **completely crossed**. The results of an Incompletely crossed ANOVA become more difficult to interpret using an ANOVA framework and are thus almost never used.

Three-Way ANOVA

Figure 1 on the next page shows in abstract form the results of a factorial experiment having two levels for the first independent variable (Variable *A*), two levels for the second independent variable (Variable *B*), and two levels for the third independent variable (Variable *C*). There are five cases per cell, and the cell, row, and column means have all been computed.

The term "**levels**" is used here generically. If the conditions of an independent variable are quantitatively different amounts of the variable, then the term "levels" appropriately means the amount of the variable. If the conditions of the independent variable are qualitatively different categories (e.g., types of teaching methods or type of drug administered), then the term "**categories**" is used.

When the categories are intact groups such as race, gender or profession, the term "**grouping variables**" is often used to distinguish the variable from true independent variables. Significant effects for such grouping variables are often ambiguous because many things are likely to be correlated with the variable. For example, the grouping variable gender frequently is significant either as a main effect or as part of an interaction (see the example below). In fact, males and females differ along so many dimensions that it's often not clear what a gender effect really means.

The diagram in **Figure 1** is restricted to two levels (categories) for each variable simply for ease of drawing! In fact, any of the variables could have more than two levels or categories.

Recall from the previous chapter that diagrams of this type show the **design of the experiment**. The levels or categories of the three independent variables (factors) are

Figure 1

Diagrammatic Representation of a Completely Crossed 2 X 2 X 2 Factorial ANOVA with Five Cases per Cell.

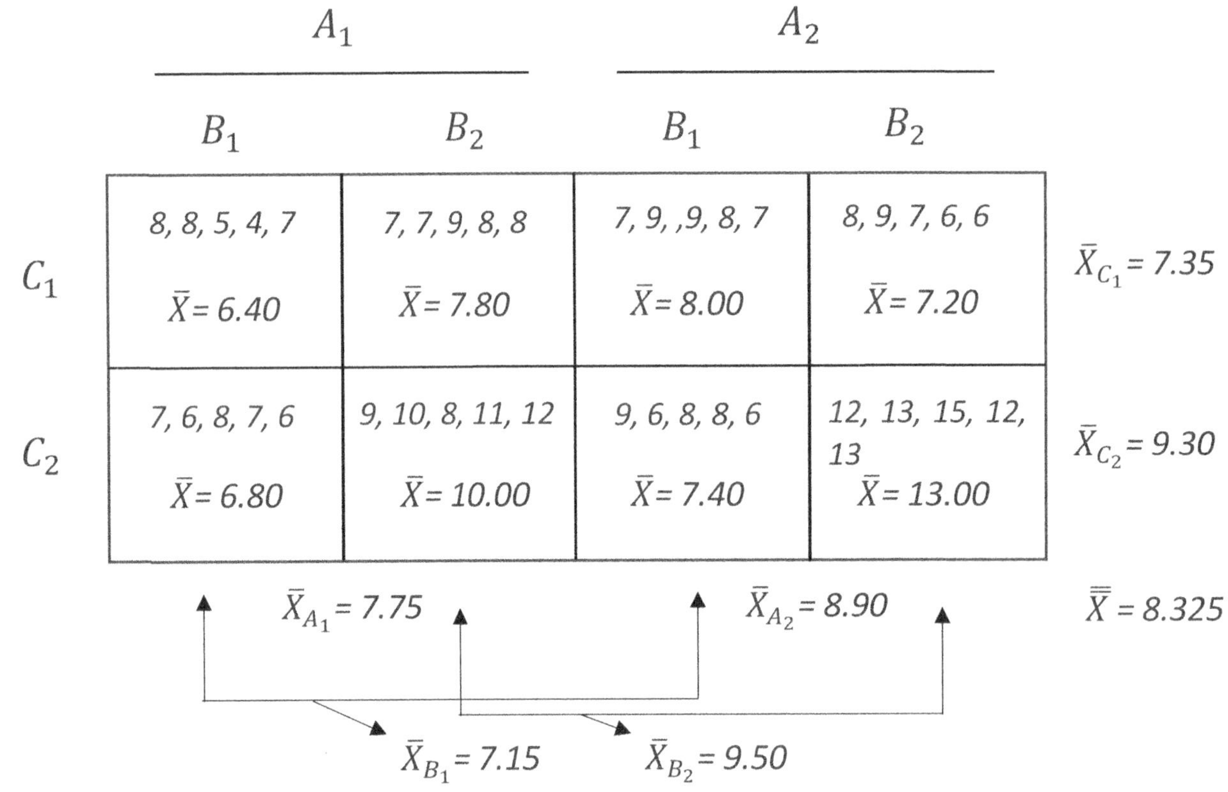

	A_1		A_2		
	B_1	B_2	B_1	B_2	
C_1	8, 8, 5, 4, 7 $\bar{X} = 6.40$	7, 7, 9, 8, 8 $\bar{X} = 7.80$	7, 9, ,9, 8, 7 $\bar{X} = 8.00$	8, 9, 7, 6, 6 $\bar{X} = 7.20$	$\bar{X}_{C_1} = 7.35$
C_2	7, 6, 8, 7, 6 $\bar{X} = 6.80$	9, 10, 8, 11, 12 $\bar{X} = 10.00$	9, 6, 8, 8, 6 $\bar{X} = 7.40$	12, 13, 15, 12, 13 $\bar{X} = 13.00$	$\bar{X}_{C_2} = 9.30$

$\bar{X}_{A_1} = 7.75$ $\bar{X}_{A_2} = 8.90$ $\bar{\bar{X}} = 8.325$

$\bar{X}_{B_1} = 7.15$ $\bar{X}_{B_2} = 9.50$

called the "experimental treatments," and the combinations of treatments are called the "**cells**" of the design. The experiment would be called a 2 X 2 X 2 design. In this example, there are five observations per cell. In more complex ANOVA designs it's an extremely good idea to have equal or near equal cell sample sizes. Markedly unequal sample sizes make the interpretation of the results (if any) more difficult.

Recall that the means computed for the cells represent the **simple effects** of the two variables. There are also means computed for the columns and rows of the design. The two column means corresponding to the two levels of the A variable are the main effect means and are labeled $\bar{X}_{A_1}$ and $\bar{X}_{A_2}$. As in a two-way ANOVA, these main effect values are computed by averaging *across* the 10 observations in the two levels of the B variable and the 10 observations in two levels of the C variable. The two means labeled $\bar{X}_{B_1}$ and $\bar{X}_{B_2}$ are the main effect means for the B variable. These are computed by averaging the 20 observations in the A and C levels. $\bar{X}_{C_1}$ and $\bar{X}_{C_2}$ are the main effect means for the C variable. The mean indicated by $\bar{\bar{X}}$ is the overall mean and the average of all 40 cases.

The three first-order interaction tables are shown in **Figures 2–4**. Each shows the 2 X 2 table for the two variables involved in the interaction collapsed across the third variable. The cell means for each table are included.

Figure 2

First order Interaction Table for the A X B Interaction.

	A_1	A_2
B_1	$\bar{X}_{A_1 B_1} = 6.60$ n = 10	$\bar{X}_{A_2 B_1} = 7.70$ n = 10
B_2	$\bar{X}_{A_1 B_2} = 8.90$ n = 10	$\bar{X}_{A_2 B_2} = 10.10$ n = 10

Figure 3

First order Interaction Table for the A X C Interaction.

	A_1	A_2
C_1	$\bar{X}_{A_1C_1} = 7.10$ n = 10	$\bar{X}_{A_2C_1} = 8.60$ n = 10
C_2	$\bar{X}_{A_1C_2} = 8.40$ n = 10	$\bar{X}_{A_2C_2} = 10.20$ n = 10

Figure 4

First order Interaction Table for the B X C Interaction.

	B_1	B_2
C_1	$\bar{X}_{B_1C_1} = 7.20$ n = 10	$\bar{X}_{B_2C_1} = 7.50$ n = 10
C_2	$\bar{X}_{B_1C_2} = 7.10$ n = 10	$\bar{X}_{B_2C_2} = 11.50$ n = 10

The three-way interaction is simply the eight cells of the design having five observations per cell.

Computing the Variances In a Three-Way ANOVA

As we noted in Chapter 12, the term "analysis of variance" literally means breaking down the variance sources in an array of data. Once the design of the experiment becomes fairly complex, the analysis is best carried out working with the sums of squares (SS) for the various components of the experiment. We'll use the data in the layout shown in **Figure 1** to show how such an analysis is carried out—either "by hand" or by a computer program like *JASP*.

1. Compute the Total Sums of Squares.

The total sums of squares (SS_{Tot}) for the data shown in **Figure 1** is simply the sum of the squared differences between every score and the grand mean:

$$(1) \sum_{i=1}^{n}(X_i - \bar{\bar{X}})^2 = 218.775$$

where: n = total number of observations (40).

2. Compute the Main Effects Sums of Squares.

The three main effect sums of squares are computed in the same way that we did for a two-way ANOVA:

The sums of squares for the *A* variable:

$$SS_A = n_a \sum_{a=1}^{A}(\bar{X}_a - \bar{\bar{X}})^2 = 13.225$$

where: n_a = number of observations in each level of the *A* variable (20) and
A = number of levels of the *A* Variable (2).

The sums of squares for the *B* variable:

$$SS_B = n_b \sum_{b=1}^{B}(\bar{X}_b - \bar{\bar{X}})^2 = 55.225$$

where: n_b = number of observations in each level of the *B* variable (20) and
B = number of levels of the *B* Variable (2).

The sums of squares for the *C* variable:

$$SS_C = n_c \sum_{c=1}^{C}(\bar{X}_c - \bar{\bar{X}})^2 = 38.025$$

where: n_c = number of observations in each level of the *C* variable (20) and
C = number of levels of the *C* Variable (2).

3. Compute the First Order Interaction Sums of Squares.

The first order (two-way) interactions are computed by first finding the sums of squares for the cells involved collapsed across the other variables, and then subtracting the sums of squares for the main effects involved in the interaction. In finding the sums of squares for cells, the cell means are taken about the overall (grand) mean.

For the A X B interaction, the sums of squares for the four cells is given by:

$$SS_{AB\ Cells} = n_{abcell} \sum_{cell=1}^{Cell} (\bar{X}_{cell} - \bar{\bar{X}})^2 = 68.475$$

where: n_{abcell} = number of observations in each AB cell (10) and
Cell = number of cells (4).

The sums of squares for the A X B interaction is thus:

$$SS_{AB} = SS_{AB\ Cells} - SS_A - SS_B$$

$$= 68.475 - 13.225 - 55.225$$

$$= .025$$

The sums of squares for the A X C and the B X C interactions are computed in exactly the same way:

$$SS_{AC\ Cells} = n_{accell} \sum_{cell=1}^{Cell} (\bar{X}_{cell} - \bar{\bar{X}})^2 = 55.475$$

$$SS_{AC} = SS_{AC\ Cells} - SS_A - SS_C$$

$$= 55.475 - 13.225 - 38.025$$

$$= 4.225$$

and,

$$SS_{BC\ Cells} = n_{bccell} \sum_{cell=1}^{Cell} (\bar{X}_{cell} - \bar{\bar{X}})^2 = 135.275$$

$$SS_{BC} = SS_{BC\ Cells} - SS_B - SS_C$$

$$= 135.275 - 55.225 - 38.025$$

$$= 42.025$$

4. Compute the Second Order Interaction Sums of Squares.

The A X B X C interaction is computed in exactly the same manner as the first order interactions—by subtracting sums of squares. The "Cells" sums of squares for the three-way interaction is simply the squared deviations of all eight cells about the grand mean:

$$SS_{ABC\ Cells} = n_{abccell} \sum_{cell=1}^{Cell} (\bar{X}_{cell} - \bar{\bar{X}})^2$$

$$= 165.975$$

where: $n_{abccell}$ = Number of observations in each ABC cell (5) and
Cell = number of cells (8).

Once the SS for the eight ABC cells has been computed, the A X B X C interaction is computed by subtracting the three main-effect SS and the three two-way interaction SS from the cells SS:

$$SS_{ABC} = SS_{ABC\ Cells} - SS_A - SS_B - SS_C - SS_{AB} - SS_{AC} - SS_{BC}$$

$$= 165.975 - 13.225 - 55.225 - 38.025 - .025 - 4.225 - 42.025$$

$$= 13.225$$

5. Compute the Within Cells Sums of Squares.

The computation of the within cells sums of squares is done by subtracting all main effect and interaction sums of squares from the total sums of squares. That's why JASP calls the within cells sums of squares the "**Residuals.**" Without going through all the computations, let's just say that the value is 52.800

Figure 5 shows the JASP ANOVA source table for the data that we've just analyzed by hand. We'll carry out a brief interpretation of the results shown in the table.

The first thing to note is that all three main effects are highly significant, with the main effects of variables A, B, and C accounting for almost half the variance in the data. Two of the interactions, B X C and A X B X C are also highly significant and that complicates the interpretation of the analysis. Generally, in complex factorial ANOVAs, if interactions involving main effect variables are also significant, one needs to carefully examine what is giving rise to the main effects. Inspection of the cell means shown in **Figures 1** and **Figure 4** makes it clear that two cell means are producing almost all of the main effects. The much larger means for $Cell_{A1,B2,C2}$ and $Cell_{A2,B2,C2}$ are producing the very strong B X C interaction and contributing to the A X B X C interaction. They are also inflating the A_2, B_2, and C_2 marginal means leading to the three main effects.

Because the data used in this analysis are purely symbolic, there is no meaningful theoretical interpretation of the interactions, so we're at a stopping point in this example. We'll make the process of interpretation clearer in the more realistic example that follows.

The two effect size statistics η^2 and ω^2 included in **Figure 5** are close in value and indicate that the main effects for Variables B and C, and the interaction of B X C account for an appreciable portion of the variation in the experiment—collectively about 59%.

Figure 5

JASP Generated Source Table for the 2 X 2 X 2 ANOVA Shown In Figure 1.

ANOVA - Score is simply the generic term for the dependent variable

Cases	Sum of Squares	df	Mean Square	F	p	η^2	ω^2
Var A	13.225	1	13.225	8.015	0.008	0.060	0.053
Var B	55.225	1	55.225	33.470	< .001	0.252	0.243
Var C	38.025	1	38.025	23.045	< .001	0.174	0.165
Var A ∗ Var B	0.025	1	0.025	0.015	0.903	0.000	0.000
Var A ∗ Var C	4.225	1	4.225	2.561	0.119	0.019	0.012
Var B ∗ Var C	42.025	1	42.025	25.470	< .001	0.192	0.183
Var A ∗ Var B ∗ Var C	13.225	1	13.225	8.015	0.008	0.060	0.053
Residuals (WC)	52.800	32	1.650				

Note. Type III Sum of Squares I added "(WC)" to indicate the nature of the Residuals.

A Three-Way ANOVA With *JASP*

Using the sums of squares formulas to compute the seven *F* ratios required for a three-way ANOVA makes manual calculations straightforward and feasible, but the labor involved is still considerable. For the following sample problem called "A Good Choice?," we'll use *JASP* for all calculations. **Read the description of the problem on the next page completely before continuing.**

The Design of the Experiment

The "A Good Choice?" experiment is classified as a 2 X 3 X 2 design, with equal sample sizes for the 12 cells of the design. There are two main effect means for the levels variable of *Implied SES*, three main effect means for the levels variable *Age*, and two main effect means for the categorical (grouping) variable *Gender* of the participant. There are three first-order interactions (*Implied SES* X *Age*, *Implied SES* X *Gender*, and *Age* X *Gender*) and one higher-order interaction (*Implied SES* X *Age* X *Gender*)

Assumptions of the Procedure

The assumptions for the three-way ANOVA are the same as those for the two-way ANOVA.

1. **The data are numerical in character.**

2. **The samples representing the cells of the design are randomly selected from the populations of interest and are independent of one another (uncorrelated samples.**

3. **The scores for the populations represented by the cell means are normally distributed.**

4. **The populations represented by the cell means all have the same variance.**

Are the assumptions Met?

The sampling strategy used in the experiment is a convenience sample, i.e., volunteers recruited from classes by virtue of monetary or extra credit inducements. The recruitment strategy is well known, and most investigators would be comfortable generalizing the findings to "college students" in general. Within gender, the subjects were randomly and independently assigned to the conditions of the experiment so there is no intrinsic bias for or against any of the conditions. The samples are too small to conclude anything about the shape of distributions represented by the cells, but most well-constructed rating scales yield at least unimodal distributions. The homogeneity of variance assumption will be tested using Levene's Test for equal variances.

In general, if the sample sizes are not too small and are equal for all cells, the independent groups ANOVA procedure will yield accurate values for the *F* tests and associated *p* values even with moderate violations of the normality and homogeneity of variance assumptions. Recall that this tendency for a statistical test to work correctly in the face of violation of its mathematical assumptions is called **robustness.**

Formatting the data

Once the number of variables gets larger, it's important format the data to maximize the flexibility and completeness of the statistical analysis. In this example, each participant has four bits of data—the first three indicating which condition of the experiment he or she was

A Good Choice?

A great deal of research in evolutionary psychology has been directed toward understanding the differences between females and males in the characteristics that they look for in a long-term partner. Susan, an up and coming young evolutionary psychologist, is working to develop a simple methodology for investigating such differences. Her approach is to devise structured descriptions of hypothetical people and ask male and female participants to evaluate the attractiveness of the people as longer-term partners. The gender of the participant (*Gender*) was the main variable of interest in the present experiment. Thirty female and 30 male students were recruited from introductory psychology and sociology classes at the University of Colorado, Boulder. All self-identified as heterosexual in orientation and had a mean age of 21.8 years. Each participant was randomly assigned to read <u>one</u> version of the following description of a hypothetical person that varied systematically in the following two dimensions: Implied socioeconomic status of the person (*Implied SES* – Medium vs. Upper) and age of the person (*Age* – 24 or 27 or 30 years). The gender of the hypothetical person was always opposite that to the participant. Immediately after reading the description, the participant rated the hypothetical person on Attractiveness, Intelligence, Health, Motivation, and Friendliness, and then indicated how interested they would be in dating the person if they were to meet at a party (potential partner rating, *P-Rating*, using an 11-point scale, ranging from (1) "Not at All" to (6) "Moderately Interested" to 11 "Extremely Interested").

Bill (or Kari) is a 24 year old (or 27 or 30 year old) resident of Boulder. He is a junior at the University majoring in Industrial Engineering. He lives alone in a small one-bedroom rental apartment (or a spacious three-bedroom condo that his parents own) just south of campus. Bill has a two-year old rescue dog named Jack whom he takes with him on long weekend hikes in the mountains. He also has a gym membership but struggles to use it more than once or twice a week. Recently Bill landed an internship at one of Denver's larger electronics research firms. To handle the commute from Boulder to his internship he treated himself to a two-year old Honda Civic (or a new BMW SUV).

Here are the *P-Rating* rating scale data for the male and female participants:

	Medium *Implied SES*			High *Implied SES*		
	24 years old	27 years old	30 years old	24 years old	27 years old	30 years old
Female Participants	3	5	3	5	7	7
	5	4	6	9	9	10
	2	3	5	8	8	9
	4	6	5	7	9	6
	3	4	3	6	10	10
Male Participants	5	3	3	7	6	5
	6	4	2	8	6	7
	6	2	1	9	7	6
	5	4	4	5	7	4
	3	4	3	7	5	5

in and the fourth the dependent variable, the rating score. **Figure 6** shows a screen capture of the first three cases of the data set entered into an Excel spreadsheet.

When running a three-way ANOVA, it's important to use a stepwise approach that begins with the computation of the ANOVA Source Table and then using the results in the table to guide additional analyses. The analysis employs the same *JASP* ANOVA procedure that we used for a two-way analysis. The only difference is that there are

Figure 6

Excel Spreadsheet Showing the Data for the First 3 Cases.

	A	B	C	D
1	Gender	SES	Age	P-Rating
2	Female	Medium	24	3
3	Female	Medium	24	5
4	Female	Medium	24	2

Figure 7

Result of the Initial Three-Way ANOVA showing Levene's Test For Homogeneity of Variance and the ANOVA Source Table.

Test for Equality of Variances (Levene's)

F	df1	df2	p
0.692	11.000	48.000	0.740

ANOVA – *P-Rating* is the dependent variable

Cases	Sum of Squares	df	Mean Square	F	p	η^2	ω^2
Gender	17.067	1	17.067	10.611	0.002	0.058	0.052
SES	160.067	1	160.067	99.523	<.001	0.543	0.534
Age	2.700	2	1.350	0.839	0.438	0.009	0.000
Gender ✶ SES	6.667	1	6.667	4.145	0.047	0.023	0.017
Gender ✶ Age	30.233	2	15.117	9.399	<.001	0.102	0.091
SES ✶ Age	1.033	2	0.517	0.321	0.727	0.004	0.000
Gender ✶ SES ✶ Age	0.033	2	0.017	0.010	0.990	0.000	0.000
Residuals (WC)	77.200	48	1.608				

Note. Type III Sum of Squares. I added "(WC)" to indicate the nature of the Residuals.

now three independent variables to be moved over to the active "Fixed Factors" box. In doing the initial analysis, the following options were selected: Estimates of effect size (η^2 and ω^2) and Homogeneity tests (Levine's test).

Figure 7 shows the results of the initial analysis. Levene's test for homogeneity of variance does not approach significance, which indicates that the assumption of homogeneity of variance of the cell scores is tenable.

Inspection of the source table indicates that the main effects for *Gender* and *Implied SES* are highly significant, with *Gender* of the participant accounting for 5% of the variance (ω^2 = .052) and the *Implied SES* of the hypothetical person accounting for 53% of the variance (ω^2 = .534) in attraction rating scores. The main effect of *Age* of the hypothetical person didn't approach significance.

Among the first-order interactions, *Gender* X *Implied SES* was significant at the .05 significance level but accounted for only about 2% of the variance in rating scores. In contrast, the *Gender* X *Age* interaction was highly significant and accounted for almost 10% of the variance in rating scores. The interaction of *Implied SES* X *Age* did not approach significance and the three-way interaction has an F ratio of effectively zero.

Figure 8 shows the descriptive statistics for the two significant main effects. For Gender, female participants rated the hypothetical person slightly, but significantly higher than the male participants. For all participants, when the hypothetical person was implied to be of upper

Figure 8

JASP Descriptive Statistics for the Main Effects of Gender and Implied SES.

Descriptive Statistics for Gender

	P-Rating	
	Female	Male
Valid	30	30
Mean	6.033	4.967
Std. Error of Mean	0.446	0.347
Std. Deviation	2.442	1.903

Descriptive Statistics for Implied SES

	P-Rating	
	Medium	Upper
Valid	30	30
Mean	3.867	7.133
Std. Error of Mean	0.243	0.310
Std. Deviation	1.332	1.697

SES the ratings were markedly higher than when the hypothetical person was of only medium implied SES.

Figure 9 shows the *Gender* X *Implied SES* interaction plot produced by the *JASP* Descriptive plots option. Post-hoc comparisons using *JASP*'s Post-Hoc tests option with the Tukey correction for multiple tests revealed that there

Figure 9

JASP Interaction Plot Showing Female and Male P-Ratings for the Two Implied SES Levels of the Hypothetical Person.

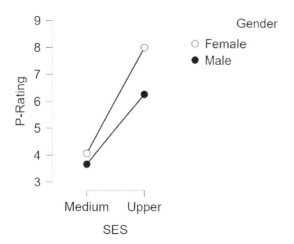

Figure 10

JASP Interaction Plot Showing Female and Male P-Ratings for the Three Stated Ages of the Hypothetical Person.

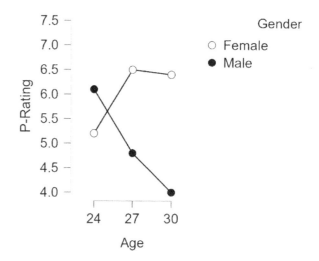

was no difference between the genders in their ratings for medium *Implied SES,* and both increased significantly from medium to upper *Implied SES*. The interaction arose from a significant difference between the genders in response to upper *Implied SES*, $t(28) = 3.74$, $p = .003$, with females showing a larger increase in ratings.

Figure 10 is the plot of the interaction between the gender of the participants and the Age of the hypothetical person. Post-hoc comparisons showed that although the females increased their ratings as the age of the hypothetical person increased, the increases did not reach significance due to the Tukey correction for the large number of comparisons made. Male participants, by contrast, showed large declines in their rating scores with increases in the hypothetical person's age, with the decline from 24 years to 30 years being highly significant $t(18) = 3.70$, $p = .007$. Although the genders did not differ significantly in their ratings for 24 years, the difference was significant for both 27 years, $t(18) = 3.00$, $p = .046$ and 30 years, $t(18) = 4.23$, $p = .001$.

Here's a sample summary of the findings of the analysis:

Thirty female and 30 male participants read a short description of a hypothetical person of the opposite sex which systematically varied the age and apparent socioeconomic status of the person. Overall, female participants rated the hypothetical persons slightly but significantly higher than male participants (**Figure 8** upper, 6.0 and 5.0), and both genders rated higher *Implied SES* persons much more favorably (7.1) than medium *SES* persons (3.9,

Figure 8 lower). Clearly, higher SES makes a person a more attractive partner for both genders.

The interaction of *Gender X Implied SES* shown in **Figure 9** was significant at $p < .05$ but accounted for only 2% of the rating variance. Female and male participants rated medium *Implied SES* persons as relatively unattractive partners (4.0 and 3.6) but both significantly increased their ratings for higher *Implied SES*, with the greater increase for females accounting for the interaction.

The interaction of *Gender X Age* (**Figure 10**) was highly significant ($p < .001$) and reflected the fact that females and males showed very different responses to increases in the age of the hypothetical person. Females gave higher ratings to older male persons, although post-hoc comparisons using Tukey's correction did not reach statistical significance. By contrast, male ratings declined markedly and significantly as the hypothetical person's age increased from 24 to 30 years. The difference in ratings between female and male participants was significant for both 27 and 30 years. This strong interaction of gender and age is likely responsible for the main effect of gender.

Overall, the results support the hypothesis that young females and males differ in the attributes that they find attractive in potential partners. Females appear to place a higher value on partner resources and are not averse to having a partner who is older than they. Males also value partner resources but respond negatively to partners older than they. Presumably increases in age may be accompanied by declines in physical attractiveness or social compatibility.

Final Comments

One of the limitations of three-way and higher ANOVA designs is the tendency for sample sizes for each cell of the design to be smaller than would be the case for one- and two-way designs (many more cells to fill). As a result, potentially important differences fail to reach conventional significance levels. This problem is exacerbated in post-hoc comparisons of simple effects where even relatively liberal corrections for multiple comparisons such as Tukey's prevent a normally significant t test value from being significant. For example, in the interaction shown in **Figure 10**, the difference between female volunteer *P-Ratings* for 23-year olds and 27-year olds yielded a t value of 2.29, which in a standard t test is significant at the .05 level. The Tukey corrected p value was .217 based upon the fact that 15 post-hoc comparisons were made in evaluating the interaction. The best solution to this problem of course is to make every effort to have adequate cell sample sizes.

In many psychological experiments, gender is often analyzed as a grouping variable—not because there is a theoretical interest in gender, but because separating males and females reduces within-group or within-cell error variance. This reflects the fact that females and males differ on many variables that may be related to the dependent variable, so mixing female and male participants adds systematic gender variation in the value of the dependent variable that is treated as "random error" in the denominator of the F ratio.

In the present hypothetical experiment, gender was the major variable of interest. Evolutionary psychologists, including our up-and-coming Susan, are prone to ascribe widespread gender differences to hard-wired behavioral tendencies which have evolved through natural selection pressures in human evolution. Mate selection differences between males and females is one such difference that has broad empirical support. But there are always alternative explanations for any given experimental finding. In the present situation, the female volunteers may have figured out the purpose of the study and are helpfully giving Susan what she was looking for!

"Those who understand others are clever, those who understand themselves are wise."

                                                  Lao Tzu

Chapter 14, Factorial ANOVA
Chapter Exercises

Some questions are purely factual and have only one correct answer. Others require greater thought on your part and may have more than one possible answer.

1. Fill in the missing term(s) that makes the statement true.

   a. Two-way ANOVA is simply a special case of _____.
   b. An A X B interaction is called a _____ interaction.
   c. A third-order interaction would involve _____ independent variables.
   d. If a 2 X 2 X 4 design has no cases in one of the cells, it is called _____.
   e. In three-way ANOVA's it is important to have _____ of cases.
   f. A three-way ANOVA has how many variance sources? _____
   g. The first thing computed in a factorial ANOVA is _____.
   h. *JASP* uses the term "Residuals" to mean the _____ sums of squares.
   i. The design of the "A Good Choice?" experiment is a _____.
   j. The major variable of interest in the "A Good Choice?" experiment is _____ because _____.
   k. Factorial ANOVA is _____ in the face of modest violation of the assumptions.
   l. The extremely significant *Gender* X *Age* interaction arose because _____.
   m. Gender or race is sometimes included in an ANOVA design simply to _____.

2. Susan, our up and coming evolutionary psychologist, enthusiastically presented her findings to her dissertation advisor, Dr. Mueller. Dr. Mueller was suitably impressed with the apparent success of the research strategy but puzzled by the lack of a three-way interaction. "There should be one you know," she told Susan. "I don't think you made your *Implied SES* manipulation strong enough." Why don't you run the experiment again with a stronger *Implied SES*. Being doggedly enthusiastic, Susan agreed to do so.

Here is her reworked hypothetical scenario:

   Bill (or Kari) is a 24 year old (or 27 or 30 year old) resident of Boulder. He is a junior at the University majoring in Industrial Engineering. He lives alone in a small one-bedroom rental apartment (or a spacious three-bedroom condo owned by his wealthy parents) just south of campus. Bill has a two-year old rescue dog named Jack whom he takes with him on long weekend hikes in the mountains. He also works out at a local gym (at the Country Club gym) but struggles to do it more than once or twice a week. Recently Bill landed an internship at one of Denver's larger electronics research firms. To handle the commute from Boulder to his internship he treated himself to a two-year old Honda Civic (or a new Mercedes EV SUV).

Here are Susan's data from the new experiment:

	Medium *Implied SES*			Very High *Implied SES*		
	24 years old	27 years old	30 years old	24 years old	27 years old	30 years old
Female Participants	4	4	2	9	9	10
	4	2	4	7	9	10
	5	3	3	7	11	11
	6	4	1	6	10	9
	5	3	2	8	9	10
Male Participants	6	4	3	7	3	5
	7	3	3	8	4	4
	5	4	2	6	3	3
	5	2	3	7	6	3
	6	4	1	9	5	2

Analyze these data using Susan's first experiment as a model. Dr. Mueller is correct about the three-way interaction being significant and that has a big effect on the interpretation of the results. Do a thorough descriptive and inferential analysis of the data, including the appropriate graphs. JASP doesn't provide for a three-way interaction to be graphed on the same axes. Instead, you will need to us the Descriptive plots option and put *Age* in the Horizontal Axis window, *Gender* in the Separate Lines window, and *Implied SES* in the Separate Plots window. Once that's done, use the JASP Edit Image option to scale the Y Axis the same for both graphs. How would you interpret the findings of the study?

3. Recall the three students in Chapter 13 who did an altruism experiment in which they asked 65 participants what portion of $10,000 they would give a cousin, brother, or sister who found themselves in financial difficulty, with closeness of the relationship as a second independent variable? Their advisor admitted they did find some interesting results but was disappointed they did not analyze the data separately for the male and female participants in the experiment (in fact, the three experimenters had forgotten to record the gender of the participants). The advisor suggests that they do the experiment again—this time omitting the cousin condition but including the gender of the participants as a grouping variable.

Encouraged by their professor's interest, the three repeated the modified experiment—this time keeping track of the gender of their volunteer participants and insuring equal sample sizes for each cell of the design. They used the same scenario as in the previous experiment, except on the professor's advice they omitted the adjective "older" in describing either the brother or sister.

Your _____ has gotten himself (herself) into a financial bind and comes to you for a loan, saying, "Anything would help." You have managed to save $10,000 by working two jobs and have plans for the money. You have always been close to your _____ growing up (OR, You were never close to your _____, although you got along well enough). Understanding that you're not likely to get it back, how much would you be willing to loan your _____?

Here are the data rounded to the nearest $1,000:

	Not Close Growing Up		Close Growing Up	
	Brother	Sister	Brother	Sister
Male Participants	4 4	6 7	5 6	10 10
	3 5	8 6	6 4	9 9
	6	8	5	8
Female Participants	3 4	3 4	10 8	10 8
	5 5	5 2	8 9	9 9
	4	2	9	6

Do a complete descriptive and inferential analysis of the data and try to interpret what is going on psychologically.

Chapter 15
Repeated Measures ANOVA

What's In This Chapter?

In **repeated measures** ANOVA cases appear in more than one condition of an experiment. This is usually done to increase the statistical power of the experiment by reducing the size of the within-group or within-cell error variance estimate. When the cases are human participants, **individual differences** often greatly inflate the within-group or within-cell estimates of random variation, thus making the resultant F Ratio smaller and less likely to be statistically significant. Repeated measures ANOVA procedures treat individual differences variation as a variable and allow its removal from the estimate of error variation.

In the simplest repeated measures ANOVA, the cases appear in all conditions of the independent variable and the total sums of squares is broken into a **"within subjects"** component which evaluates effect of the independent variable and a **"between subjects"** component which is simply the random individual differences variation.

More complex repeated measures designs include both repeated measures independent variables and between subjects variables like those described in Chapters 12–14. These are **called mixed design** ANOVAs.

Repeated measures ANOVAs have more complex statistical assumptions about population characteristics than random assignment ANOVAs. In addition to random sampling and normality of the populations, repeated measures variables must show **homogeneity of covariance** and **homogeneity of variance** for all repeated measures conditions AND the <u>differences between conditions</u>. The homogeneity of variance requirement is referred to as **sphericity**, which is easily violated and which produces unpredictable changes in Type I and II Errors.

Mauchly's W is a statistic used to detect violations of the sphericity assumptions. The **Geenhouse-Geisser and Huynh-Feldt corrections** modify the analysis to reduce the likelihood of Type I Errors.

Introduction

This chapter introduces what is called "repeated measures" ANOVA. Recall that in the ANOVA procedures discussed in the previous chapters, cases were randomly assigned to only one condition of the experiment or to one treatment combination in the case of two-way or higher-order ANOVAs. In repeated measures ANOVA, the same cases appear in more than one condition of the experiment.

In psychology where the cases are usually human participants, the procedure is often called "within subjects" ANOVA. The reason for having the same cases appear in multiple conditions of the experiment is the same as that seen in correlated t-test situations, i.e., making the experiment more powerful. In a correlated t test this is achieved by making the sampling distribution of differences between means narrower—by reducing the size of $s_{\bar{x}_1 - \bar{x}_2}$. In the repeated measures ANOVA, random factors that make cases vary in their response to the independent variable ("individual differences" in the case of human participants) are partitioned out, leaving a smaller estimate of random variation in the denominator of the F ratio.

Here's a bit of history. A long time ago in psychology and medicine, human volunteers for research were referred to as "Subjects." The term derived from the fact that animals used in research (rats, cats, pigeons, etc.) were called subjects, so it became common to refer to human participants as subjects. When it finally dawned on researchers that the term "subjects" was degrading when applied to humans, it was replaced by more neutral terms like "participant" or "volunteers." The change, however, somewhat bypassed statistics terminology, hence such terms as "within subjects," "between subjects," "subjects X treatment" interaction and the like.

The Logic of Repeated Measures ANOVA

We'll begin with the simplest example of a repeated measures design, in which the procedure is used simply to remove individual difference variation between participants from the random variability estimate in the denominator of the F ratio.

Table 1 shows the results of a hypothetical experiment involving solving anagrams. As you know, an anagram is a word in which the letters are scrambled, and the person's task is to rearrange the letters to recover the actual word. Psychologists have long studied the variables that influence anagram solving in the attempt to understand how the brain processes language—specifically how words are "looked up" from what is called semantic memory.

In this experiment, participants are given a list of 20 anagrams, each of five to seven letters, with instructions to solve as many as possible in three minutes. The anagrams yield only one solution, and all the words are of equal frequency in normal discourse. In the first condition, the 20 words are unrelated. In condition 2 all of the words are drawn from a single category (e.g., animals). In the third condition all of the words are drawn randomly from two categories (e.g., animals and furniture).

Traditional One-Way ANOVA

To see the advantage of a repeated measures design, we'll compare two ways to do the experiment. First, we'll begin by assuming that the data in Table 1 were derived from a simple one-variable experiment involving 15 participants randomly assigned to one of the three conditions of the experiment. Inspection of the sample means makes it clear that there is a substantial advantage in having the words drawn from a single category and a smaller advantage for two categories.

The data would be analyzed using a standard one-way ANOVA, as described in Chapter 12. Figure 1 shows the result of that analysis. Levene's test and inspection of the

Table 1

Number of Anagrams Out of 20 Solved By the End of Three Minutes by 15 Participants.

	Unrelated	1 Category	2 Categories
	6	10	8
	12	19	15
	3	7	8
	11	13	10
	8	15	10
$\bar{X} =$	8.0	12.8	10.2

within-group variation seen in Table 1 suggests that the homogeneity of variance assumption is tenable. Disappointingly, the F Ratio of 2.019 is not statistically significant (p = .176). Notice that the value for ω^2 is markedly smaller than that for η^2, indicative of the latter's tendency toward overestimation bias with small samples.

The explanation for the non-significant F Ratio is obvious from inspection of the within-group variability. There is enormous variation within each group in the number of anagrams solved. Clearly, regardless of the experimental condition, some participants are simply much better than others at solving anagrams (e.g., better language skills, prior practice, etc.). Those individual differences make the within-group variance estimate (s_{WG}^2) very large relative to the between-group estimate (s_{BG}^2).

Repeated Measures ANOVA

Now let's conceptualize the experiment differently. Instead of using 15 participants, we'll use five and each

Figure 1

JASP Analysis of the Anagram Experiment Including Levene's Test for Homogeneity of Variance and the One-Way Source Table.

Test for Equality of Variances (Levene's)

F	df1	df2	p
0.638	2.000	12.000	0.546

ANOVA – Number of anagrams solved

Cases	Sum of Squares	df	Mean Square	F	p	η^2	ω^2
Condition (*BG*)	57.733	2	28.867	2.019	0.176	0.252	0.120
Residuals (*WG*)	171.600	12	14.300				

Note. Type III Sum of Squares   Note: I've added the *BG* and *WG* abbreviations

participant will serve in all three conditions (*Cond*) of the experiment. Naturally, this introduces things like practice effects from being in a previous condition of the experiment and potential interference effects from finding words in a new and unfamiliar category. These are called **carryover effects.** Psychologists have developed methods to deal with such matters, like using different orders of the conditions for different participants, which is called **counterbalancing,** and giving extensive prior practice on words not used in the experiment. These issues can make repeated measures designs rather complex! But we'll ignore those issues for the purpose of this example.

Table 2 shows what the new repeated measures design would look like—with the data remaining the same. The table now includes the code numbers of the participants and the mean number of anagrams solved by each participant across the three conditions ($\bar{X}_{Par}$). The mean of 10.33 is the overall mean of the 15 observations.

The repeated measures analysis is going to treat the mean scores of the five participants across the three conditions as a second variable called the between participants variable. When conceptualized this way, the design looks like a 3 (conditions) X 5 (participants) two-way ANOVA, with one observation per cell. Let's see how the analysis would proceed, again expressing the variance estimates as sums of squares (*SS*).

The total *SS* is simply the variability of all 15 scores about the grand mean of 10.33, i.e.,

$$SS_{Tot} = (1) \sum_{i=1}^{n}(X_i - \bar{\bar{X}})^2 = 229.333$$

where: n = total sample size (15) and
$\bar{\bar{X}}$ = overall mean of the 15 observations.

Note that in the *JASP* Source Table (**Figure 1** above), the sum of the Condition *SS* and the Residual *SS* equals the Total *SS*, i.e., 57.733 + 171.600 = 229.333.

The main effect for experimental condition is unchanged and is given by:

$$SS_{Cond} = n_C \sum_{c=1}^{C}(\bar{X}_c - \bar{\bar{X}})^2 = 57.733$$

where: $\bar{X}_c$ = mean of each condition,
C = the number of conditions (3), and
n_C = number of observations per condition (5).

What is new in the analysis is finding the sums of squares for the between-participants variation (SS_{Par}). It is computed in exactly the same way the experimental conditions sums of squares is computed:

Table 2

Number of Anagrams Solved Out of 20 By the End of Three Minutes by Five Participants.

Particip.	No Cat.	1 Cat.	2 Cat.	$\bar{X}_{Par}$
1	6	10	8	8.0
2	12	19	15	15.33
3	3	7	8	6.0
4	11	13	10	11.33
5	8	15	10	11.0
$\bar{X}$ =	8.0	12.8	10.2	10.33

$$SS_{Par} = n_{Par} \sum_{par=1}^{Par}(\bar{X}_{par} - \bar{\bar{X}})^2 = 151.881$$

where: $\bar{X}_{par}$ = mean of each participant,
Par = the number of participants (5), and
n_{Par} = Num of observations per participant (3).

What exactly is this variability estimate? Basically, it's an estimate of random variation between individual participants. *JASP* refers to it as *Residuals*, but it's really what we have called within-group (WG) random variation in a one-way ANOVA. It is placed in the lower portion of the *JASP* source table, **Figure 2** on the next page.

We're going to subtract it and the *Cond* sums of squares from the total sums of squares, leaving an estimate of error variation **free of individual differences variation.** *JASP* also refers to this estimate as "Residuals" which is confusing, but we'll do so as well, abbreviating it SS_{Resid}.

$$SS_{Resid} = SS_{Tot} - SS_{Cond} - SS_{Par}$$
$$= 229.333 - 57.733 - 151.881 = 19.719$$

If the above looks like the formula for computing a two-way interaction, it turns out that that's exactly what it is! It's the interaction of Participants X Condition, with one observation per "cell" of the design. **In words, it's the unpredicted variation in how participants respond to the different conditions of the experiment.** Notice in **Table 2** that the first participant increases the number of solved anagrams from 6 to 10 (+4) in going from the Unrelated to the one-category condition, while Participant 2 increases 7 solutions, and Participant 4 only 2 solutions. The non-parallel changes reflect the unpredictable "interaction" of Participants X Condition. For this reason, another abbreviation for this SS_{Resid} is $SS_{Par\ X\ Cond}$.

This repeated measures analysis thus yields three variance estimates:

$$s_{Cond}^2 = \frac{SS_{Cond}}{C-1} = \frac{57.733}{2} = 28.867$$

$$s_{Par}^2 = \frac{SS_{Par}}{P-1} = \frac{151.881}{4} = 37.970$$

$$s_{Resid}^2 = \frac{SS_{Resid}}{(P-1)(C-1)} = \frac{19.719}{8} = 2.465$$

The F ratio test for the Condition variable is given as:

$$F = \frac{s_{Cond}^2}{s_{Resid}^2} = \frac{28.867}{2.465} = 11.710$$

Figure 2 shows how *JASP* carried out this analysis. From the upper table labeled "Within Subjects Effects," it is apparent that switching to a repeated measures (within subjects) design enormously increased the power of the experiment. The F ratio increased from 2.019 with a p value of .176 to 11.782 with a p value of .004. Notice that there is no statistical test of the significance for the between-subjects variance. It is simply given to indicate the magnitude of individual difference variation.

The small discrepancies in the values of the various statistics reflects the greater precision with which *JASP* carried out the computations.

In repeated measures designs the number and types of effect size statistics increases. **Figure 2** includes three effect size statistics--η^2, η_G^2 ("general" η^2) and ω^2. Plain η^2 reflects just the within participants sources of variance, i.e.,

$$\eta^2 = \frac{SS_{Cond}}{SS_{Cond} + SS_{Resid}} = \frac{57.773}{77.33} = .747$$

The statistic η_G^2 includes both the within and between components, i.e.,

$$\eta_G^2 = \frac{SS_{Cond}}{SS_{Cond} + SS_{Resid} + SS_{WG}} = \frac{57.773}{229.33} = .252$$

where: SS_{WG} = the between participants "Residuals."

An obvious question is which effect size statistic to use in this situation? Given that individual differences in anagram solution will always be present, η_G^2 may better reflect for this design how potent the independent variable is in real life. However, as we'll see below when there is a between-participants variable in the design, η_G^2 does not include that variable and thus tends to overstate the magnitude of the effect.

An Alternative to a Pure Repeated Measures ANOVA Design

It's apparent that the gain in power by using a repeated measures design in this situation comes at the cost of increased complexity in actually carrying out the experiment. Counterbalancing the order of presentation of the experimental conditions usually entails statistically evaluating the effectiveness of the counterbalance. **Asymmetrical transfer effects** occur when Condition A followed by Condition B produces a different effect than

Figure 2

JASP Repeated Measures Analysis of the Anagram Solution Data.

Within Subjects Effects

Cases	Sum of Squares	df	Mean Square	F	p	η^2	η_G^2	ω^2
Condition	57.733	2	28.867	11.782	0.004	0.747	0.252	0.198
Residuals (*Par X Cond*)	19.600	8	2.450					

Note. Type III Sum of Squares   Note: I have added the (*Par X Cond*) interaction to indicate what the error term actually is.

Between Subjects Effects

Cases	Sum of Squares	df	Mean Square
Residuals (*WG*)	152.000	4	38.000

Notes. Type III Sum of Squares. I've added the (*WG*) annotation.

when Condition B is followed by Condition A. In fact, the analysis and interpretation of counterbalanced designs can become quite messy.

An alternative strategy that maintains some of the gain in power from the design is to use **matched groups** in place of individual participants. Recall that in the correlated *t* test, an alternative to testing the same cases under two conditions was to use pairs of cases matched on a variable highly correlated with the dependent variable. The same is true for repeated measures ANOVA. What we would do in this situation is to test a larger group of volunteers for their ability to solve anagrams and make up groups or triads of equal ability. The three-person groups would take the place of individual participants, with each member of the triad assigned to one of the three experimental conditions at random. No counterbalancing required!

This design would technically be called a **Randomized Blocks Design,** where the "blocks" are matched triads. The statistical analysis would proceed exactly as that described above, with the between-subjects *SS* replaced by a between-blocks *SS*. The key here is that the gain in power through matching is a function of the effectiveness of the matching strategy. In this situation, matching on a pretest of anagram solving would probably be quite effective as long as the initial pool of potential participants was large enough to make the matching relatively precise.

Assumptions of Repeated Measures ANOVA

The assumptions for the repeated measures ANOVA are similar to those for between groups ANOVA, **with two very important differences.** In addition to the usual random sampling and normality assumptions, data from the repeated measures variable have to possess two important characteristics.

1. The covariance between scores in all of the repeated measures conditions must be the same. This is called the **homogeneity of covariance assumption**. The term "covariance" refers to the numerator in the formula for Pearson *r* (i.e., $\sum z_X z_Y$), so in the ANOVA situation, homogeneity of covariance means basically that the correlation of scores among <u>all</u> pairs of repeated measures conditions is the same.

2. The repeated measures conditions AND the **differences between scores** between the various conditions must have the same variance. This is an expanded and very restricting version of the homogeneity of variance assumption and is technically called **sphericity**.

These two assumptions are very difficult to meet and the sphericity assumption is frequently violated in practice. In fact, violations of this assumption are sufficiently common that all statistical packages, including *JASP*, automatically test for violations of sphericity and note when they are present. Moreover, under the option Assumption Checks, *JASP* provides a **statistical test for *non-sphericity*** called **Mauchly's *W***, which allows you to see how close you have come to violating the sphericity assumption. In addition, *JASP* provides two possible procedures for correcting for non-sphericity— **Greenhouse-Geisser** and **Huynh-Feldt**. These procedures adjust the *df* for *F* ratios, thereby making it more difficult to claim a significant effect when it might actually be due to Type I error.

This fairly threatening discussion of the assumptions underlying the repeated measures ANOVA should alert you to the fact that **violating those assumptions has very negative outcomes.** Unlike independent groups ANOVA, the repeated measures ANOVA is **NOT robust** in the face of assumption violations. Departures from sphericity produce unpredictable changes in Type I and Type II Errors.

For the hypothetical anagram experiment data, Mauchly's test has a *p* value of .99, indicating that the assumption of sphericity is completely tenable.

An important moral to this extended discussion of assumptions is to use adequate sample sizes in your research, so that an aberrant observation or two won't completely invalidate your results. This is particularly important in repeated measures ANOVA because one of the attractions of the procedure is the gain in statistical power while ostensibly using fewer participants.

A More Complex Repeated Measures Design

The above example is the simplest application of the repeated measures ANOVA, where the design is used only to control individual differences variation. Most often, however, the repeated measures component is part of a more complex design that also has one or more between-subjects variables, in which participants are randomly assigned to different experimental conditions. These are referred to as **mixed designs**. We'll develop this idea in the context of a hypothetical experiment to better teach female police cadets to shoot their service handguns. Read through the following experiment called, "A Necessary Evil" on the next page before continuing.

A Necessary Evil

In one large Police Training Academy, it has been policy to use female firearms instructors (FIs) in an effort to make female cadets more comfortable in learning to use the commonly issued 9-millimeter pistol. As a full-size duty pistol, this handgun is relatively large for many women's' hands and, like all guns, is loud and violent when fired, with recoil kicking the gun back and up. Use of women FIs has been only partially successful in getting women cadets qualified with the handgun.

One FI believes that the solution is not just to use female FIs, but to match the size of the instructor to the size of the trainee—particularly in terms of hand size. Loaded pistols are heavy and awkward, and safe handling and accurate firing require a firm grip and learning many counter-intuitive coordination skills. The instructor reasons that a small-statured woman will learn best from a small-statured FI, regardless of gender, who has mastered handling and shooting with smaller hands.

To test this theory, the instructor recruits the smallest male FI and explains her reasoning to him. That instructor is assigned to six randomly selected incoming female recruits for the first week of firearm instruction. Six other randomly selected female recruits are assigned as usual to the two large-statured female FIs and are used as a comparison group. Following two days of instruction in safe handling, loading, and cleaning of their issued pistols, all 12 recruits go through an initial week of six 1.5-hour firing exercises where they fire 50 rounds at 9" targets from 5 to 10 meters away. Their daily score is the number of hits in the target out of the 50 rounds fired.

Below are the daily hit data for the 12 female recruits broken down by Type of Instructor. Marginal and group means have been included because they will be used to show how the analysis is carried out.

Small Male FI	Recruit	Day 1	Day 2	Practice Days Day 3	Day 4	Day 5	Day 6	$\bar{X}_{Par}$	
	1	8	12	13	16	22	29	16.67	
	2	12	15	19	24	30	34	22.33	
	3	13	20	23	25	32	41	25.67	$\bar{X}_{SMFI} = 21.94$
	4	10	13	18	26	30	34	21.83	
	5	9	12	17	24	29	38	21.50	
	6	10	15	22	27	32	36	23.67	
$\bar{X}_{RM_{SMFI}} =$		10.33	14.50	18.67	23.67	29.17	35.33		

Regular Female FI's	Recruit	Day 1	Day 2	Practice Days Day 3	Day 4	Day 5	Day 6	$\bar{X}_{Par}$	
	7	7	9	13	18	23	26	16.00	
	8	10	14	16	20	22	26	18.00	
	9	5	8	11	16	20	23	13.83	$\bar{X}_{RFFI} = 15.14$
	10	6	9	12	12	15	20	12.33	
	11	8	10	13	12	20	22	14.17	
	12	8	11	14	17	23	26	16.50	
$\bar{X}_{RM_{RFFI}} =$		7.33	10.17	13.17	15.83	20.50	23.83		
$\bar{X}_{RM_{All}} =$		8.83	12.33	15.92	19.75	24.83	29.58	$\bar{\bar{X}}_{Grand} = 18.54$	

Note: RM = repeated measures; SMFI = small male firearms instructor; RFFI = regular female firearms instructor

In addition to the raw data which will be entered into the *JASP* REPEATED MEASURES ANOVA routine, the problem also shows the marginal means for the rows (12 participants), columns (six repeated measures), conditions (two instructor conditions), and the grand mean of all 72 observations. We'll use those computed means to show the logic of analyzing a mixed design experiment.

The analysis begins by computing to total sums of squares (SS_{Tot}) as follows:

$$SS_{Tot} = (1)\sum_{i=1}^{n}(X_i - \bar{\bar{X}})^2 = 5153.876$$

where: X_i = individual scores,
n = Total number of observations, and
$\bar{\bar{X}}$ = mean of all 72 observations.

Between-Subjects Variance Components

The next step is to compute the two between-subjects sums of squares components. We'll begin with the main effect sums of squares for type of instructor (*Cond*):

$$SS_{Cond} = (n_C)\sum_{c=1}^{C}(\bar{X}_c - \bar{\bar{X}})^2 = 833.680$$

where: $\bar{X}_c$ = mean for each condition,
C = number of instructor conditions (2),
n_C = number of cases in each condition (6).

In words, subtract the grand mean from each instructor condition mean, square the difference, sum the squared differences, add them together for the two conditions, and multiply by the number of observations in each of the two conditions.

Next, we'll compute the between-subjects error sums of squares, i.e., within-groups random variation between participants. The sums of squares is abbreviated SS_{WG}:

$$SS_{WG} = (n_P)\sum_{c=1}^{C}\sum_{par=1}^{P}(\bar{X}_{par} - \bar{X}_c)^2 = 398.03$$

where: $\bar{X}_{par}$ = mean for each participant,
$\bar{X}_c$ = mean of the condition,
P = number of participants in each condition (6),
C = number of instructor conditions (2),
n_P = number of observations per participant 6).

In words, for each condition, subtract the condition mean from each participant's mean, square the difference, and sum the squared differences. Add the sums of squares for the two conditions and multiply by the number of observations in each participant's mean.

The between conditions variance estimate is:

$$s_{Cond}^2 = \frac{SS_{Cond}}{df_{Cond}} = \frac{833.68}{1} = 833.68$$

The within-groups variance estimate is:

$$s_{WG}^2 = \frac{SS_{WG}}{df_{WG}} = \frac{SS_{WG}}{2(P-1)} = \frac{398.03}{10} = 39.80$$

where: P = number of participants in each Condition (6).

The *F* ratio test for the between participants variable of type of instructor main effect is given by:

$$F = \frac{s_{Cond}^2}{s_{WG}^2} = \frac{833.68}{39.80} = 20.947, \quad p = .001$$

Clearly, overall, the matching of hand size of the instructor to the cadet exerts a significant effect on the shooting performance of female cadets.

Within-Subjects Variance Components.

The next step in the analysis is to compute the within-subjects repeated measures variance components. We'll begin with the main effect sums of squares for days of practice (*Days*). This uses the *Days* means at the bottom of the problem data array, i.e. the means averaged across all 12 participants in the experiment.

$$SS_{Days} = (n_D)\sum_{d=1}^{D}(\bar{X}_d - \bar{\bar{X}})^2 = 3631.464$$

where: $\bar{X}_d$ = mean for each day averaged across all 12 participants,
D = number of practice days(6), and
n_D = number of observations per day (12).

The next sums of squares is the interaction of *Days* X *Cond*. Recall that in Chapter 14 we computed interactions by finding the cell sums of squares and subtracting out the main effects sums of squares. We'll do the same thing here. The cells are the six participants' *Days* means for each of the two Instructor conditions, i.e., $\bar{X}_{D1,SMFI}$, $\bar{X}_{D2,SMFI}$, + ... + $\bar{X}_{D6,RFFI}$, yielding 12 cells. The formula would be:

$$SS_{Cells} = (n_{Cells})\sum_{cells=1}^{Cells}(\bar{X}_{cells} - \bar{\bar{X}})^2 = 4612.134$$

where: $\bar{X}_{cells}$ = mean for each of the 12 cells,
Cells = number of cells (12), and
n_{Cells} = number of observations for each cell (6).

So, the sums of squares for the interaction of *Days* X *Cond* would be:

$$SS_{DC} = SS_{Cells} - SS_{Days} - SS_{Cond}$$

$$= 4612.134 - 3631.464 - 833.680$$

$$= 146.990$$

The last within-subjects component to be computed is the remaining random variation that participants show across days within the two conditions of the experiment. Technically, this is the *Par* X *Days* interaction, but again *JASP* calls it the "Residuals," which we'll abbreviate "Resid." It's computed by finding the cell sums of squares and subtracting out main effects of *Days*, *Cond*, the *Days* X *Cond* interaction and the WG error term. The SS_{Cells} in this case is simply SS_{Tot}, so:

$$SS_{Resid} = SS_{Tot} - SS_{Days} - SS_{Cond} - SS_{DC} - SS_{WG}$$

$$= 5153.88 - 3631.46 - 833.68 - 146.990 - 398.03$$

$$= 143.71$$

The three within-subjects variance estimates for this mixed design are as follows:

$$s^2_{Days} = \frac{SS_{Days}}{df_{Days}} = \frac{3631.464}{(6-1)} = 726.293$$

$$s^2_{DC} = \frac{SS_{DC}}{df_{DC}} = \frac{146.990}{(6-1)(2-1)} = 29.398$$

$$s^2_{Resid} = \frac{SS_{Resid}}{df_{Resid}} = \frac{143.71}{2(P-1)(6-1)(2-1)} = 2.874$$

The *F* Ratios for the main effect of *Days* and the *Days* X *Cond* interaction would thus be:

$$F_{Days} = \frac{s^2_{Days}}{s^2_{Resid}} = \frac{726.293}{2.874} = 252.712, \quad p < .001$$

$$F_{DC} = \frac{s^2_{DC}}{s^2_{Resid}} = \frac{29.398}{2.874} = 10.229, \quad p < .001$$

The enormous *F* Ratio for the main effect of Days of training make it clear that practice of any sort helps, regardless of instructor. The significant interaction of Days X Condition indicates that the type of instructor influences the extent of improvement across days of training (see below for the interaction plot).

Doing the Problem Using *JASP*

As usual, actually doing a mixed design repeated measures problem by hand is a lot of work! So, we'll now see how it's done by *JASP*. The procedure is ANOVA and the sub procedure is Repeated Measures ANOVA.

Begin with the formatting of the data. **Figure 3** shows a screen capture of the Excel data file for the first eight cases. The between participants condition of the experiment (Cond) is in the first column, the Cadet's ID number in the second, and the six days of practice in columns 3 – 8.

Figure 3

Screen Shot of Excel Data File Showing the Data for Eight Cases In the Necessary Evil Experiment.

	A	B	C	D	E	F	G	H
1	Cond	Cadet	Day 1	Day 2	Day 3	Day 4	Day 5	Day 6
2	Small Fl	1	8	12	13	16	22	29
3	Small Fl	2	12	15	19	24	30	34
4	Small Fl	3	13	20	23	25	32	41
5	Small Fl	4	10	13	18	26	30	34
6	Small Fl	5	9	12	17	24	29	38
7	Small Fl	6	10	15	22	27	32	36
8	Larger Fl's	7	7	9	13	18	23	26
9	Larger Fl's	8	10	14	16	20	22	26

Figure 4 shows a screen shot of the Statistics Panel in *JASP* once the sub procedure REPEATED MEASURES ANOVA procedure has been selected. The data for the experiment have already been retrieved and are in the left-hand list-of-variables window. Setting up the analysis is a bit tricky, so we'll go through the steps carefully.

Figure 4

Screen Shot of JASP REPEATED MEASURES ANOVA Statistics Panel Before Variables Have Been Named and Moved.

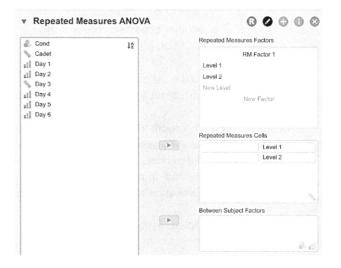

The top window on the right-hand side of the Statistics Panel ("Repeated Measures Factors") is used to define the repeated measures data. It already shows three levels for the first repeated measures variable ("RM Factor 1") defined generically as "Level 1," "Level 2, and "new Level."

First, we'll rename "RM Factor 1" to "Days of Training." Then, we'll change the Levels names as we indicate that there are six levels of the repeated measures variable. Instead of "Level 1," we'll name the first level as "D1." The short name will make *JASP*-produced graphs and tables "cleaner" and less cluttered.

The next window down ("Repeated Measures Cells") is where we'll move the actual variables over to their soon-to-be defined slots using the small "move variable" arrow.

Finally, we'll move the between participants variable *Cond* into the next window down labeled "Between Subject Factors."

Figure 5 shows a screen shot of the three right-hand windows when all of the variables have been named and the active windows filled.

After all of the variables have been moved to the active windows, *JASP* will carry out a basic mixed design ANOVA. This is shown in **Figure 6**. Satisfy yourself that the values in the table agree with those computed by hand except for small rounding errors with manual calculations.

Not including η^2_p, there are three effect size statistics available for the mixed model ANOVA, η^2, η^2_G, and ω^2. The first, η^2, gives the proportion of variance for the two main effects and the interaction as a proportion of the total sums of squares in the analysis. For Days of Training, η^2 equals:

Figure 5

Screen Shot of JASP Statistics Panel After All of the Variables Have Been Named and Moved into the Active Variables Windows.

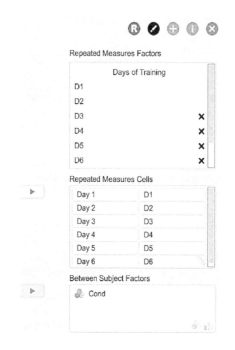

$$\eta^2 = \frac{SS_{Days}}{SS_{Days} + SS_{DC} + SS_{Resid} + SS_{Cond} + SS_{WG}}$$

$$= \frac{3631.79}{5153.87} = .705$$

Figure 6

JASP Mixed Model ANOVA of the Necessary Evil Data.

Within Subjects Effects

Cases	Sum of Squares	df	Mean Square	F	p	η^2	η^2_G	ω^2
Days of Training	3631.792	5	726.358	252.549	< .001	0.705	0.870	0.859
Days of Training ✶ Cond	146.569	5	29.314	10.192	< .001	0.028	0.213	0.182
Residuals (Par X Days)	143.806	50	2.876					

Note. Type III Sum of Squares

Between Subjects Effects

Cases	Sum of Squares	df	Mean Square	F	p	η^2	η^2_G	ω^2
Cond	833.681	1	833.681	20.945	0.001	0.162	0.606	0.476
Residuals (*WG*)	398.028	10	39.803					

Note. Type III Sum of Squares   Note: I've added the (Par X Days) and (*WG*) annotations.

The second, η^2_G, expresses the main effects and interactions relative to the random variation present in the data, very similarly to η^2_p. For Days of Training, η^2_G equals:

$$\eta^2_G = \frac{SS_{Days}}{SS_{Days} + SS_{Resid} + SS_{WG}}$$

$$= \frac{3631.79}{4173.63} = .870$$

Finally, ω^2 is the value of η^2_G corrected for its overestimation bias based on the sample sizes for the within and between groups components of the analysis. The correction is much less for the Within-Subjects main effect and interaction because they are based on larger numbers of observations than the between-subjects main effect for Condition.

So, given the wealth of effect size statistics available, which one should you use? Here is where it gets complicated. The statistic η^2_G reflects just the effect of the variable (or interaction) relative to the total error present in the experiment. It is useful when you want to compare different experiments in which that variable is included. Such comparisons are technically called **meta-analyses** and are beyond the scope of this text. My preference is now η^2, simply because it has the most intuitive interpretation as the proportion of total variation in the experiment accounted for by the experimental variables in real world situations.

Figure 7 gives the descriptive statistics for the main effect of days of training. The large day-to-day increases in mean hits account for the enormous size of the F ratio and the very large value for η^2. Also notice that the variability as measured by the standard deviation increases dramatically across the six days—clearly violating the assumption of sphericity.

Figure 7

JASP Descriptives Table for the Main Effect of Days of Training.

Descriptive Statistics for Days of Training

	Day 1	Day 2	Day 3	Day 4	Day 5	Day 6
Valid	12	12	12	12	12	12
Mean	8.83	12.33	15.92	19.75	24.83	29.58
Std. Deviation	2.33	3.37	3.94	5.34	5.56	6.83
Minimum	5.00	8.00	11.00	12.00	15.00	20.00
Maximum	13.00	20.00	23.00	27.00	32.00	41.00

Figure 8 shows the means and standard deviations for the main effect of the condition variable. The difference of 6.8 in the mean number of hits averaged

Figure 8

JASP Descriptive Statistics for the Main Effect of Condition.

Descriptive Statistics

	Overall	
	Larger FI's	Small FI
Valid	6	6
Mean	15.139	21.944
Std. Deviation	2.064	3.001
Minimum	12.333	16.667
Maximum	18.000	25.667

across six training days is relatively large and consistent with the fact that Condition accounts for 16% of the variance in accuracy.

Figure 9 is the *JASP* interaction plot of the *Days X Cond* interaction. Post-hoc analyses using the Post Hoc Tests sub procedure with Tukey's correction shows that for the interaction, the difference between the small male FI and the lager female Fi's conditions was not significant for the first three days of practice but reached statistical significance in Days 4–6 ($p < .01$ or greater).

Figure 9

JASP Interaction Plot of Significant Days X Cond Interaction.

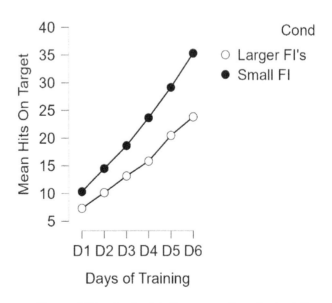

The next thing to check is the assumption of sphericity. JASP automatically flags an analysis if the assumption is clearly violated ($p < .05$). This didn't happen in the example data, but checking the Sphericity tests under the Assumption Checks option reveals that the data are very close to violating the assumption (Mauchy's $W = .057$, $p = .067$).

Figure 10

JASP Mixed Design ANOVA of Necessary Evil Data with Greenhouse-Geisser Corrections Applied.

Within Subjects Effects

Cases	Sphericity Correction	Sum of Squares	df	Mean Square	F	p	η²
Days of Training	None	3631.792	5.000	726.358	252.549	< .001	0.705
	Greenhouse-Geisser	3631.792	**2.338**	1553.529	252.549	< .001	0.705
Days of Training ✶ Cond	None	146.569	5.000	29.314	10.192	< .001	0.028
	Greenhouse-Geisser	146.569	**2.338**	62.696	10.192	< .001	0.028
Residuals	None	143.806	50.000	2.876			
	Greenhouse-Geisser	143.806	**23.378**	6.151			

Note. Type III Sum of Squares

Given the near violation of sphericity, it's prudent to rerum the analysis with the conservative Greenhouse-Geisser corrections applied. **Figure 10** above shows the recomputed analysis. Notice that the *df*'s have been reduced markedly (indicated in boldface), which sets a higher bar for the *F* Ratios to reach significance. Happily, our hypothetical data have yielded such large *F* Ratios that the significance levels haven't been reduced.

Another Approach to Correcting Sphericity Assumption Violations.

Real-world violations of the sphericity assumptions often can't be solved by using the Geenhouse-Geisser corrections simply because the results are only marginally significant to begin with and application of the corrections renders the results nonsignificant. When faced with that situation, the high-minded researcher will, of course, want to solve the problem either by redesigning the experiment or by increasing sample size. Sadly, In the research game, that's a solution of last resort. There's another, possibly cheaper solution. If the violations are a result of too many repeated measures levels or categories, one possible solution is collapsing the repeated measures levels or categories in an effort to increase the stability of the cell means and/or reducing the opportunities for violations to occur.

We'll see how that works in the present data by averaging the scores for Days 1 and 2, 3 and 4, and 5 and 6, and then rerunning the analyses. **Figure 11** shows the new analysis. Notice that the *F* ratio for the main effect of Days has been enormously increased along with a modest

Figure 11

JASP Mixed Design ANOVA of the Necessary Evil Data with Days Variable Collapsed into Three Levels.

Within Subjects Effects

Cases	Sum of Squares	df	Mean Square	F	p	η²
Days of Training	1667.375	2	833.687	377.994	< .001	0.698
Days of Training ✶ Cond	61.847	2	30.924	14.021	< .001	0.026
Residuals	44.111	20	2.206			

Note. Type III Sum of Squares

Between Subjects Effects

Cases	Sum of Squares	df	Mean Square	F	p	η²
Cond	416.840	1	416.840	20.945	0.001	0.174
Residuals	199.014	10	19.901			

Note. Type III Sum of Squares

increase in *F* for the Days X Condition interaction. The between-subjects SS_{Cond} and SS_{WG} have both been reduced by half because each Condition mean is now based on 18 observations and each Participants mean on three observations. That leaves the *F* Ratio for Condition unchanged.

So, what happened to the near violation of sphericity? Mauchly's *W p* value went from .067 to .185, which most researchers would classify as a "near escape" and seek to publish the results.

A Final Comment

Repeated measures ANOVA is a powerful option in many areas of research and, indeed, for some research questions it's a requirement. But as we've seen, it's a lot "pickier" than pure between-participants ANOVAs. Sphericity assumption violations are most likely to occur in those very situations where repeated measures ANOVA is most needed, i.e., repeatedly measuring participants over time and/or across too many treatments. Generally, however, the problems of sphericity violations can be minimized by thinking through the experimental procedures before actually collecting data and by using the largest samples feasible.

Chapter 15, Repeated Measures ANOVA
Chapter Exercises

Some questions are purely factual and have only one correct answer. Others require greater thought on your part and may have more than one possible answer.

1. Fill in the missing term(s) that makes the statement true.

    a. In a repeated measures ANOVA cases appear _____.
    b. Where cases are humans or animals, repeated measures variables are often called _____ variables.
    c. One important benefit of repeated measures designs is _____.
    d. In the "bad old days" in psychology and medicine, people were called _____.
    e. Non-significant F ratios are due either to _____ or _____.
    f. The problem with many repeated measures experiments is the possibility of _____ effects.
    g. In repeated measures designs, the within-subjects SS_{Resid} really means _____.
    h. To avoid the problems associated with repeated measures designs, an alternative is _____.
    i. In repeated measures designs, homogeneity of variance means that _____ AND _____.
    j. This requirement is called _____.
    k. _____ is a test for this requirement.
    l. The Greenhouse-Geisser correction reduces the _____, making it more difficult to _____.
    m. When an experiment has both repeated measures and random assignment variables it is called _____.
    n. Such a design has both _____ and _____ components.
    o. In the "A Necessary Evil" experiment, I preferred simple η^2 over η^2_G because _____.
    p. I suggested that another approach to correcting for _____ violations was to _____.

2. Here is a classical application of the randomized blocks design. A university researcher secures a large grant to investigate effective ways of incorporating artificial intelligence (AI) into high school mathematics teaching. She devises an experiment to compare AI instructional approaches to teaching Algebra I. There are three conditions: (a) the current teacher-homework model (control or reference condition), (b) teacher instruction with AI-assisted homework tutoring after school, and (c) pure on-line AI with an Avatar to present concepts during the school day and assist with homework after school.

    She secures the cooperation of six large high schools in her state and 12 Algebra teachers to participate in the experiment. Upcoming first and second-year students who want to take Algebra I are invited to participate in the experiment. With parental consent and appropriate waivers, students agree to be randomly assigned to be in one of the three conditions. Students in the first two conditions meet in regular classrooms at the same time. Students in the third condition meet in a computer laboratory for the same time period. The dependent variable of the experiment is the class average on the standardized achievement test at the end of the first semester.

    Because the high schools differ widely in such things as location, funding, ethnic composition and family income, schools are blocked for the purpose of analysis. Here are the data:

	Current Approach	AI – Aided Homework	Avatar Instruction
School A	3.12	3.65	2.89
School B	2.72	3.19	2.67
School C	3.70	3.82	3.69
School D	3.11	3.67	3.50
School E	2.45	2.94	2.09
School F	3.96	3.78	3.79

Use *JASP* to do a complete descriptive and inferential analysis of the data and write a short summary of your findings.

3. The ability to read effectively is an enormously difficult cognitive skill. It requires that words or groups of words in a text activate their neural representations, which have to remain activated long enough for the entire sentence to be parsed

and a meaning assembled and stored. Here is an example of the type of experiment used to assess people's ability to hold an activated meaning (word) in working memory.

Participants watch a computer screen as a series of individual words appear on the screen for some short period of time (say one second) to be replaced by the next word in the series, and so on. The participant's task is simply to press a key if they have seen the word before. Sounds simple, right. Actually, the task is grueling. A given word may be preceded by itself with 0, 1, 2, 3, 4, 5, 6, or 7 interpolated words, e.g., "rabbit, car, lazy, rabbit" or "rabbit, car, lazy, coat, rabbit." And, to make things worse, lazy might also be a target word, as in "rabbit, car, lazy, coat, rabbit, horse, bicycle, stream, lazy", so you never know what word will be repeated. The idea is to see how long an activated word remains available in memory before it is replaced by other words. The word strings are often hundreds of words in length. It sucks to participate in one of these experiments.

Two graduate students decide to conduct such an experiment to see whether the task differentiates between proficient and not-so proficient readers. Thiry-five undergraduates are recruited for the experiment. They do a standardized reading comprehension test and are then divided into the top 5 and bottom 5 scorers on the test. All 10 then do the serial recognition task for a great deal of money. The data are the percentage of correct recognition of repeated words with 1, 3, 5, and 7 interpolated words. The score is adjusted for false positives by subtracting the number of false positives from the number of correct responses.

		Number of Interpolated Words			
		1	3	5	7
Top Readers	K. L.	96	86	68	32
	J. N.	100	88	72	38
	A. H.	86	88	74	34
	D. A	94	90	66	30
	T. S.	98	86	58	38
Poor Readers	L. K.	100	82	54	18
	P. A.	98	86	54	20
	J. R.	100	90	48	22
	G. D.	98	82	46	20
	R. W.	98	86	48	28

Carry out a mixed model 2 X 4 repeated measures ANOVA with these data. Do all the necessary tests and corrections if necessary and present your findings using the appropriate statistics and graphs.

4. Here's an example of a mixed design having two within-subjects variables and one between-subjects variable. It is based on an experiment that a research group carried out in one of my research seminars in evolutionary psychology. The purpose of the experiment was to try out a new approach to measuring people's levels of altruism as a function of social closeness and genetic relationship. In previous studies volunteers read and responded to only one hypothetical scenario in a between-subjects design, e.g., end of chapter Problem 2 in Chapter 13 and Problem 3 in Chapter 14. In this experiment, the students wanted to see what would happen if participants got to read all of the scenarios and had the opportunity to adjust their responses after seeing how they responded to the various possibilities.

Volunteers in an Introduction to Evolutionary Psychology class and a Learning and Cognition class read four scenarios presented on separate pages in a random order for each participant. The scenarios were similar to those used in Problem 3 in Chapter 14 in which students were asked how much of their saved $10,000 they would loan (realistically, give) to another person who was experiencing financial problems. In this experiment the potential recipient was either a first cousin (average 12.5% shared genes) or sibling (average 50% shared genes) and either "not very close but got along OK" (Not Close) or "very close and shared many things" (Very Close). Female participants read scenarios in which the potential recipient was female and male participants read about male recipients. The participants were instructed to read all four scenarios and to adjust their responses as they saw fit.

The design was thus a 2 (*Gender* of the participant) X 2 (*Closeness*) X 2 (Genetic *Relationship*) mixed design, with the dependent variable the amount they would "lend" the person who was in financial trouble. Here are the data rounded to the nearest $1,000):

	Partic	Relationship Not Close		Relationship Very Close	
		Cousin	Sibling	Cousin	Sibling
Female Participants	1	2	5	3	8
	2	0	3	2	9
	3	3	5	4	5
	4	4	6	5	10
	5	2	5	4	9
	6	5	8	6	9
	7	2	4	4	6
	8	1	1	3	6
	9	4	6	6	7
	10	4	5	6	8
Male Participants	11	4	6	5	8
	12	0	4	2	7
	13	5	8	7	10
	14	4	6	4	8
	15	2	4	3	6
	16	4	6	4	10
	17	0	3	0	7
	18	2	5	3	8
	19	2	4	3	6
	20	3	4	3	8

Whenever you have more than one repeated measures variable, the set-up is a bit tricky, so here are some hints. First, the data should be entered into Excel or the *JASP* Data Editor just as they appear above, i.e.,

	A	B	C	D	E	F
1	Partic	Gender	NC-Cousin	NC_Sib	C-Cousin	C-Sib
2	1	Female	2	5	3	8
3	2	Female	0	3	2	9
4	3	Female	3	5	4	5

Second, the repeated measures variable definition windows in *JASP* should look like this:

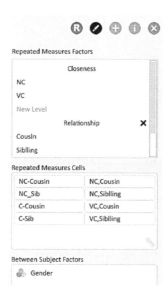

The *JASP* Source Table is going to look a little strange because there are different "Residuals" error terms for different within-subject main effects and interactions. Basically, they are different Participants X Variables interactions that are treated as random error. Just interpret the *F* ratios and effect size statistics in the usual way. Do a complete descriptive and inferential analysis of the data and write an insightful interpretation of the results.

Chapter 16
Chi Square

What's In This Chapter?

The statistic χ^2 (**Greek chi squared**) **is defined in terms of z^2 for continuous variables**, but the family of theoretical χ^2 distributions can be used to test hypotheses about the frequencies of categorical data. A **goodness-of-fit** χ^2 tests the H_0 that the distribution of frequencies among two or more categories follows a prescribed pattern. A significant χ^2 means that the frequencies for at least two categories differ significantly from those specified in the H_0.

A **contingency table** χ^2 tests the H_0 that two categorical variables are independent or uncorrelated. In a contingency table χ^2 analysis, a significant χ^2 means that the two categorical variables are correlated with one another. The strength of the association between categorical variables is assessed using **Cramer's V**.

Introduction

Most of the statistical procedures that we have covered thus far in the book are called **parametric statistics**. There are two reasons for this. First, these procedures are concerned with testing hypotheses about or making estimates of specific population parameters, like μ and p. Second, these procedures make restrictive assumptions about parametric characteristics of the population or populations—things like homogeneity of variance and normality of the distributions.

Some statistical procedures are considered **non-parametric** because they (a) test hypotheses about shapes of populations rather than values of specific parameters, and/or (b) make fewer assumptions about the specific characteristics of the populations under consideration. In the chapter on *t* tests, we encountered a **non-parametric** statistic called the Wilcoxon Matched-Pairs Signed-Ranks test as an alternative to the correlated *t* test.

This chapter is concerned with a non-parametric statistical procedure called χ^2 (Greek "chi" squared). Chi square is a statistical procedure for testing hypotheses about categorical data. Recall that a categorical variable is one in which the values of the variable are discrete, often unordered categories. The data for such a variable are the number of cases that fall into each category. The first application that we will consider tests whether the sample distribution of cases among the categories of a variable is consistent with some hypothetical distribution in the population from which the sample was selected. This application is called a goodness-of-fit χ^2 test. Another name for this procedure is a **one-way** χ^2, because only one variable is involved.

The second application that you will learn about is called a contingency table χ^2 test. The purpose of this procedure is to determine whether two categorical variables are correlated or "contingent" upon one another. Because two variables are involved, some people call this a **two-way** χ^2 **test**.

The Chi Square Statistic

Although we will use χ^2 to test hypotheses about categorical data, in the most fundamental sense, χ^2 is related to old friends of ours, the *z* score and the normal distribution curve. Imagine a normally distributed population of scores with mean μ and standard deviation σ. For such a population, χ^2 is defined as follows:

$$\chi^2 = z^2 = \frac{(X - \mu)^2}{\sigma^2}$$

If you were to select at random a score from the population, translate the score to a *z* score, and then square it, you would have the χ^2 value of the score. For example, imagine that $\mu = 50$ and $\sigma = 10$. You select a score at random whose value is 42. Translating 42 to a χ^2 score yields:

$$\chi^2 = z^2 = \frac{(42 - 50)^2}{10^2} = \frac{64}{100} = .64$$

Notice that although *z* scores can be negative, χ^2 values are always positive.

Suppose that we drew an infinitely large number of scores from the population and computed the χ^2 value of each score. This is equivalent to drawing an infinitely large number of samples of size *n* = 1. If we then put the values of χ^2 into a continuous probability distribution, we would have constructed the sampling distribution of χ^2 for sample size $n = 1$. The df for this distribution would also be 1, because no parameters were estimated in the computation of χ^2. What would such a distribution look like? First, χ^2 cannot be negative because the z scores are squared. But χ^2 potentially can be substantial in value, because squaring even a modest number produces rather large numbers. **Figure 1** shows what the theoretical sampling distribution of χ^2 looks like for this situation.

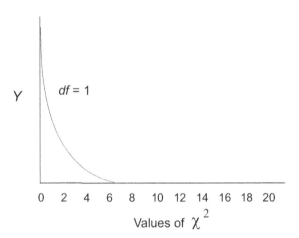

Figure 1

Sampling Distribution of χ^2 for n = df = 1.

Notice that χ^2 is highly positively skewed in this case. Most of the area under the curve is very close to 0, because most of the z scores on which it is based tend to be near 0 in a normally distributed population. Also, recall that squaring values less than 1.00 makes the resulting value *smaller*, not larger. Thus, in our example, the z score equaling -.80, when squared, became .64. To get a χ^2 of 4 or larger, you must draw a member of the population that lies ± 2.00 standard deviations away from μ. In a normally distributed population, such scores are relatively rare.

One of the interesting things about χ^2 values is that you can add them up. Suppose we repeat our hypothetical sampling experiment and draw an infinitely large number of samples of size $n = 2$ from our normally distributed population. For each sample, we compute the z score equivalent for each of the two scores, square the z scores to get χ^2 values, and then add the two values of χ^2 together. The resulting sum of χ^2's is also called a χ^2, i.e.,

$$\chi^2 = \chi_1^2 + \chi_2^2$$

where: χ_1^2 = value of χ^2 for first score and
χ_2^2 = χ^2 for second score in sample.

What would the sampling distribution of the summated χ^2 look like? Obviously, adding together two χ^2 values will tend to yield larger numbers, with fewer values very close to zero. **Figure 2** shows the theoretical sampling distribution for χ^2 for the $n = 2$ example.

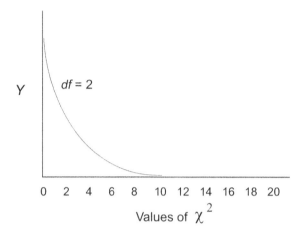

Figure 2

Sampling Distribution of χ^2 for n = df = 2.

Notice that the distribution is still highly skewed, although not as severely as was the case when each sample consisted of a single score.

Suppose that we increase sample size to $n = 6$ and repeat the process. Now for each sample we will have an overall value of χ^2 based on the sum of the χ^2 values for the six scores of the sample. What would the sampling distribution of these summated values of χ^2 look like? Clearly, the distribution should "move over" to the right, because the sum of a set of six χ^2's can get sizeable in value. **Figure 3** shows what the sampling distribution looks

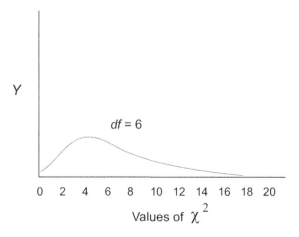

Figure 3

Sampling Distribution of χ^2 for n = df = 6.

like.

Notice that although the distribution is still necessarily positively skewed, it is now, at least, unimodal. It can be shown mathematically (and you can see by inspection) that for samples of $n \geq 2$, the sampling distribution will have its mode at $n - 2$ and its mean will be equal to n.

We will boost sample size one more time to $n = 10$ and repeat the whole process. **Figure 4** shows what the sampling distribution looks like for a summated χ^2 based on samples of size $n = 10$. The distribution has shifted further to the right ($Mo = 8$, $\mu = 10$), and is now fairly symmetrical in shape. As sample size continues to increase, these trends will continue. The limiting shape of the χ^2 sampling distribution when n becomes very large is (you guessed it) a normal distribution with $\mu = n$.

Figure 4

Sampling Distribution of χ^2 for n = df = 10.

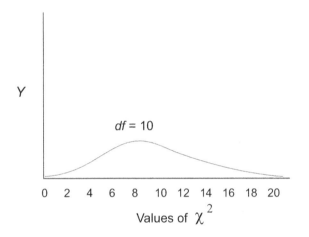

By now you should be wondering what all of this has to do with using χ^2 to analyze categorical data. It turns out that **we can compute χ^2 for frequency data and use the theoretical sampling distributions described above to evaluate such χ^2's.** Let's show how this is done in the context of a simple hypothesis testing situation. Recall yet again the "(Perhaps) Crooked Coin" example introduced in Chapter 3. In case you have forgotten, the crux of the problem in that example was to determine whether a particular quarter was biased to come up "heads." We flipped the coin 16 times and counted the number of times that heads occurred. The sample of 16 flips yielded 12 heads. The nondirectional H_0 and H_A that we tested in that situation could be formulated as:

H_0: $p(H) = .50$
H_A: $p(H) \neq .50$

To apply χ^2 to this problem, imagine the sampling distribution for the statistic "Number of Heads" that would be generated across an infinite number of 16-flip samples of the coin. It can be shown that the discrete distribution of the statistic "Number of Heads" would be approximately normal in shape with mean μ_H and variance σ_H^2 given by the following:

$$\mu_H = np = 16(.50) = 8.00$$

$$\sigma_H^2 = npq = 16(.50)(.50) = 4.00$$

where: n = sample size,
p = probability of Heads, and
q = $1 - p$ = probability of Tails.

Our sample outcome was 12 heads. What is the χ^2 value for the sample result of 12? Recalling the definition of χ^2 given earlier, we have:

$$\chi^2 = z^2 = \frac{(X - \mu)^2}{\sigma^2} = \frac{(\#H - \mu_H)^2}{\sigma_H^2}$$

$$= \frac{(12 - 8.00)^2}{4.00} = \frac{4^2}{4} = 4.00$$

What does this value of χ^2 tell us? Refer back to **Figure 1**, where we displayed the sampling distribution of χ^2 for $n = df = 1$. Notice that relatively few χ^2 values are as large as 4.00. In fact, fewer than 5% of the samples of $n = 1$ would yield a χ^2 of 4.00 or greater. Seemingly, our result is a very unlikely one if H_0 were true and $p(H) = .50$. If we were using a .05 α level to make our decision, we would conclude that H_0 is false and accept H_A that $p(H) \neq .50$. (You might also note that because $\chi^2 = z^2 = 4.00$, z in this case equals 2.00, which exceeds the $\pm$ 1.96 required to reject H_0 at the .05 α level.)

We can make the connection between χ^2 and frequency data even more explicit by recasting the definitional formula for χ^2 in terms of frequencies rather than means and standard deviations. The μ of the sampling distribution of outcomes of 16 flips is called the "expected value" of the statistic. In frequency terms, we will call the mean of the distribution the **expected frequency** and abbreviate it "f_E." We will call the value of the sample statistic the **observed frequency** and abbreviate it "f_O." Recalling that the variance of the population is equal to npq, we can rewrite the formula for χ^2 as

$$\chi^2 = \frac{(\#H - \mu_H)^2}{\sigma_H^2} = \frac{(f_O - f_E)^2}{npq}$$

The np in the denominator can be rewritten as f_E because np equals the mean of the sampling distribution, which we are calling f_E. This results in:

$$\chi^2 = \frac{(f_O - f_E)^2}{f_E q}$$

In our example, $q = .5$, so

$$\chi^2 = \frac{(f_O - f_E)^2}{f_E(.50)} = \frac{2(f_O - f_E)^2}{f_E}$$

This gives us the fundamental frequency definition of χ^2 for a two-category situation. We can compute χ^2 for our perhaps crooked coin problem using this definition as well:

$$\chi^2 = \frac{2(f_O - f_E)^2}{f_E} = \frac{2(12-8)^2}{8}$$

$$= \frac{2(4)^2}{8} = \frac{2(16)}{8} = \frac{32}{8} = 4.00$$

We need to do one more thing and then we'll be finished with the development of the definition of χ^2 in terms of frequency data. The "2" in the formula doubles the value of $(f_O - f_E)^2/f_E$. There's another way to think of this. There are actually two frequencies observed in the coin-tossing experiment, the frequency of heads and the frequency of tails. The "2" in the formula means that we are summing across two equal discrepancies, i.e., across two categories. This can be diagrammed in **Table 1**.

Let's compute $(f_O - f_E)^2/f_E$ for each discrepancy (category) and add up the result.

For Heads:

$$\chi_H^2 = \frac{(f_O - f_E)^2}{f_E} = \frac{(12-8)^2}{8} = \frac{4^2}{8} = \frac{16}{8} = 2.00$$

For tails:

$$\chi_T^2 = \frac{(f_O - f_E)^2}{f_E} = \frac{(4-8)^2}{8} = \frac{-4^2}{8} = \frac{16}{8} = 2.00$$

The sum of the values of χ^2 for the two categories is 4.00, which is exactly what we got from our original frequency definition of χ^2. This leads us to the final form of the frequency definition of χ^2.

$$\chi^2 = \sum \left(\frac{(f_O - f_E)^2}{f_E} \right)$$

According to this formula, χ^2 is the sum of the values of $(f_O - f_E)^2/f_E$ for each category, i.e., the sum of the χ^2's

Table 1
Observed and Expected Frequencies for Outcome of Coin-Tossing Experiment.

	Categories	
	Heads	Tails
Observed Frequency (f_O)	12	4
Expected Frequency (f_E)	8	8

computed for each of the categories. This formula is applicable to any number of categories, because χ^2 is an additive statistic. Technically, χ^2 defined in this way is called **Pearson chi square**, because the statistician, Karl Pearson, was the person who invented the application of chi square to the analysis of frequency or categorical data.

There is one puzzling thing about this definition of χ^2. If the computed value of χ^2 is based on what appears to be two separate χ^2's (one for each category), why did we originally evaluate the result with respect to the sampling distribution of χ^2 based on a sample size of $n = 1$? Why not for an n of 2, given that there appear to be two χ^2 values summed up? The answer lies in the concept of degrees of freedom, which we introduced in Chapter 10 in connection with the t distribution. Although there are two categories, the discrepancies $(f_O - f_E)$ for the two categories are not independent of one another. In fact, the value of one $(f_O - f_E)$ discrepancy exactly determines the other, i.e., if you have $12 - 8 = +4$ heads, there must be $4 - 8 = -4$ tails. For purposes of using the theoretical sampling distribution of χ^2 to evaluate frequency data, you must refer to the distribution corresponding to the **number of independent discrepancies**—which is to say, the df for the particular χ^2 problem.

When the $df = 1$ for a χ^2 analysis, many statisticians recommend that an adjustment made in the way that χ^2 is computed to improve the fit of the discrete frequency values to the underlying continuous distribution. This adjustment is called **Yate's correction for continuity** and involves the following modification in the formula for computing χ^2:

$$\chi^2 = \sum \left(\frac{(|f_O - f_E| - .5)^2}{f_E} \right)$$

where: $|f_O - f_E|$ = the absolute (unsigned) discrepancy between the observed and expected frequencies.

If we apply Yate's correction to the computation of χ^2 for our coin-flipping example, the obtained value of χ^2 = 3.063, which falls in the upper 8% of the area under the sampling distribution of χ^2 for $n = df = 1$.

Assumptions for the Procedure

Certain assumptions must be met to use the theoretical sampling distributions of χ^2 to evaluate frequency data. We cleverly slipped those assumptions into our coin-tossing example without drawing attention to them. It's now time to make them explicit.

1. **The sample of cases on which the frequency data are collected is a random sample of the population of interest.** This assumption was met in the coin-flipping example but is frequently violated insofar as most samples are not truly random.

2. **The frequencies comprising the data set must be completely independent of one another.** Clearly, the outcome associated with each flip of the coin was independent of every other flip. More broadly, this assumption requires that **each case on which the study is conducted contribute *only one* frequency to the data set.**

3. **The discrepancies must be normally distributed in the population of interest.** Recall that χ^2 is fundamentally defined in terms of sampling from a normally distributed population of scores. If the $(f_O - f_E)$ discrepancies are not normally distributed, the theoretical sampling distribution of χ^2 will not be appropriate for evaluating a sample value of χ^2 obtained on frequency data. If sampling has been random *and* the sample size is large, this assumption tends to be met. Problems arise, however, when the sample size is small. In small samples with low f_E's, the $(f_O - f_E)^2$ discrepancies tend to be skewed in a positive direction. Moreover, in small samples, the $(f_O - f_E)$ discrepancies are markedly discontinuous, because frequencies change in increments of 1.00. Highly discontinuous data preclude an adequate fit to the smooth, continuous theoretical distribution of χ^2. The moral of the story is to use adequately sized samples to make it likely that this assumption will be met. A couple of "rules of thumb" are offered below concerning sample sizes for different applications of χ^2 for frequency data. The sample size in the coin-tossing experiment was large enough to meet this assumption.

Goodness of Fit χ^2

The goodness-of-fit application of χ^2 is used to test the H_0 that the distribution of observations among two or more categories in a population follows some specified pattern. The H_A says that they are not distributed in the manner stated under H_0. **Because most populations are assumed to be indefinitely large, the hypotheses must be stated in terms of proportions rather than absolute frequencies.** The coin-tossing example through which we developed the logic of χ^2 was a goodness-of-fit test of the H_0 that $p(H) = p(T) = .50$. We'll now apply the procedure to a more complicated goodness-of-fit problem in which there are more than two categories. **The problem is concerned with testing a hypothesis about the distribution of grades and is titled, "The Embarrassing Mode." Carefully read through the description of this problem below before continuing.**

The Dean wants to test the H_0 that the distribution of grades given at her college is the same as that reported for the prestigious colleges and universities. The H_0 and H_A could be stated as follows:

H_0: $p(A) = .315$, $p(B) = .398$, $p(C) = .124$, $p(D) = .092$, $p(F) = .071$

H_A: The proportions are not as stated in H_0.

Computational Procedures for Goodness-of-Fit χ^2

This test of the H_0 is a goodness-of-fit χ^2 test involving the distribution of frequencies among five categories. From the description of the problem, we know that the Dean's

---

The Embarrassing Mode

In Chapter 9 ("The Dean's Dilemma"), the Dean of the college learned that students on average work significantly fewer hours per week on each of their courses than do students at more prestigious colleges and universities. The same issue of the *Chronicle of Higher Education* also reported the average distribution of letter grades awarded to students at these prestigious institutions. The distribution was A's = 31.5%, B's = 39.8%, C's = 12.4%, D's = 9.2%, and F's = 7.1%. The Dean wonders how her institution compares to the more prestigious ones in this regard. Fearing the worst, she secures a random sample of 150 student grade slips for the most recent semester. She records the *first* grade listed on each sheet, and then tabulates the frequency of A's, B's, C's, D's, and F's. Here are her data: A's = 62, B's = 44, C's = 25, D's = 13, and F's = 6.

sample is random (Assumption 1) and the frequencies are independent of one another because the Dean used only one grade from each student (Assumption 2). We don't know for sure whether the $(f_O - f_E)$ discrepancies are normally distributed in the population (Assumption 3), but with a sample size of 150, the assumption is certainly plausible.

We'll work through the manual solution of this problem and then show how it would be analyzed using *JASP*. We will begin by describing the sample data for the college. This is easy because all that you can do with frequency data is to convert them to proportions or percentages. For our problem, the proportion of the sample earning each grade is obtained by dividing the observed frequency for the grade by the sample size. **Table 2** gives the proportion distribution of grades at the institution, along with the distribution reported for the prestigious colleges and universities.

Table 2

Relative Frequency Distribution of a Sample of 150 Grades at the Deans's College Compared with the Distribution of Grades at Prestigious Colleges and Universities.

Grade	Dean's College	Prestigious Colleges
A	.413	.315
B	.293	.398
C	.167	.124
D	.087	.092
F	.040	.071

As can be seen in **Table 2**, 41.3% of the sample of grades given at the college are A's, while at the more prestigious institutions, the figure is 31.5%. In contrast, faculty members at the college seem to give proportionately fewer B's than the comparison group. **The question, of course, is whether the sample distribution is *significantly different* from the distribution for the prestigious institutions**, or whether the departures could be due simply to sampling fluctuations. That is where the χ^2 goodness-of-fit test comes in.

Our first task is to "lay out" the problem in a way that will facilitate analyzing the data. Such a layout is given in **Table 3**. The observed frequencies are those obtained by the Dean. The expected frequencies are computed by multiplying her sample size (n = 150) times the decimal fraction of each grade reported for the prestigious colleges, i.e., the distribution stated in the H_0. The fact that the expected frequencies are not whole numbers is of no

Table 3

Observed and Expected Frequencies for the Goodness-of-Fit χ^2 Analysis of the "Embarrassing Mode" Study.

Grade	f_O		f_E
A	62	150 X .315 =	47.25
B	44	150 X .398 =	59.70
C	25	150 X .124 =	18.60
D	13	150 X .092 =	13.80
F	6	150 X .071 =	10.65
n =	150		150

concern, because we are approximating a continuous distribution anyway.

Table 4 on the next page shows the computation of χ^2 for these data. The obtained value of χ^2 for our data is 13.01. The question is whether this value is sufficiently large to cause us to reject H_0. Because we are computing χ^2 by hand, it is necessary for us to compare this result to a table that lists the critical values of χ^2 for different α levels. **Table C** in the Appendix is used for this purpose. It consists of six columns. The first is labeled *df* and refers to the number of degrees of freedom for the χ^2 distribution. The remaining five columns are the critical values of χ^2 for α levels of .10, .05, .025, .01, and .005. The *df* for our test is 4, because of the five $(f_O - f_E)$ discrepancies, only four are independent of one another. Once any four discrepancies have been computed, the remaining one is constrained by the fact that the sum of the discrepancies must equal 0. You can satisfy yourself of this by consulting **Table 4**, where you will note that the sum of the $(f_O - f_E)$ discrepancies must equal 0. **The general formula for determining the *df* for a goodness-of-fit χ^2 test is *df* = number of categories minus one, i.e., (C – 1)**.

Consulting **Table C** in the Appendix for an α level of .05, we find that the critical value of χ^2 defining the area of rejection is 9.49. Our obtained value of χ^2 exceeds this, so we can reject H_0 that the distribution of grades given at the college is the same as that of the reference group of more prestigious institutions.

Where, exactly, does the significant difference arise? Because the obtained χ^2 is summed across the separate χ^2's for each grade category, it reflects an "overall" difference, much like that of the *F* ratio. There is no simple procedure for making "pairwise" comparisons between categories, but usually the source of the major part of the

Table 4

Manual Computation of χ^2 Goodness-of-Fit Analysis for the "Embarrassing Mode" Study.

Grade	f_O	f_E	$f_O - f_E$	$(f_O - f_E)^2$	$(f_O - f_E)^2 / f_E$
A	62	47.25	14.75	217.56	4.60
B	44	59.70	-15.70	246.49	4.13
C	25	18.60	6.40	40.96	2.20
D	13	13.80	-.80	.64	.05
F	6	10.65	-4.65	21.62	2.03

$$\chi^2 = 13.01$$

total χ^2 is apparent. Inspection of the summary information in **Table 2** suggests the major difference between the institution and those described in the *Chronicle* is a surplus of A's and a shortage of B's. The distribution of the other grades is relatively similar for the college and the reference institutions.

There is one other feature of the data analysis to be discussed. Notice that the χ^2 value computed for the "F" category is fairly large (2.03) in spite of the fact that the $(f_O - f_E)$ discrepancy was only -4.65. This occurred because the expected frequency for that category was fairly small, 10.65. Low expected frequencies can sometimes produce very large χ^2 values and make the overall χ^2 significant, even when the $(f_O - f_E)$ discrepancies for the categories having most of the frequencies are relatively small. This leads to a rule of thumb concerning expected frequencies: **For goodness-of-fit χ^2 tests having 1 *df*, the f_E for both categories should be greater than 5.0. For tests having more than 1 *df*, no more than 1/5 of the expected frequencies should be less than 5.0.** Our data meet this rule of thumb.

Before proceeding, let's consider how one might write a narrative summary of the results of the analysis for the grade distribution study.

Sample data from 150 students were used to determine whether the distribution of letter grades assigned at the college differed significantly from the average grade distribution reported in a national survey of prestigious colleges and universities. A χ^2 goodness-of-fit test with α set at .05 compared the obtained frequencies for the sample across the five grade categories with expected frequencies generated from the proportions reported in the national survey. The resulting χ^2 was statistically significant, χ^2 (4, n = 150) = 13.012, p < .011, indicating that the college's distribution differed significantly from that reported in the survey.

A comparison of the college's grade distribution with that reported in the national survey suggests that the college seems to award more A grades and fewer B grades than the more prestigious institutions. There appears to be no difference in the frequency of C, D, or F grades.

JASP Analysis of the Data

For a goodness-of-fit test *JASP* offers two ways to input the data. The first requires that the data be in the form of raw scores entered into a data file. For the Embarrassing Mode data, this would require entering the value of "4" (the numerical coding of a grade of A) 64 times, "3" (for B) 44 times, and so on. **This is not the way to do it unless the frequency data are part of a larger data set.**

The preferred approach allows you to enter the frequency counts directly as data. **Figure 5** shows what the data set looks like for *JASP*.

Figure 5

Data Formatted for JASP Goodness-of-Fit Test.

	grades	Observed	Expected
1	A	62	47.25
2	B	44	59.7
3	C	25	18.6
4	D	13	13.8
5	F	6	10.65

The general analysis routine is called FREQUENCIES and the test that is used for goodness-of-fit is called the Multinominal Test. **Figure 6** shows a screen capture of the format of the test, indicating where the three variables are entered and the option of generating a Descriptive statistics table.

Figure 7 shows the results of the chi square analysis of the data. Note that the value of χ^2 of 13.01 is the same as we obtained by hand calculation.

Figure 7

Value of χ^2 Computed by the Multinominal Test.

Multinomial Test

	χ^2	df	p
H₀ (a)	13.012	4	0.011

Descriptives

grades	Observed	Expected: H₀ (a)
A	62	47.250
B	44	59.700
C	25	18.600
D	13	13.800
F	6	10.650

Contingency Table χ^2

We will develop our understanding of the contingency table χ^2 using the following example problem called "Sad to Say:" Read the description of the problem on the next page carefully before going on.

The purpose of the contingency table χ^2 analysis is to decide whether two categorical variables are "contingent" upon one another. The word "contingent" is used in this context to mean "correlated." Because we are already familiar with correlation, we'll use "correlation" instead of contingent. The H_0 and H_A tested take the following form:

H_0: The two variables are uncorrelated.
H_A: The two variables are correlated.

The H_0 and H_A for the *"Sad to Say"* study would be formulated as follows:

H_0: Grades and gender are uncorrelated.
H_A: Grades and gender are correlated.

Figure 6

JASP Screen Capture of Multinominal Test Analysis Window.

As usual, we will set α at .05.

Let's begin by evaluating how well the assumptions underlying χ^2 are met in this situation. First, the 98 students do not constitute a random sample of any population (Assumption 1). At best, we might hope that they are a representative sample of students who have taken statistics and research design in the past and who will take it in the future. Second, the frequencies in the table are independent of one another, because a particular student appears only once in the table (Assumption 2). This is because the instructor recorded only the *first* grade for students who failed the course and had to repeat it. Had the instructor not done so, then those students who failed the course and repeated it would contribute at least two frequencies to the table. These frequencies would not be independent of one another. The third assumption of normality of the $(f_O - f_E)$ discrepancies in the population is completely unknown. In view of the lack of random sampling, we may be on slippery ground here, particularly given the fact that some of the f_E's for the table are going to be small (see below). We should be cautious in interpreting the results of this analysis.

Obtaining f_E's and *df* for a Contingency Table χ^2

Once you have computed one χ^2 by hand, you probably have a "feel" for how it works Instead, we'll concentrate on explaining one feature of contingency table χ^2 analysis that many students have trouble understanding—getting the expected frequencies. We will

Sad to Say

A common observation at many colleges is that women students seem to do better in the psychology major program than do the male students. To document this, the instructor who teaches the Statistics & Research Design course consulted his grade books for that course for the past five years and tabulated the gender and grade in the course for 98 students who completed the course. Here are the data placed in a contingency table. Notice that many more women than men major in psychology!

	\multicolumn{5}{c}{Grade Received}
	A	B	C	D	F
Males	3	5	8	2	9
Females	15	26	20	4	6

use the data layout in **Table 5** to explain where the expected frequencies come from in a contingency table χ^2.

Table 5

Data Layout for a χ^2 Contingency Table Analysis of Grade Distributions for Male and Female Students.

			\multicolumn{5}{c}{Grade}	Row
			A	B	C	D	F	Totals
Males	f_O	=	3	5	8	2	9	27
	f_E	=	?					
Females	f_O	=	15	26	20	4	6	71
	f_E	=						
Column Totals	=		18	31	28	6	15	98

For all types of χ^2 analyses, the expected frequencies are *always* derived from the H_0. In a goodness-of-fit χ^2 this connection between the H_0 and the expected frequencies is obvious. It is less so in a contingency table χ^2. What, for example, is the expected frequency for the upper left-hand "cell" of the table (Males, Grade of A)? It is not the case that all grades are equally likely to occur, so we don't expect 1/5 of all males to get A's.

The solution to this problem is to look at the distribution of grades for the entire sample (Males plus Females) indicated by the column totals. Assuming that H_0 is true, and there is no relationship between gender and grades, the distribution of frequencies for the entire sample should provide the best guess concerning how grades are distributed in the population. We will use this distribution to generate the f_E's for the Male and Female students.

For the entire sample, 18/98 or 18.4% of the students received A's. Assuming that gender is unrelated to grades (H_0), we would expect that 18.4% of the males should receive A's. Thus, 18.4% of the 27 males should fall in the A category, i.e., .184 x 27 = 4.97. Similarly, 18.4% of the 71 females is .184 x 71 = 13.06. These are the f_E's for the Male and Female, Grade of A categories. For the entire sample, 31/98 = 31.6% of the students received B's, so we expect that 31.6% of the 27 males will fall into the B category and 31.6% of the 71 females will fall into the B category, and so on. The necessary computed expected frequencies are shown in **Table 6**.

Table 6

Summary of the χ^2 Contingency Table Analysis of Grade Distributions for Male and Female Students.

		\multicolumn{5}{c}{Grade}
		A	B	C	D	F
Males		3	5	8	2	9
		11%	19%	30%	7%	33%
f_E	=	4.97	8.53	7.72	1.65	4.13
Females		15	26	20	4	6
		21%	37%	28%	6%	8%
f_E	=	13.06	22.44	20.31	4.33	10.86
Total		18.4%	31.6%	28.6%	6.1%	15.3%

Doing the analysis by hand, once the expected frequencies had been computed in this way, you would proceed to compute χ^2 for each cell using the standard formula. The sum of the χ^2 values for the 10 ($f_O - f_E$) discrepancies would be your obtained value of χ^2. You would compare the obtained value with the critical value of χ^2 for the .05 α level from **Table C** for the theoretical distribution of χ^2 having **4 Degrees of freedom**.

The analysis of the contingency table by hand is straightforward but somewhat laborious. Using the formula:

$$\chi^2 = \sum \left(\frac{(f_O - f_E)^2}{f_E} \right)$$

$$= \frac{(3 - 4.97)^2}{4.97} = \frac{-1.97^2}{4.97} = \frac{3.88}{4.97} = .78$$

$$= \frac{(5 - 8.53)^2}{8.53} = \frac{-3.53^2}{8.53} = \frac{12.46}{8.53} = 1.46$$

+
.
.
.
+

$$= \frac{(6 - 10.86)^2}{10.86} = \frac{-4.86^2}{10.86} = \frac{23.62}{10.86} = 2.17$$

The sum of the 10 χ^2 values = 11.12, which with 4 *df* is significant at *p* = .025.

Why are there only 4 df for this contingency table? The answer again lies in the concept of independent $(f_O - f_E)$ discrepancies. Consider the upper left-hand cell in **Table 6** (Males, Grade of A). Once f_E has been computed for this cell, the f_E is also determined for the lower left-hand cell (Females, Grade of A). This is because the sum of the observed and expected frequencies for the Grade of A column must sum to 18. Thus, the $(f_O - f_E)$ discrepancy for the lower left-hand cell gives the same information that the upper left-hand cell does.

Move over one cell to the right in the upper row (Males, Grade of B). The $(f_O - f_E)$ discrepancy for this cell is independent of that for the previous cell, which moves the *df* up to 2. But the cell below it (Females, Grade of B) is not independent for the same reason that the $(f_O - f_E)$ discrepancy for the Females. This continues until you reach grade of F for the Males, which is now constrained by the row total. Thus, this last row $(f_O - f_E)$ is not independent of the first four. So, total *df* = 4. **More generally, you can figure the *df* for any contingency table by multiplying (R – 1) times (C – 1).**

Where does the significant chi square come from? Clearly, males are earning far too few A's and B's and way too many F's compared to their female counterparts!

Notice in **Table 6** that three of the cells have expected frequencies below 5.00, and that the expected frequency for the Males, Grade of D cell is particularly low (f_E = 1.65). **Recall that the rule of thumb with contingency tables is that no more than 1/5 of the cells should have expected frequencies lower than 5**. We have just barely violated this rule of thumb, mostly because the number of males is so small (*n* = 27).

A good solution would be to try to increase sample size by including another year's worth of data in the analysis. An alternative solution favored by some investigators is to "collapse" categories to increase the f_E's. You could, for example, collapse the D and F categories to a new category called "D or F." There would be 11 males in that category and the expected frequency would be 5.78 (i.e., .214 x 27). This approach has hazards. Manipulating categories like this is analogous to carrying out all possible comparisons among a set of three or more sample means. As a result, your α level doesn't stay put, increasing quickly.

In our situation, we had already rejected H_0 based on the overall table as originally constructed, so we might be justified in collapsing the two categories to see if the results were still significant. If they were (and they are, χ^2 = 9.60, *df* = 3, *p* value = .022), we would be confident that the original significant χ^2 did not arise because of capitalizing on low f_E's.

Analysis of the Data Using *JASP*

It's clear that computing χ^2 for a sizable contingency table is a good deal of work, which is best left to computers. To make most efficient use the *JASP* Contingency Tables procedure under **Frequencies**, the data need to be entered in a particular manner. **Figure 8** shows data format for the first six cases in an *Excel* spreadsheet.

Figure 8

Data *Layout For the Contingency Table*

	A	B	C
1	Gender	Grade	Count
2	m	a	3
3	m	b	5
4	m	c	8
5	m	d	2
6	m	f	9
7	f	a	15

Figure 9 shows the way the variables are entered in the Statistics Panel for the Contingency Tables procedure.

Figure 9

Variable Entry Into the Three Windows of the Contingency Table Procedure.

In analyzing the data, under the Statistics option, χ^2 is selected by default and we'll also select Cramer's V as a measure of the strength of the association between the two variables. Under the Cells option we'll select Expected Counts and Row Percentages. **Figure 10** shows the *JASP* summary Contingency Table, the χ^2 and its *p* value, and the Cramer's *V* statistic (see below). Satisfy yourself that this agrees with the table and value for χ^2 that we computed by hand. Females come before Males in the table because *JASP* uses alphabetical order in laying out tables.

A Measure of the Strength of Association

The significant value for χ^2 allows us to reject the H_0 of no correlation, but by itself, it doesn't tell us how strong the correlation is. This is analogous to the situation with independent *t* tests and ANOVAs, where an effect size statistic is used to objectively quantify the strength of the effect or association. A statistic called Cramer's *V* is appropriate for contingency tables.

Figure 10

JASP Contingency Table, Value of Chi Square, and Contingency Coefficient.

Contingency Table of Gender X Grade for 98 Students Who Took Statistics & Research Design

Gender		Grade					
		A	B	C	D	F	Total
Female	Count	15.00	26.00	20.00	4.00	6.00	71.00
	Expected count	13.04	22.46	20.29	4.35	10.87	71.00
	% within row	21.13 %	36.62 %	28.17 %	5.63 %	8.45 %	100.00 %
Male	Count	3.00	5.00	8.00	2.00	9.00	27.00
	Expected count	4.96	8.54	7.71	1.65	4.13	27.00
	% within row	11.11 %	18.52 %	29.63 %	7.41 %	33.33 %	100.00 %
Total	Count	18.00	31.00	28.00	6.00	15.00	98.00
	Expected count	18.00	31.00	28.00	6.00	15.00	98.00
	% within row	18.37 %	31.63 %	28.57 %	6.12 %	15.31 %	100.00 %

Note: I added information to the title and for clarity inserted blank rows between the Female, Male, and Total sections of the table.

Chi-Squared Tests

	Value	df	p
X²	11.12	4	0.03
N	98		

Nominal	
	Value[a]
Cramer's V	0.34

Cramer's *V* is a measure of the magnitude of the correlation between the two categorical variables. It can range in value from 0 (no association) to 1.00 (perfect correlation). The formula for *V* is:

$$V = \sqrt{\frac{\chi^2}{n * min(R-1, C-1)}}$$

where: *n* = total sample size,
 R = number of row categories,
 C = number of column categories, and
 min(*R*–1,*C*–1) = whichever is smaller of *R*–1 and *C*–1.

For our data, *V* is computed as:

$$V = \sqrt{\frac{11.122}{98 * min(2-1, 5-1)}}$$

$$= \sqrt{\frac{11.122}{98 * 1}} = \sqrt{.113} = .337$$

It is important to note that *V* is not the same as Pearson *r* and cannot be interpreted in the same way. You cannot, for example, square *V* to get predicted variance. Just consider *V* an index of the strength of association between the two variables. Values less than .2 indicate a weak relationship, between .2 and .4, a moderate relationship, and between .4 and .6, a strong relationship.

Let's finish our discussion of contingency table analysis by considering a narrative summary of the data analysis for the study of the relationship between gender and grades in Statistics & Research Design.

The study compared the distribution of letter grades earned by 27 male students in Statistics & Research Design with the distribution of grades earned by 71 female students in the same course. A 2 (Gender) X 5 (Grade) χ^2 contingency table analysis was used to determine whether the distribution of grades differed significantly for males and females. The obtained value of χ^2 was significant at the .05 α level, χ^2 (4, *n* = 98) = 11.122, *p* < .025, indicating that the distributions of grades for the two sexes differed significantly.

The relative frequency distributions of grades for the males and female students are shown in **Figure 10**. Comparison of the distributions suggests that female students are much more likely than male students to receive A's and B's in the course and much less likely to receive F's. There seemed to be no difference between the sexes in the proportion of students who received C and D grades.

Notice that in the above description, we stated that the purpose of using the contingency table χ^2 was to determine whether the distributions of grades for male and female students *differed significantly*. This may seem like an odd way to state the purpose, because the H_0 and H_A for a contingency table given at the beginning of this section focused on determining whether a correlation or contingency exists between two variables. There is no real inconsistency, however. The statement that the "distribution of grades is correlated with sex" is equivalent to saying, "the distribution of grades differs for the two sexes."

Final Comments

Congratulations! If you're reading this you made it through the entire course. You've learned some extremely valuable knowledge and skills and you should look for opportunities to keep practicing those skills. Volunteer as a statistics tutor or offer to help others with statistical problems. *JASP* is an incredibly useful tool which you should hang onto and periodically update. And by all means, when you put your professional resume together, include Statistics and Data Analysis as one of your competencies!

Chapter 14, Chi Square
Chapter Exercises

Some questions are purely factual and have only one correct answer. Others require greater thought on your part and may have more than one possible answer.

1. Fill in the missing term(s) that makes the statement true.

   a. ANOVA is a _____ statistic, while χ^2 is a _____ statistic.
   b. χ^2 is used to analyze _____ data.
   c. χ^2 is fundamentally defined as _____.
   d. The sampling distribution for χ^2 is heavily _____ skewed.
   e. When df > 2, the sampling distribution of χ^2 has a mode at _____ and its mean at _____.
   f. As df increases, the limiting shape of the χ^2 distribution is the _____.
   g. The assumption of independence requires that each case contribute _____.
   h. Another name for a χ^2 goodness of fit test is a _____.
   i. The fe's for a goodness of fit test come from the _____.
   j. The sum of (fo – fe)'s for a given χ^2 analysis always = _____.
   k. A χ^2 with 6 df = 6.75. According to **Table C**, can you reject H_0 with α =.05? _____
   l. Another term for "contingent" is _____.
   m. A 4 X 3 contingency table has _____ df.
   n. In a contingency table, the fe's are derived by assuming H_0 is true and summing across _____.
   o. If H_0 in a contingency table analysis is rejected, a measure of the strength of association is _____.

2. A new inexpensive multiple choice I.Q. test has been developed for classroom use in middle schools. The makers of the test claim that when used with unselected 6th, 7th, or 8th graders, the instrument will yield a normal distribution of scores having a mean equal to 100 and a standard deviation of 15. A large urban school district is interested in using the test. To verify that the test behaves as advertised, a psychologist for the district administers the test to all 6th grade students at one of the district's larger middle schools (n = 403 students). The sample mean and standard deviation are close to the advertised values of 100 and 15. The distribution of scores for the sample, however, appears somewhat non-normal in shape. Below are the scores organized into a frequency distribution where the class intervals are expressed in terms of the sample $\bar{X}$ and $\hat{s}$: Is it reasonable to believe that the test yields a normal distribution of IQ scores?

Interval	No. of Scores
Greater than +1.00 $\hat{s}$	37
Between +.51 and +1.00 $\hat{s}$	106
Sample $\bar{X}$ ± .50 $\hat{s}$	153
Between -.51 and -1.00 $\hat{s}$	88
Less than -1.00 $\hat{s}$	19

State H_0 and H_A and then derive the expected frequencies assuming H_0 is true. Use *JASP* to test the H_0.

3. Many people, particularly parents of teenage children, believe that the full moon contributes to "crazy" behavior. Indeed, the old term "lunatic" (*Luna* meaning moon) makes this connection explicit. Mary Meyer, a junior at Evansdale High School, has heard her mom blame full moons for everything from incontinent dogs to mass murders. Desperate for a quick-and-dirty Science Fair project, Mary has hit upon the idea of seeing whether the phases of the moon are associated with different frequencies of disruptive behavior in school. She persuades Mr. Snargle, the High School Principal, to allow her access to "sanitized" (names removed) discipline records. Mr. Snargle, although an indifferent educator, keeps meticulous discipline records, and Mary is able to compile two years' worth of infractions. By cross checking the date of the infraction against the *Farmer's Almanac*, Mary is able to identify the phase of the moon at the time of the event. Below are the number of discipline incidents that occurred during each phase of the moon for the past two years. (Note: A lunar month is approximately 29.5 days in length. Mary arbitrarily defined each phase to be 7 days in length, centered symmetrically about the precise phase day.)

Phase of Moon	# of incidents
Full Moon	76
2nd Quarter	46
3rd Quarter	50
4th Quarter	62

Do a complete descriptive and inferential analysis of the data, including an appropriate graph. How would you interpret the results of this study?

4. A particular species of moth shows a normal genetic variation in wing color ranging from light green to dark green. The species inhabits mountainous regions encompassing both open meadows and dense coniferous forests, and individuals tend not to travel very far from where they were hatched. The process of natural selection implies that light winged moths should predominate in meadow regions because their wing coloration blends in with the flowers and grasses and makes them less subject to predation. In contrast, the darker winged members should predominate in forest regions because they are less conspicuous in the heavy shadows. To test this prediction, a biology student systematically traps moths from both areas. Below are the data (1 = light winged individual; 2 = intermediate coloration; 3 = dark winged individual).

Sample trapped from meadows:

1	3	1	2	1	2	1	2	3
1	2	1	3	2	1	1	1	2
1	2	1	3	1	1	3	2	1
1	3	1	1	2	2	1	1	3
3	2	3	1	1	1	2	1	2
2	1	2	3					

Sample trapped from forest:

3	1	2	2	3	2	2	1	3
3	3	3	2	3	3	2	3	3
1	3	3	1	2	3	2	3	2
3	1	2	3	3	3	1	3	3
1	2	2	1					

State the H_0 and H_A and then use *JASP* to generate a descriptive statistics table and test the H_0. Write a professional interpretation of the data.

Appendix
Statistical Tables

This appendix contains the following statistical tables:

Table A	Areas Under the Unit Normal Curve corresponding to Different Values of *z*.
Table B	Student's Distribution of *t*.
Table C	The χ^2 Distribution.

These tables are less comprehensive than those normally included in statistics textbooks. The reason for this is that the *JASP* program handles all the necessary computations that normally require the use of statistical tables. These tables are intended for instructional purposes, rather than routine statistical use.

Appendix Table A
Normal Curve Table

Column 1   Values of z
Column 2   Area Under the Curve Between μ and z
Column 3   Area Under the Curve Beyond z

z	μ to z	> z	z	μ to z	> z	z	μ to z	> z
0.00	.0000	.5000	0.35	.1368	.3632	0.70	.2580	.2420
0.01	.0040	.4960	0.36	.1406	.3594	0.71	.2611	.2389
0.02	.0080	.4920	0.37	.1443	.3557	0.72	.2642	.2358
0.03	.0120	.4880	0.38	.1480	.3520	0.73	.2673	.2327
0.04	.0160	.4840	0.39	.1517	.3483	0.74	.2704	.2296
0.05	.0199	.4801	0.40	.1554	.3446	0.75	.2734	.2266
0.06	.0239	.4761	0.41	.1591	.3409	0.76	.2764	.2236
0.07	.0279	.4721	0.42	.1628	.3372	0.77	.2794	.2206
0.08	.0319	.4681	0.43	.1664	.3336	0.78	.2823	.2177
0.09	.0359	.4641	0.44	.1700	.3300	0.79	.2852	.2148
0.10	.0398	.4602	0.45	.1736	.3264	0.80	.2281	.2119
0.11	.0438	.4562	0.46	.1772	.3228	0.81	.2910	.2090
0.12	.0478	.4522	0.47	.1808	.3182	0.82	.2939	.2061
0.13	.051	.4483	0.48	.1844	.3156	0.83	.2967	.2033
0.14	.0557	.4443	0.49	.1879	.3121	0.84	.2995	.2005
0.15	.0596	.4404	0.50	.1915	.3085	0.85	.3023	.1977
0.16	.0636	.4364	0.51	.1950	.3050	0.86	.3051	.1949
0.17	.0675	.4325	0.52	.1985	.3015	0.87	.3078	.1922
0.18	.0714	.4286	0.53	.2019	.2981	0.88	.3106	.1894
0.19	.0753	.4247	0.54	.2054	.2946	0.89	.3133	.1867
0.20	.0793	.4207	0.55	.2088	.2912	0.90	.3159	.1841
0.21	.0832	.4168	0.56	.2123	.2877	0.91	.3186	.1814
0.22	.0871	.4129	0.57	.2157	.2843	0.92	.3212	.1788
0.23	.0910	.4090	0.58	.2190	.2810	0.93	.3238	.1762
0.24	.0948	.4052	0.59	.2224	.2776	0.94	.3264	.1736
0.25	.0987	.4013	0.60	.2257	.2743	0.95	.3289	.1711
0.26	.1026	.3974	0.61	.2291	.2709	0.96	.3315	.1685
0.27	.1064	.3936	0.62	.2324	.2676	0.97	.3340	.1660
0.28	.1103	.3897	0.63	.2357	.2643	0.98	.3365	.1635
0.29	.1141	.3859	0.64	.2389	.2611	0.99	.3389	.1611
0.30	.1179	.3821	0.65	.2422	.2578	1.00	.3413	.1587
0.31	.1217	.3783	0.66	.2454	.2546	1.01	.3438	.1562
0.32	.1255	.3745	0.67	.2486	.2514	1.02	.3461	.1539
0.33	.1293	.3707	0.68	.2517	.2483	1.03	.3485	.1515
0.34	.1331	.3669	0.69	.2549	.2451	1.04	.3508	.1492

z	μ to z	> z	z	μ to z	> z	z	μ to z	> z
1.05	.3531	.1469	1.50	.4332	.0668	1.95	.4744	.0256
1.06	.3554	.1446	1.51	.4345	.0655	1.96	.4750	.0250
1.07	.3577	.1423	1.52	.4357	.0643	1.97	.4756	.0244
1.08	.3599	.1401	1.53	.4370	.0630	1.98	.4761	.0239
1.09	.3621	.1379	1.54	.4382	.0618	1.99	.4767	.0233
1.10	.3643	.1357	1.55	.4394	.0606	2.00	.4772	.0228
1.11	.3665	.1335	1.56	.4406	.0594	2.01	.4778	.0222
1.12	.3686	.1314	1.57	.4418	.0582	2.02	.4783	.0217
1.13	.3708	.1292	1.58	.4429	.0571	2.03	.4788	.0212
1.14	.3729	.1271	1.59	.4441	.0559	2.04	.4793	.0207
1.15	.3749	.1251	1.60	.4452	.0548	2.05	.4798	.0202
1.16	.3770	.1230	1.61	.4463	.0537	2.06	.4803	.0197
1.17	.3790	.1210	1.62	.4474	.0526	2.07	.4808	.0192
1.18	.3810	.1190	1.63	.4484	.0516	2.08	.4812	.0188
1.19	.3830	.1170	1.64	.4495	.0505	2.09	.4817	.0183
1.20	.3849	.1151	1.65	.4505	.0495	2.10	.4821	.0179
1.21	.3869	.1131	1.66	.4515	.0485	2.11	.4826	.0174
1.22	.3888	.1112	1.67	.4525	.0475	2.12	.4830	.0170
1.23	.3907	.1093	1.68	.4535	.0465	2.13	.4834	.0166
1.24	.3925	.1075	1.69	.4545	.0455	2.14	.4838	.0162
1.25	.3944	.1056	1.70	.4554	.0446	2.15	.4842	.0158
1.26	.3962	.1038	1.71	.4564	.0436	2.16	.4846	.0154
1.27	.3980	.1020	1.72	.4573	.0427	2.17	.4850	.0150
1.28	.3997	.1003	1.73	.4582	.0418	2.18	.4854	.0146
1.29	.4015	.0985	1.74	.4591	.0409	2.19	.4857	.0143
1.30	.4032	.0968	1.75	.4599	.0401	2.20	.4861	.0139
1.31	.4049	.0951	1.76	.4608	.0392	2.21	.4864	.0136
1.32	.4066	.0934	1.77	.4616	.0384	2.22	.4868	.0132
1.33	.4082	.0918	1.78	.4625	.0375	2.23	.4871	.0129
1.34	.4099	.0901	1.79	.4633	.0367	2.24	.4875	.0125
1.35	.4115	.0885	1.80	.4641	.0359	2.25	.4878	.0122
1.36	.4161	.0869	1.81	.4649	.0351	2.26	.4881	.0119
1.37	.4147	.0853	1.82	.4645	.0344	2.27	.4884	.0116
1.38	.4162	.0838	1.83	.4664	.0336	2.28	.4887	.0113
1.39	.4177	.0823	1.84	.4671	.0329	2.29	.4890	.0110
1.40	.4192	.0808	1.85	.4678	.0322	2.30	.4893	.0107
1.41	.4207	.0793	1.86	.4686	.0314	2.31	.4896	.0104
1.42	.4222	.0778	1.87	.4693	.0307	2.32	.4898	.0102
1.43	.4236	.0764	1.88	.4699	.0301	2.33	.4901	.0099
1.44	.4251	.0749	1.8	.4706	.0294	2.34	.4904	.0096
1.45	.4265	.0735	1.90	.4713	.0287	2.35	.4906	.0094
1.46	.4279	.0721	1.91	.4719	.0281	2.36	.4909	.0091
1.47	.4292	.0708	1.92	.4726	.0274	2.37	.4911	.0089
1.48	.4306	.0694	1.93	.4732	.0268	2.38	.4913	.0087
1.49	.4319	.0681	1.94	.4738	.0262	2.39	.4916	.0084

z	μ to z	> z
2.40	.4918	.0082
2.45	.4929	.0071
2.50	.4938	.0062
2.55	.4946	.0054
2.60	.4953	.0047
2.65	.4960	.0040
2.70	.4965	.0035
2.75	.4970	.0030
2.8	.4974	.0026
2.85	.4978	.0022
2.90	.4981	.0019
2.95	.4984	.0016
3.00	.4987	.0013
3.05	.4989	.0011
3.10	.4990	.0010
3.15	.4992	.0008
3.20	.4993	.0007
3.30	.4995	.0005
3.45	.4997	.0003
3.50	.4998	.0002
3.60	.4998	.0002
3.70	.4999	.0001

Table B
Critical Values of *t*

Column 1 Degrees of Freedom (*df*) for Test
Columns 2 – 5 Critical Values of *t* for the Indicated Alpha Level

	1-Tail Probabilities			
	.05	.025	.01	.005
	2-Tail Probabilities			
df	.10	.05	.02	.01
1	6.314	12.706	31.821	63.657
2	2.920	4.303	6.965	9.925
3	2.353	3.182	4.541	5.841
4	2.132	2.776	3.747	4.604
5	2.015	2.571	3.365	4.032
6	1.943	2.447	3.143	3.707
7	1.895	2.365	2.998	3.500
8	1.860	2.306	2.896	3.355
9	1.833	2.262	2.821	3.250
10	1.182	2.228	2.764	3.169
11	1.796	2.201	2.718	3.106
12	1.782	2.179	2.681	3.054
13	1.771	2.160	2.650	3.012
14	1.761	2.145	2.624	2.977
15	1.753	2.132	2.602	2.947
16	1.746	2.120	2.584	2.921
17	1.740	2.110	2.567	2.898
18	1.734	2.101	2.552	2.878
19	1.729	2.093	2.540	2.861
20	1.725	2.086	2.528	2.845
21	1.721	2.080	2.518	2.831
22	1.717	2.074	2.508	2.819
23	1.714	2.069	2.500	2.807
24	1.711	2.064	2.492	2.797
25	1.708	2.060	2.485	2.787
26	1.706	2.056	2.479	2.779
27	1.703	2.052	2.473	2.771
28	1.701	2.048	2.467	2.763
29	1.699	2.045	2.462	2.756
30	1.697	2.042	2.457	2.750
31	1.696	2.040	2.453	2.744
32	1.694	2.037	2.449	2.738
33	1.692	2.034	2.445	2.733

	1-Tail Probabilities			
	.05	.025	.01	.005
	2-Tail Probabilities			
df	.10	.05	.02	.01
34	1.691	2.032	2.441	2.728
35	1.690	2.030	2.438	2.724
36	1.688	2.028	2.434	2.720
37	1.687	2.026	2.431	2.715
38	1.686	2.024	2.429	2.712
39	1.685	2.023	2.426	2.708
40	1.684	2.021	2.423	2.704
45	1.679	2.014	2.412	2.690
50	1.676	2.009	2.403	2.678
55	1.673	2.004	2.396	2.668
60	1.671	2.000	2.390	2.660
70	1.667	1.994	2.381	2.648
80	1.664	1.990	2.374	2.639
90	1.662	1.987	2.368	2.632
100	1.660	1.984	2.364	2.626

Table C
Critical Values of χ^2

Column 1   Degrees of Freedom (*df*) for Test
Columns 2 – 6   Critical Values of χ^2 for the Indicated Alpha Level

Area in Upper Tail of Distribution (α Level)

df	.10	.05	.025	.01	.005
1	2.71	3.84	5.02	6.63	7.88
2	4.61	5.99	7.38	9.21	10.60
3	6.25	7.81	9.35	11.34	12.84
4	7.78	9.49	11.14	13.28	14.86
5	9.24	11.07	12.83	15.09	16.75
6	10.64	12.59	14.45	16.81	18.55
7	12.02	14.07	16.01	18.48	20.28
8	13.36	15.51	17.53	20.09	21.96
9	14.68	16.92	19.02	21.67	23.59
10	15.99	18.31	20.48	23.21	25.19
11	17.28	19.68	21.92	24.72	26.76
12	18.55	21.03	23.34	26.22	28.30
13	19.81	22.36	24.74	27.69	29.82
14	21.06	23.68	26.12	29.14	31.32
15	22.31	25.00	27.49	30.58	32.80
16	23.54	26.30	28.85	32.00	34.27
17	24.77	27.59	30.19	33.41	35.72
18	25.99	28.87	31.53	34.81	37.16
19	27.20	30.14	32.85	36.19	38.58
20	28.41	31.41	34.17	37.57	40.00
21	29.62	32.67	35.48	38.93	41.40
22	30.81	33.92	36.78	40.29	42.80
23	32.01	35.17	38.08	41.64	44.18
24	33.20	36.42	39.36	42.98	45.56
25	34.38	37.65	40.65	44.31	46.93

Made in the USA
Middletown, DE
20 April 2024

53212777R00152